Cristina Tonasellu

D0206430

Student Solutions Manual

Essential University Physics
Second Edition

Volume 2: Chapters 20–39

Richard Wolfson
Middlebury College

Brett Kraabel
PhD, Physics, University of California, Santa Barbara

Michael Schirber
PhD, Astrophysics, Ohio State University

Addison-Wesley

Boston   Columbus   Indianapolis   New York   San Francisco   Upper Saddle River
Amsterdam   Cape Town   Dubai   London   Madrid   Milan   Munich   Paris   Montréal   Toronto
Delhi   Mexico City   São Paulo   Sydney   Hong Kong   Seoul   Singapore   Taipei   Tokyo

Executive Editor: Nancy Whilton

Project Editor: Martha Steele

Director of Development: Michael Gillespie

Development Editor: Ashley Eklund

Editorial Assistant: Peter Alston

Managing Editor: Corinne Benson

Production Supervisor: Beth Collins

Production Management and Compositor: PreMediaGlobal

Manufacturing Buyer: Jeffrey Sargent

Senior Marketing Manager: Kerry Chapman

Cover Photo Credit: Andrew Lambert Photography / Science Photo Library

Copyright © 2012, 2007 Pearson Education, Inc., publishing as Addison-Wesley, 1301 Sansome Street, San Francisco, CA 94111. All rights reserved. Manufactured in the United States of America. This publication is protected by Copyright and permission should be obtained from the publisher prior to any prohibited reproduction, storage in a retrieval system, or transmission in any form or by any means, electronic, mechanical, photocopying, recording, or likewise. To obtain permission(s) to use material from this work, please submit a written request to Pearson Education, Inc., Permissions Department, 1900 E. Lake Ave., Glenview, IL 60025. For information regarding permissions, call (847) 486-2635.

Many of the designations used by manufacturers and sellers to distinguish their products are claimed as trademarks. Where those designations appear in this book, and the publisher was aware of a trademark claim, the designations have been printed in initial caps or all caps.

1 2 3 4 5 6 7 8 9 10—BR—13 12 11 10 09

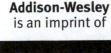

www.pearsonhighered.com

ISBN 10: 0-3217-1205-6
ISBN 13: 978-0-3217-1205-9

CONTENTS

Volume 1 contains chapters 1–19

Volume 2 contains chapters 20–39

Preface   iv

PART FOUR   Electromagnetism
Chapter 20   Electric Charge, Force, and Field   20-1
Chapter 21   Gauss's Law   21-1
Chapter 22   Electric Potential   22-1
Chapter 23   Electrostatic Energy and Capacitors   23-1
Chapter 24   Electric Current   24-1
Chapter 25   Electric Circuits   25-1
Chapter 26   Magnetism: Force and Field   26-1
Chapter 27   Electromagnetic Induction   27-1
Chapter 28   Alternating-Current Circuits   28-1
Chapter 29   Maxwell's Equations and Electromagnetic Waves   29-1

PART FIVE   Optics
Chapter 30   Reflection and Refraction   30-1
Chapter 31   Images and Optical Instruments   31-1
Chapter 32   Interference and Diffraction   32-1

PART SIX   Modern Physics
Chapter 33   Relativity   33-1
Chapter 34   Particles and Waves   34-1
Chapter 35   Quantum Mechanics   35-1
Chapter 36   Atomic Physics   36-1
Chapter 37   Molecules and Solids   37-1
Chapter 38   Nuclear Physics   38-1
Chapter 39   From Quarks to the Cosmos   39-1

PREFACE

This *Student Solutions Manual* to *Essential University Physics,* Second Edition, by Richard Wolfson, is designed to increase your skill and confidence in solving physics problems—the key to success in your physics course. It coaches you through a range of helpful and effective problem-solving techniques and provides solutions to all the odd-numbered problems in your text. By working carefully through these techniques and solutions, you will master the proven IDEA four-step problem-solving approach used in the text (Interpret, Develop, Evaluate, Assess) and learn to successfully

- interpret problems and identify the key physics concepts involved.
- develop a plan and draw figures for your solution.
- evaluate any mathematical expressions.
- assess your solution to check that it makes sense and see how it adds to your broader understanding of physics.

Do your best to solve each problem *before* reading the solution. When you need to refer to the solution, focus on the reasoning—make sure you understand how and why each step is taken. By pushing yourself, you'll develop and hone your problem-solving skills. Don't fall into the trap of just passively reading the solutions.

Through multiple reviews involving many instructors, we have made every effort to ensure these solutions are accurate and correct. If you find any errors or ambiguities, we would be very grateful to hear from you. Please ask your professor to contact her/his Pearson Education/Addison-Wesley sales representative, who will pass on any errors to the appropriate person.

EXERCISES

Section 20.1 Electric Charge

13. **INTERPRET**   We'll estimate the charge your body would carry if the electron charge slightly differed from the proton charge.

DEVELOP   Since the human body is about 60% water, let's assume that the number of protons/electrons per kilogram in your body is the same as that of a water molecule. Water is 2 hydrogen atoms (with one proton and one electron each) and one oxygen atom (with 8 protons and 8 electrons). If the charges on the electrons and protons are different by one part in a billion, then the net charge on a water molecule would be

$$|\Delta q| = 10|q_{proton} - q_{electron}| = 10(10^{-9}e) = 10(1.60 \times 10^{-28} \text{C}) = 1.60 \times 10^{-27} \text{C}$$

The mass of a water molecule is the mass of two hydrogens and one oxygen: $1u + 1u + 16u = 18u$, where $1 \text{ u} = 1.66 \times 10^{-27} \text{ kg}$. Therefore, the net charge to mass ratio for a water molecule would be:

$$\left(\frac{|\Delta q|}{m}\right)_{H_2O} = \frac{1.60 \times 10^{-27} \text{C}}{18(1.66 \times 10^{-27} \text{kg})} = 0.05 \text{ C/kg}$$

EVALUATE   We'll assume your mass is 65 kg. If we approximate this mass as pure water, then the total charge on your body would be approximately

$$|\Delta q| \approx m\left(\frac{|\Delta q|}{m}\right)_{H_2O} = (65 \text{ kg})(0.05 \text{ C/kg}) \approx 3 \text{ C}$$

ASSESS   This is a huge amount of charge. Imagine half of the charge was in your head/chest and the other half was in your legs, about a meter away. Then the magnitude of the force of repulsion between the upper and lower parts of your body would be

$$F_{12} = \frac{kq_1q_2}{r^2} = \frac{(9.0 \times 10^9 \text{ N} \cdot \text{m}^2/\text{C}^2)(\frac{1}{2}3 \text{ C})^2}{(1 \text{ m})^2} = 2 \times 10^{10} \text{ N}$$

This would rip you apart!

15. **INTERPRET**   This problem asks us to find the combination of u and d quarks needed to make a proton, which has a positive unit change e, and a neutron, which has zero charge.

DEVELOP   The u quark has charge $2e/3$ and the d quark has charge $-e/3$, so we can combine these so the charges sum to unity (for the proton) or zero (for the neutron).

EVALUATE   (a) Two u quarks and a d quark make a total charge of $2(2/3) - 1/3 = 1$, so the proton can be constructed from the quark combination uud.

(b) Two d quarks and a u quark make a total charge of $2(-1/3) + 2/3 = 0$, so the neutron can be constructed from the quark combination udd.

ASSESS   Because of a phenomenon called color confinement, quarks can only be found in hadrons, of which protons and neutrons are the most stable examples.

Section 20.2 Coulomb's Law

17. **INTERPRET** We want to know how far a proton must be from an electron to exert an electrical attraction equal to the electron's weight on Earth.

    DEVELOP The force between a proton and electron has a magnitude given by Coulomb's law (Equation 20.1): $F = ke^2/r^2$, where $1\ e = 1.60 \times 10^{-19}\,\mathrm{C}$. We'll solve this for the distance, r, at which $F = m_e g$.

    EVALUATE The distance at which the electrical force is equal to the gravitational force is

    $$r = \sqrt{\frac{ke^2}{m_e g}} = \sqrt{\frac{\left(9.0 \times 10^9\,\mathrm{N \cdot m^2/C^2}\right)\left(1.60 \times 10^{-19}\,\mathrm{C}\right)^2}{\left(9.11 \times 10^{-31}\,\mathrm{kg}\right)\left(9.8\,\mathrm{m/s^2}\right)}} = 5.1\,\mathrm{m}$$

    ASSESS On the molecular scale, protons and electrons are roughly $10^{-10}\,\mathrm{m}$ apart. Since the Coulomb attraction scales as $1/r^2$, electric forces will clearly overwhelm any gravity effects.

19. **INTERPRET** This problem involves finding the unit vector associated with the electrical force one charge exerts on another, given the coordinates of the positions of both charges.

    DEVELOP A unit vector from the position of charge q at $\vec{r}_q = (1\,\mathrm{m},\ 0)$, to any other point $\vec{r} = (x, y)$ is

    $$\hat{n} = \frac{\left(\vec{r} - \vec{r}_q\right)}{\left|\vec{r} - \vec{r}_q\right|} = \frac{(x - 1\,\mathrm{m}, y)}{\sqrt{(x - 1\,\mathrm{m})^2 + y^2}}$$

    EVALUATE **(a)** When the other charge is at position $\vec{r} = (1\,\mathrm{m}, 1\,\mathrm{m})$, the unit vector is

    $$\hat{n} = \frac{(0, 1\,\mathrm{m})}{\sqrt{0 + (1\,\mathrm{m})^2}} = (0, 1) = \hat{j}$$

    (b) When $\vec{r} = (0, 0)$,

    $$\hat{n} = \frac{(-1\,\mathrm{m}, 0)}{\sqrt{(-1\,\mathrm{m})^2 + 0}} = (-1, 0) = -\hat{i}$$

    (c) Finally, when $\vec{r} = (2\,\mathrm{m}, 3\,\mathrm{m})$, the unit vector is

    $$\hat{n} = \frac{(1\,\mathrm{m}, 3\,\mathrm{m})}{\sqrt{(1\,\mathrm{m})^2 + (3\,\mathrm{m})^2}} = \frac{(1, 3)}{\sqrt{10}} = 0.316\hat{i} + 0.949\hat{j}$$

    The sign of q doesn't affect this unit vector, but the signs of both charges do determine whether the force exerted by q is repulsive or attractive; that is, in the direction of $+\hat{n}$ or $-\hat{n}$.

    ASSESS The unit vector always points away from the charge q located at $(1\,\mathrm{m}, 0)$.

Section 20.3 The Electric Field

21. **INTERPRET** This problem is about calculating the electric field strength due to a source, when the force experienced by the electron is known.

    DEVELOP Equation 20.2a shows that the electric field strength (magnitude of the field) at a point is equal to the force per unit charge that would be experienced by a charge at that point: $E = F/q$.

    EVALUATE With $q = |e|$, we find the field strength to be

    $$E = \frac{F}{|e|} = \frac{0.61 \times 10^{-9}\,\mathrm{N}}{1.60 \times 10^{-19}\,\mathrm{C}} = 3.8 \times 10^9\,\mathrm{N/C}$$

    ASSESS Since the charge of electron is negative, the electric field will point in the opposite direction as the force.

23. **INTERPRET** This problem involves calculating the electric field needed to produce the given force on the given charge, and then find the force experienced by a second charge is the same electric field.

    DEVELOP Equation 20.2a shows that the electric field strength (magnitude of the field) at a point is equal to the force per unit charge that would be experienced by a charge at that point:

    $$E = \frac{F}{q}$$

The equation allows us to calculate E give that $F = 150$ mN and $q = 68$ nC. For part **(b)**, the force experienced by another charge q' in the same field is given by Equation 20.2b: $F' = q'E$.

EVALUATE   **(a)** With $q = 68$ nC, we find the field strength to be

$$E = \frac{F}{|e|} = \frac{150 \text{ mN}}{68 \text{ nC}} = 2.2 \times 10^6 \text{ N/C}$$

(b) The force experienced by a charge $q' = 35$ μC in the same field is

$$F' = q'E = (35 \ \mu C)(2.21 \times 10^6 \text{ N/C}) = 77 \text{ N}$$

ASSESS   The force a test charge particle experiences is proportional to the magnitude of the test charge. In our problem, since $q' = 35 \ \mu C > q = 68$ nC, we find $F' > F$.

25.   **INTERPRET**   This problem is similar to Problem 20.23, in that we are given the force exerted on a given charge by an unknown electric field, and we are to find the force exerted by this field on another charge (a proton, in this case).

DEVELOP   Apply Equation 20.2b $\vec{E} = \vec{F}/q$ to find the electric field, with $q = -1$ μC. The force on the proton ($q_p = 1.6 \times 10^{-6}$ C) is given by Equation 20.2a, $\vec{F}_p = q_p \vec{E}$. Inserting the result of Equation 20.2b gives

$$\vec{F}_p = q_p \vec{E} = q_p \frac{\vec{F}}{q}$$

EVALUATE   Inserting the given quantities into the expression above for the force on the proton gives

$$\vec{F}_p = (1.6 \times 10^{-19} \text{ C}) \frac{10\hat{i} \text{ N}}{-1.0 \times 10^{-6} \text{ C}} = -1.6\hat{i} \text{ pN}$$

ASSESS   The force on the proton acts in the opposite direction compared to the force on the original charge, because the two charges have opposite signs.

Section 20.4 Fields of Charge Distributions

27.   **INTERPRET**   For this problem, we are to find the electric field at several locations in the vicinity of two given point charges. We can apply the principle of superposition to solve this problem.

DEVELOP   Take the origin of the x-y coordinate system to be at the midpoint between the two charges, as indicated in the figure below, and use Equation 20.4 to find the electric field at the given points. Let $q_1 = +2.0$ μC and $q_2 = -2.0$ μC. Let $\vec{r}_{\pm} = \pm(2.5 \text{ cm})\,\hat{j}$ denote the positions of the charges and $\vec{r}$ denote that of the field point (i.e., the point at which we are calculating the electric field). A unit vector from charge i to the field point is $(\vec{r} - \vec{r}_{\pm})/|\vec{r} - \vec{r}_{\pm}|$ [where the plus (minus) sign corresponds to the positive (negative) charge)]. Thus, the spatial factors in Coulomb's law are $\hat{r}_i/r_i^2 = \vec{r}_i/r_i^3 = (\vec{r} - \vec{r}_{\pm})/|\vec{r} - \vec{r}_{\pm}|^3$. By the principle of superposition (Equation 20.4), the total electric field at any point is

$$\vec{E} = k\left(\frac{q_1 \vec{r}_1}{r_1^3} + \frac{q_2 \vec{r}_2}{r_2^3}\right)$$

EVALUATE   (a) For the point at 5.0 cm to the left of P, we have

$$\vec{r} = (-5.0 \text{ cm})\,\hat{i}$$

$$\vec{r}_1 = \vec{r} - \vec{r}_+ = (-5.0 \text{ cm})\,\hat{i} - (-2.5 \text{ cm})\,\hat{i} = (-2.5 \text{ cm})\,\hat{i}$$

$$\vec{r}_2 = \vec{r} - \vec{r}_- = (-5.0 \text{ cm})\,\hat{i} - (2.5 \text{ cm})\,\hat{i} = (-7.5 \text{ cm})\,\hat{i}$$

so the electric field is

$$\vec{E} = k\left(\frac{q_1 \vec{r}_1}{r_1^3} + \frac{q_2 \vec{r}_2}{r_2^3}\right) = k\left(-\frac{q_1 \hat{i}}{r_1^2} - \frac{q_2 \hat{i}}{r_2^2}\right)$$

$$= \left(9.0 \times 10^9\,\frac{\text{N} \cdot \text{m}^2}{\text{C}^2}\right)\left[-\frac{(2.0 \times 10^6\,\text{C})\hat{i}}{(0.0025\,\text{m})^2} - \frac{(-2.0 \times 10^{-6}\,\text{C})\hat{i}}{(0.0075\,\text{m})^2}\right] = (-2.6\,\text{GN/C})\hat{i}$$

(b) For the point at 5.0 cm directly above P, we have

$$\vec{r} = (5.0 \text{ cm})\,\hat{j}$$

$$\frac{\vec{r}_1}{r_1^3} = \frac{\vec{r} - \vec{r}_+}{|\vec{r} - \vec{r}_+|^3} = \frac{-(-2.5 \text{ cm})\,\hat{i} + (5.0 \text{ cm})\,\hat{j}}{\left[(5.0 \text{ cm})^2 + (-2.5 \text{ cm})^2\right]^{3/2}}$$

$$\frac{\vec{r}_2}{r_2^3} = \frac{\vec{r} - \vec{r}_-}{|\vec{r} - \vec{r}_-|^3} = \frac{-(2.5 \text{ cm})\,\hat{i} + (5.0 \text{ cm})\,\hat{j}}{\left[(5.0 \text{ cm})^2 + (2.5 \text{ cm})^2\right]^{3/2}}$$

so the electric field is

$$\vec{E} = \left(9.0 \times 10^9\,\frac{\text{N} \cdot \text{m}^2}{\text{C}^2}\right)\left(\frac{2.0 \times 10^{-6}\,\text{C}}{\text{m}^2}\right)\left\{\frac{-(-0.0025)\,\hat{i} + 0.0050\,\hat{j}}{\left[(-0.0025)^2 + (0.0050)^2\right]^{3/2}} - \frac{-(0.0025)\,\hat{i} + 0.0050\,\hat{j}}{\left[(0.0025)^2 + (0.0050)^2\right]^{3/2}}\right\}$$

$$= \left(9.0 \times 10^9\,\frac{\text{N} \cdot \text{m}^2}{\text{C}^2}\right)\left(\frac{2.0 \times 10^{-6}\,\text{C}}{\left[(0.0025)^2 + (0.0050)^2\right]^{3/2}\,\text{m}^2}\right)(0.0050\,\hat{i}) = (0.52\,\text{GN/C})\hat{i}$$

(c) For $\vec{r} = 0$, we have

$$\frac{\vec{r}_1}{r_1^3} = \frac{\vec{r} - \vec{r}_+}{|\vec{r} - \vec{r}_+|^3} = \frac{-(2.5 \text{ cm})\,\hat{i}}{(2.5 \text{ cm})^3} = \frac{-\hat{i}}{(2.5 \text{ cm})^2}$$

$$\frac{\vec{r}_2}{r_2^3} = \frac{\vec{r} - \vec{r}_-}{|\vec{r} - \vec{r}_-|^3} = \frac{(2.5 \text{ cm})\,\hat{i}}{(2.5 \text{ cm})^3} = \frac{\hat{i}}{(2.5 \text{ cm})^2}$$

so the electric field is

$$\vec{E} = \left(9.0 \times 10^9\,\frac{\text{N} \cdot \text{m}^2}{\text{C}^2}\right)\left(\frac{2.0 \times 10^{-6}\,\text{C}}{\text{m}^2}\right)\left[\frac{-\hat{i}}{(0.0025)^2} - \frac{\hat{i}}{(0.0025)^2}\right] = (-5.8\,\text{GN/C})\hat{i}$$

ASSESS   The electric field for part (b) is much weaker because the fields from the two charges largely cancel.

29.   INTERPRET   We are given a long wire with a uniform charge density and are asked to find the electric field strength 38 cm from the wire. We can assume that the wire length is much, much greater than 38 cm.

DEVELOP   For a very long wire ($L \gg 38$ cm), Example 20.7 shows that the magnitude of the electric field falls off like $1/r$. Therfore, the electic field is simply scaled by the ratio of the distances, or

$$E_2 = E_1 \frac{r_1}{r_2}$$

EVALUATE   Inserting the given quantities into the expression above gives

$$E_2 = (1.9 \text{ kN/C})\frac{22}{38} = 1.1 \text{ kN/C}$$

ASSESS   The electric field gets weaker the farther we are from the wire, as expected.

31.   INTERPRET   We will use Coulomb's law and the definition of electric field to find the electric field at a point on the axis of a charged ring.

DEVELOP   From Example 20.6, which is done for a general distance x, we see that

$$E = \frac{kQx}{(x^2 + a^2)^{3/2}}$$

We want to know the field E at position $x = a$.

EVALUATE   Inserting $x = a$ into the expression above gives

$$E = \frac{kQx}{(x^2 + a^2)^{3/2}} = \frac{kQa}{(2a^2)^{3/2}} = \frac{kQ}{\sqrt{8}a^2}$$

ASSESS   The units are $kQ/(\text{distance})^2$, which are correct for an electric field.

Section 20.5 Matter in Electric Fields

33.   INTERPRET   This problem involves kinematics, Newton's second law, and Coulomb's law. We can use these concepts to find the electric field strength necessary to accelerate an electron from rest to $c/10$ within 5.0 cm.

DEVELOP   If the electric filed is constant in space, the force applied to the charge will be constant, so we can use Equation 2.11, which applied for constant acceleration, to find the necessary acceleration for the electron. This gives

$$v^2 = \overset{=0}{\overbrace{v_0^2}} + 2a(x - x_0)$$

where $v = c/10$, $x - x_0 = 5.0$ cm, and $v_0 = 0$ because the electron starts from rest. Thus the acceleration is

$$a = \frac{v^2}{2(x - x_0)} = \frac{c^2}{200(5.0 \text{ cm})}$$

The force needed to provide this acceleration is given by Newton's second law which, combined with Coulomb's law, gives

$$F = Eq = ma = \frac{mc^2}{10 \text{ m}}$$

which we can solve for E.

EVALUATE   Inserting $|q| = 1.6 \times 10^{-19}$ C for the electron's charge (since we are only concerned with the strength of the electric field, not its direction), we find

$$E = \frac{mc^2}{q(10 \text{ m})} = \frac{(9.11 \times 10^{-31} \text{ kg})(3.0 \times 10^8 \text{ m/s})^2}{(1.6 \times 10^{-19} \text{ C})(10 \text{ m})} = 5.1 \times 10^4 \text{ N/C}$$

ASSESS   Because the electron has as negative charge, it would move opposite to the direction of this electric field.

35.   INTERPRET   This problem involves an electrostatic analyzer like that in Example 20.8, so we will use the results of that example. We are to find the coefficient E_0 in the expression for the electric field strength in the analyzer that will permit protons to exit the analyzer.

DEVELOP   From the analysis of Example 20.8, we know that the coefficient E_0 of the analyzer is related to the particle mass m, its velocity v, and it charge q by

$$E_0 = \frac{mv^2}{qb}$$

Given that $m = 1.67 \times 10^{-27}$ kg for a proton, $q = 1.6 \times 10^{-19}$ C, and v and b are given, we can find E_0.

EVALUATE Inserting the given quantities in the expression for E_0 gives

$$E_0 = \frac{mv^2}{eb} = \frac{\left(1.67 \times 10^{-27} \text{ kg}\right)\left(84 \times 10^3 \text{ m/s}\right)^2}{\left(1.6 \times 10^{-19} \text{ C}\right)\left(0.075 \text{ m}\right)} = 980 \text{ N/C}$$

to two significant figures.

ASSESS Note that the proton exits the analyzer with the same speed with which it entered that analyzer, because the force is always perpendicular to the proton's trajectory (i.e., it's a centripetal force). Thus, the force does no work on the proton, but the proton's velocity has changed direction.

PROBLEMS

37. **INTERPRET** This problem involves Coulomb's law, which we can use to relate the force experienced by the two particles to their charges.

    DEVELOP Coulomb's law (Equation 20.1) gives the force between charged particles 1 and 2 as

    $$F = \frac{kq_1 q_2}{r^2}$$

    We are given that $q_1 = 2q_2$, and the $r = 15$ cm, so we can solve for the magnitude of the larger charge q_1.

    EVALUATE Substituting for q_2 in Coulomb's law and solving for q_1 gives

    $$q_1 q_2 = \frac{Fr^2}{k}$$

    $$q_1 \left(\frac{q_1}{2}\right) = \frac{Fr^2}{k}$$

    $$q_1 = \pm r \sqrt{\frac{2F}{k}} = \pm (0.15 \text{ m}) \sqrt{\frac{2(95 \text{ N})}{9.0 \times 10^9 \text{ N} \cdot \text{m}^2/\text{C}^2}} = \pm 22 \text{ } \mu\text{C}$$

    ASSESS Because the force is repulsive, the charges must have the same sign. However, from the information given in the problem statement, we cannot tell whether the sign is positive or negative.

39. **INTERPRET** This involves finding the electric force on one charge from another charge when both are located in the x-y plane.

    DEVELOP Denote the positions of the charges by $\vec{r}_1 = 15\hat{i} + 5.0\hat{j}$ cm for $q_1 = 9.5$ μC, and $\vec{r}_2 = 4.4\hat{i} + 11\hat{j}$ cm for $q_2 = -3.2$ μC. The vector from q_1 to q_2 is $\vec{r} = \vec{r}_2 - \vec{r}_1$, from which we can find the charge separation, r, that goes into Coulomb's law, $F_{12} = kq_1 q_2 / r^2$, for the force q_1 exerts on q_2. The direction of this force can be designated with the unit vector, $\hat{r} = \vec{r} / r$.

    EVALUATE First, we find the vector displacement:

    $$\vec{r} = \vec{r}_2 - \vec{r}_1 = \left(4.4\hat{i} + 11\hat{j} \text{ cm}\right) - \left(15\hat{i} + 5.0\hat{j} \text{ cm}\right) = -10.6\hat{i} + 6.0\hat{j} \text{ cm}$$

    The magnitude of this vector is $r = \sqrt{10.6^2 + 6.0^2}$ cm $= 12.2$ cm. So the electric force has a magnitude of

    $$F_{12} = \frac{kq_1 q_2}{r^2} = \frac{\left(9.0 \times 10^9 \text{ N} \cdot \text{m}^2/\text{C}^2\right)\left(9.5 \text{ } \mu\text{C}\right)\left(-3.2 \text{ } \mu\text{C}\right)}{\left(0.122 \text{ m}\right)^2} = -18.4 \text{ N}$$

    Using the unit vector, we can write the force in component form:

    $$\vec{F}_{12} = F_{12}\hat{r} = \left(-18.4 \text{ N}\right)\left(\frac{-10.6\hat{i} + 6.0\hat{j} \text{ cm}}{12.2 \text{ cm}}\right) = 16\hat{i} - 9.0\hat{j} \text{ N}$$

    ASSESS Without calculating the exact directions, we can see that $\vec{r}$ points to the left of the positive y-axis, whereas $\vec{F}_{12}$ points to the right of the negative y-axis. This seems to agree with the fact that the force between oppositely charged particles is attractive, pointing in the opposite direction as the vector that separates their positions.

41. **INTERPRET**   This problem involves Coulomb's law and the principle of superposition, which we can use to find the position of the three given charges so that all of them experience zero net force.

    DEVELOP   By symmetry, the negative charge must be at the midpoint between the two positive charges, as shown in the figure below. With this positioning, the attractive force between the $-q$ charge and each $+4q$ charge cancels the repulsive forces between the two $+4q$ charges. The $-q$ charges experiences no net force because each $+4q$ charge generates a force on it that is equal in magnitude but in the opposite direction.

    EVALUATE   To verify that we have the correct positioning, we calculate the net force on the left-hand charge. Coulomb's law and the superposition principle give

    $$k\frac{4q^2}{a^2}\left(-\hat{i}\right)+k\frac{16q^2}{\left(2a\right)^2}\left(\hat{i}\right)=0$$

    The expression for the force on the right-hand charge is the same, except that the sign of the unit vectors is reversed. The force on the central charge is

    $$k\frac{4q^2}{a^2}\left(-\hat{i}\right)+k\frac{4q^2}{a^2}\left(\hat{i}\right)=0$$

    Thus, all three charges experience zero net force.

    ASSESS   The equilibrium is unstable, since if $-q$ is displaced slightly toward one charge, the net force on it will be in the direction of that charge.

43. **INTERPRET**   This problem is similar to the preceding one, only the magnitude of the charges has changed. Therefore, we can use the same strategy to solve this problem.

    DEVELOP   The position of the charges are again denoted by $\vec{r}_1=\left(1\text{ m}\right)\hat{j}$, $\vec{r}_2=\left(2\text{ m}\right)\hat{i}$, and $\vec{r}_3=\left(2\text{ m}\right)\hat{i}+\left(2\text{ m}\right)\hat{j}$ The unit vector pointing from q_3 toward q_1 is

    $$\hat{r}_{31}=\frac{\left(\vec{r}_1-\vec{r}_3\right)}{\left|\vec{r}_1-\vec{r}_3\right|}=\frac{\left(1\text{ m}\right)\hat{j}-\left(2\text{ m}\right)\hat{i}-\left(2\text{ m}\right)\hat{j}}{\sqrt{\left(-2\right)^2+\left(-1^2\right)}\text{ m}}=-\frac{2}{\sqrt{5}}-\frac{1}{\sqrt{5}}\hat{j}$$

    Similarly, the unit vector pointing from q_2 toward q_1 is

    $$\hat{r}_{21}=\frac{\left(\vec{r}_1-\vec{r}_2\right)}{\left|\vec{r}_1-\vec{r}_2\right|}=\frac{\left(1\text{ m}\right)\hat{j}-\left(2\text{ m}\right)\hat{i}}{\sqrt{1^2+\left(-2\right)^2}\text{ m}}=-\frac{2}{\sqrt{5}}\hat{i}+\frac{1}{\sqrt{5}}\hat{j}$$

    The vector form of Coulomb's law and the superposition principle give the net electric force on q_3 as

    $$\vec{F}_1=kq_1\left[\frac{q_2\hat{r}_{21}}{r_{21}^2}+\frac{q_3\hat{r}_{31}}{r_{31}^2}\right]=kq_1\left[\frac{q_2\left(-2\hat{i}+\hat{j}\right)}{5^{3/2}\text{ m}^2}+\frac{q_3\left(-2\hat{i}-\hat{j}\right)}{5^{3/2}\text{ m}^2}\right]$$

    We are told that the force on q_1 is in the $\hat{i}$ direction, so the $\hat{j}$ component of $\vec{F}_1$ must be zero. This gives

    $$q_2-q_3=0$$

    EVALUATE   (a) From the equation above, we find that $q_3=-q_2$, or $q_3=20$ μC.
    (b) Inserting the result from part (a) into the expression for $\vec{F}_1$ gives

    $$\vec{F}_1=\left(9.0\times10^9\text{ N}\cdot\text{m}^2/\text{C}^2\right)\left(25\times20\text{ μC}^2\right)\left(-4.0\hat{i}\right)\left(5.0\right)^{-3/2}=\left(-1.6\right)\hat{i}\text{ N}.$$

    ASSESS   Because the charges q_2 and q_3 are positioned symmetrically above and below q_1, the result that $q_2=-q_3$ is expected.

45. **INTERPRET** We're asked to calculate the electric field from a point charge at several locations.

DEVELOP The electric field from a point charge is given by Equation 20.3: $\vec{E} = kq\hat{r}/r^2$. The charge is at the origin, so the vector, $\vec{r}$, has the same coordinate values as the points in x-y plane that we're asked to consider. Since the unit vector is $\hat{r} = \vec{r}/r$, we can write the electric field in a more compact form: $\vec{E} = kq\vec{r}/r^3$.

EVALUATE (a) At $x = 50$ cm and $y = 0$ cm, the electric field is

$$\vec{E} = \frac{kq}{r^3}\vec{r} = \frac{\left(9.0\times10^9 \ \frac{N\cdot m^2}{C^2}\right)(65 \ \mu C)}{(50 \ cm)^3}(50\hat{i} \ cm) = 2.3\hat{i} \ MN/C$$

(b) At $x = 50$ cm and $y = 50$ cm, the electric field is

$$\vec{E} = \frac{\left(9.0\times10^9 \ \frac{N\cdot m^2}{C^2}\right)(65 \ \mu C)}{\left(\sqrt{50^2+50^2} \ cm\right)^3}(50\hat{i}+50\hat{j} \ cm) = 0.82\hat{i}+0.82\hat{j} \ MN/C$$

(c) At $x = 25$ cm and $y = -75$ cm, the electric field is

$$\vec{E} = \frac{\left(9.0\times10^9 \ \frac{N\cdot m^2}{C^2}\right)(65 \ \mu C)}{\left(\sqrt{25^2+75^2} \ cm\right)^3}(25\hat{i}-75\hat{j} \ cm) = 0.30\hat{i}-0.89\hat{j} \ MN/C$$

ASSESS The magnitude of the electric field in parts (a), (b), (c) is 2.3 MN/C, 1.2 MN/C, and 0.9 MN/C, respectively. This shows that the field decreases as one moves further away from the point charge, as we would expect.

47. **INTERPRET** This problem involves two source charges, a proton at $x = 0$ and an ion at $x = 5.0$ nm. We can apply the principle of superposition to find the electric field at $x = -5$ nm.

DEVELOP The field at the point $x = -5$ nm due to the proton is

$$\vec{E}_p = \frac{ke}{(5 \ nm)^2}(-\hat{i})$$

The field at the same point due to the ion is

$$\vec{E}_i = \frac{kq_i}{(10 \ nm)^2}(-\hat{i})$$

The total electric field is the sum of these two, and is zero at $x = -5.0$ nm, so we can solve for q_i.

EVALUATE Solving for the ion's charge gives

$$\frac{ke}{(5 \ nm)^2}(-\hat{i}) + \frac{kq_i}{(10 \ nm)^2}(-\hat{i}) = 0$$

$$q_i = -e\frac{(10 \ nm)^2}{(5 \ nm)^2} = -4e$$

ASSESS Note that the field due to the ion was defined for a positive charge, so the final charge is negative, as indicated.

49. **INTERPRET** This problem is that same as Example 20.5, except that it is rotated by 90°. There are two source particles: a proton at (0, 0.60 nm) and an electron at (0, −0.60 nm), so the system is dipole.

DEVELOP We can use the result of Example 20.5, with y replaced by x, and x by −y (or, equivalently, $\hat{j}$ by $\hat{i}$, and $\hat{i}$ by $-\hat{j}$). The electric field on the x axis is then

$$\vec{E}(x) = 2kqa\hat{j}\left(a^2+x^2\right)^{-3/2}$$

where $q = e = 1.6 \times 10^{-19}$ C and $a = 0.60$ nm (see Figure 20.12 rotated 90° clockwise). The constant $2kq = 2(9.0 \times 10^9 \text{ N} \cdot \text{m}^2/\text{C}^2)(1.6 \times 10^{-19} \text{ C}) = (2.88 \text{ GN/C})(\text{nm})^2$.

EVALUATE   (a) Midway between the two charges (at $x = 0$), the electric field is

$$\vec{E}(0,0) = \frac{2kq\hat{j}}{a^2} = \frac{(2.88 \text{ nm}^2 \cdot \text{GN/C})\hat{j}}{(0.60 \text{ nm})^2} = (8.0 \text{ GN/C})\hat{j}$$

(b) for $x = 2$ nm,

$$\vec{E}(2.0 \text{ nm}, 0) = (2.88 \text{ nm}^2 \cdot \text{GN/C})(0.60 \text{ nm})(0.60^2 + 2.0^2)^{-3/2}(\text{nm})^{-3}(\hat{j}) = (190 \text{ MN/C})\hat{j}$$

(c) For $x = 20$ nm,

$$\vec{E}(-20 \text{ nm}, 0) = (2.88 \text{ nm}^2 \cdot \text{GN/C})(0.60 \text{ nm})\left[0.60^2 + (-20)^2\right]^{-3/2}(\hat{j}) = (215.71 \text{ kN/C}) = \hat{j}(220 \text{ kN/C})\hat{j}.$$

to two significant figures.

ASSESS   For part (c), because $x \gg a$, we can apply Equation 20.6a, which gives

$$\vec{E}(-20 \text{ } nm, 0) = -\frac{kp}{x^3}\hat{j} = -\frac{2kqa}{x^3}\hat{j} = -\frac{2(2.88 \text{ nm}^2 \cdot \text{GN/C})(-20 \text{ nm})}{(-20 \text{ nm})^3} = 216.0 \text{ GN/C}$$

which differs by only 0.1% from the more precise result of part (c).

51. **INTERPRET**   This problem involves finding the net charge or an unknown charge distribution. We are given the behavior of the electric field as a function of distance from the charge, for distances much, much greater than the size of the charge distribution.

DEVELOP   Taking the hint, we suppose that the field strength varies with an inverse power of the distance, $E(r) \sim r^n$. Under this hypothesis, $282/119 = (1.5/2.0)^n$, or $n = \ln(282/119)/\ln(0.75) = -3.0$.

EVALUATE   A dipole field falls off like r^{-3} for $r \gg a$, so the charge distribution must be a dipole, whose net charge is zero.

ASSESS   Note that this result is only valid if the dipole separation $a \ll 1$ m. Because this is normally the case, the result is appears valid.

53. **INTERPRET**   This problem involves Coulomb's law, which generates the given forces between two charged metal spheres. The spheres initially experience an attractive force, but when the charge on the spheres is equilibrated, the force becomes repulsive.

DEVELOP   The charges initially attract, so $q_1 = -q_2$, and, by Coulomb's law (Equation 20.1), we have

$$2.5 \text{ N} = -\frac{kq_1 q_2}{1 \text{ m}^2}$$

When the spheres are brought together, they share the total charge equally, each acquiring $\frac{1}{2}(q_1 + q_2)$. The magnitude of their repulsion is

$$2.5 \text{ N} = k\frac{(q_1 + q_2)^2}{4 \text{ m}^2}$$

Because the forces have the same magnitude, we can equation them to find the original charges q_1 and q_2.

EVALUATE   Equating these two forces, we find a quadratic equation $\frac{1}{4}(q_1 + q_2)^2 = -q_1 q_2$, or $q_1^2 + 6q_1 q_2 + q_2^2 = 0$, with solutions $q_1 = (-3 \pm \sqrt{8})q_2$. Both solutions are possible, but since $3 + \sqrt{8} = (3 - \sqrt{8})^{-1}$, they merely represent a relabeling of the charges. Since $-q_1 q_2 = (2.5 \text{ N} \cdot \text{m}^2)/(9.0 \times 10^9 \text{ N} \cdot \text{m}^2/\text{C}^2) = (1.67 \text{ μC})^2$, the solutions are $q_1 = \pm\sqrt{3 + \sqrt{8}}(16.7 \text{ μC}) = \pm 40 \text{ μC}$ and $q_2 = \mp(40.2 \text{ μC})/(3 + \sqrt{8}) = \mp 6.9 \text{ μC}$, or the same values with q_1 and q_2 interchanged.

ASSESS   The results are reported to two significant figures, which is the precision to which the data is known.

55.   INTERPRET   We will calculate the electric field magnitude a distance x from either end of a uniformly charged rod of charge Q and length L.

DEVELOP   We will use the integral form for electric field, Equation 20.7: $\vec{E} = \int \frac{k\,dq}{r^2}\hat{r}$. Because the charge is distributed uniformly along the length, the differential charge element is $dq = (Q/L)\,dx'$, where x' is the location of this charge along the rod, see figure below.

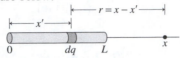

We are only interested in the electric field along the x-axis at points with $x > L$, in which case the distance to dq is just $r = x - x'$.

EVALUATE   The integral for the electric field strength is

$$E = \int_0^L \frac{k\,dq}{r^2} = \frac{kQ}{L}\int_0^L \frac{dx'}{(x-x')^2}$$

We can change variables: $u = x - x'$, $dx' = -du$, so the integral becomes

$$E = \frac{kQ}{L}\int_x^{x-L}\frac{-du}{u^2} = \frac{kQ}{L}\left(\frac{1}{(x-L)} - \frac{1}{x}\right) = \frac{kQ}{x(x-L)}$$

ASSESS   For $x \gg L$, the electric field reduces to $E = kQ/x^2$, which is what we'd expect since the rod will appear as a point charge from a great distance.

57.   INTERPRET   You want to check if a patent for an isotope separator will work. Since different isotopes have the same charge but different mass, the device can work if it discriminates between objects with different charge-to-mass ratios.

DEVELOP   You can assume the accelerating field, $\vec{E}_1$, is constant and that the plates in the figure are separated by a distance x. Therefore, if an atom stripped of its electrons starts at rest at the bottom plate, it will be accelerated by the field to a final speed of

$$v = \sqrt{2ax} = \sqrt{\frac{2qE_1 x}{m}}$$

So, it is true that isotopes of different charge-to-mass rations (q/m) will leave the first half of the device with different speeds. For example, an isotope with a relatively large charge-to-mass ratio will attain a higher speed in the accelerating field than another isotope with a lower charge-to-mass ratio. But the question is: can the second half of the device select just one of these speeds so that only one type of isotope emerges?

EVALUATE   The second half of the device is an electrostatic analyzer, as described in Example 20.8. It has a curved field, $\vec{E}_2 = E_0(b/r)\hat{r}$, which points toward the center of curvature. The parameters E_0 and b are constants with units of electric field and distance, respectively. It was shown in the text that particles entering the device from below will only emerge from the horizontal outlet if their speed satisfies:

$$v = \sqrt{\frac{qE_0 b}{m}}$$

If you equate this speed with the speed from the accelerating field, you find that the charge-to-mass ratio cancels out. This means there's no discrimination between isotopes. If one type of isotope can emerge, then they all can. The device doesn't work.

ASSESS   The problem with this device is that both the accelerating field and the curving field depend on the charge-to-mass ratio in the same way. One way to get around this is to accelerate the isotopes by heating them to high temperature. The speeds in this case won't depend on the charge. Another way is to use a magnetic field to curve the path of the isotopes. In this case, the charge-to-mass ratio doesn't cancel out, as we'll see in Example 26.2.

59.   **INTERPRET**   The charge orbiting the wire undergoes circular motion, and the centripetal force (Chapter 3) is provided by the Coulomb force. We are asked to find the line charge density, given the particle's orbital speed. This problem is similar to Problem 20.56.

DEVELOP   The electric field of the wire is radial and falls off like $1/r$ ($E = 2k\lambda/r$, see Example 20.7). For an attractive force (positive charge encircling a negatively charged wire), this is the same dependence as the centripetal acceleration (Equation 3.9, $a_c = v^2/r$). For circular motion around the wire, the Coulomb force provides the centripetal acceleration. Thus, Newton's second law gives

$$F = qE = ma_c$$

$$a_c = \frac{v^2}{r} = \frac{qE}{m} = -\frac{2kq\lambda}{mr}$$

The equation can be used to deduce the line charge density λ give the speed.

EVALUATE   The above equation gives

$$\lambda = -\frac{mv^2}{2kq} = -\frac{\left(6.8\times10^{-9}\ \text{kg}\right)\left(280\ \text{m/s}\right)^2}{2\left(9.0\times10^9\ \text{N}\cdot\text{m}^2/\text{C}^2\right)\left(2.1\times10^{-9}\ \text{C}\right)} = -14\ \mu\text{C/m}$$

ASSESS   For the force to be attractive, the line charge density must be negative.

61.   **INTERPRET**   This problem is about an electric dipole aligned with an external electric field. We are to find the dipole moment given the electric field and the amount of work needed to reverse the dipole's orientation.

DEVELOP   Using Equation 20.10, the energy required to reverse the orientation of such a dipole is

$$\Delta U = -pE\left(\cos\theta_f - \cos\theta_i\right) = -pE\left[\cos\left(180°\right) - \cos\left(0°\right)\right] = 2pE$$

EVALUATE   Using the equation above, the electric dipole moment is

$$p = \frac{\Delta U}{2E} = \frac{3.1\times10^{-27}\ \text{J}}{2\left(1.2\times10^3\ \text{N/C}\right)} = 1.3\times10^{-30}\ \text{C}\cdot\text{m}$$

ASSESS   An electric dipole tends to align itself in the direction of the external electric field. Thus, energy is required to change its orientation.

63.   **INTERPRET**   This problem is about the interaction between a dipole and the electric field due to a source charge.

DEVELOP   With the x axis in the direction from Q to $\vec{p}$ and the y axis parallel to the dipole in Figure 20.30, we have $\vec{p} = (2qa)\hat{j}$ and $E = (kQ/x^2)\hat{i}$. In the limit $x \gg a$, the torque on the dipole is given by Equation 20.9, $\vec{\tau} = \vec{p}\times\vec{E}$, where $\vec{E}$ is the field from the point charge Q, at the position of the dipole.

EVALUATE   **(a)** Using Equation 20.9, we find the torque to be

$$\vec{\tau} = \vec{p}\times\vec{E} = (2qa\hat{j})\times\left(\frac{kQ}{x^2}\hat{i}\right) = -\frac{2kQqa}{x^2}\hat{k}$$

The direction is into the page, or clockwise, to align $\vec{p}$ with $\vec{E}$.

(b) The Coulomb force obeys Newton's third law. The field of the dipole at the position of Q is (Example 20.5 adapted to new axes)

$$\vec{E}_{\text{dipole}} = -\frac{2kqa}{x^3}\hat{j}$$

Thus, the force on Q due to the dipole is

$$\vec{F}_{\text{on }Q} = Q\vec{E}_{\text{dipole}} = -\frac{2kQqa}{x^3}\hat{j}$$

The force on the dipole due to Q is the opposite of this:

$$\vec{F}_{\text{on dipole}} = -\vec{F}_{\text{on }Q} = \frac{2kQqa}{x^3}\hat{j}$$

The magnitude of $\vec{F}_{\text{on dipole}}$ is $2kQqa/x^3$.

(c) The direction of $\vec{F}_{\text{on dipole}}$ is in $+\hat{j}$, or parallel to the dipole moment.

ASSESS   The net force $\vec{F}_{\text{on dipole}}$ will cause the dipole to move in the $+\hat{j}$ direction. In addition, there is a torque that tends to align p with E. So, the motion of the dipole involves both translation and rotation.

65. **INTERPRET**   You're asked to estimate the charge distribution on two different molecules given their dipole moments.

DEVELOP   Your given the dipole moment, p, for H_2O and CO in debyes. A debye (D) is a unit with dimensions of "charge" times "distance." You can look in an outside reference and see that $1\,\text{D} = 3.34 \times 10^{-30}\,\text{C}\cdot\text{m}$. You can model these dipole molecules as two opposite charges, $\pm q$, separated by a distance d. Since the atomic separation for many molecules is about an Angstrom $(1\,\text{Å} = 10^{-10}\,\text{m})$, we can estimate how the charge is distributed on each molecule: $q = p/d$.

EVALUATE   Let's first convert the dipole moments to SI units:

$$p_{H_2O} = (1.85\,\text{D})\left(\frac{3.34 \times 10^{-30}\,\text{C}\cdot\text{m}}{1\,\text{D}}\right) = 6.18 \times 10^{-30}\,\text{C}\cdot\text{m}$$

$$p_{CO} = (0.12\,\text{D})\left(\frac{3.34 \times 10^{-30}\,\text{C}\cdot\text{m}}{1\,\text{D}}\right) = 4.00 \times 10^{-31}\,\text{C}\cdot\text{m}$$

Assuming the atoms in the molecules are separated by about 1 Angstrom, the amount of charge on each "atom" is

$$q_{H_2O} = \frac{p}{d} = \frac{6.18 \times 10^{-30}\,\text{C}\cdot\text{m}}{10^{-10}\,\text{m}}\left(\frac{1\,e}{1.6 \times 10^{-19}\,\text{C}}\right) = 0.4\,e$$

$$q_{CO} = \frac{p}{d} = \frac{4.00 \times 10^{-31}\,\text{C}\cdot\text{m}}{10^{-10}\,\text{m}}\left(\frac{1\,e}{1.6 \times 10^{-19}\,\text{C}}\right) = 0.03\,e$$

We've written the result in terms elementary charge.

ASSESS   In the case of water, the oxygen atom partially "steals" the electrons from the hydrogen atoms. That results in a negative fractional charge $(q \approx -0.4e)$ on the oxygen atom, and a correspondingly positive fractional charge on the hydrogen atoms. A similar situation occurs with carbon monoxide, but this time the carbon atom is the more electrophilic ("electron-loving") species in the covalent bond, so it will have the negative fractional charge and the oxygen atom will have the positive one.

67. **INTERPRET**   This problem involves three source charges positioned on a line. We are to find the electric field on this line to the right of the right-most charge and show that this expression reduces to κ/x^4 for $x \gg a$, where κ is a constant and a is the characteristic size of the source-charge distribution.

DEVELOP   The electric field of a single charge is give by Equation 20.3. Apply the principle of superposition to find the electric field due to three charges. This gives

$$\vec{E}(x) = k\hat{i}\left[\frac{q}{(x-a)^2} - \frac{2q}{x^2} + \frac{q}{(x+a)^2}\right]$$

EVALUATE   (a) Simplifying the expression above for the electric field gives

$$\vec{E}(x) = 2kqa^2\,\frac{(3x^2 - a^2)}{x^2(x^2 - a)^2}(\hat{i})$$

(b) For $x \gg a$, we can neglect the a compared to x^2. This gives

$$\vec{E}(x) \approx \frac{6kqa^2}{x^4}(\hat{i})$$

ASSESS   The quadrupole moment of this "linear quadrupole" is $Q_{xx} = 4qa^2$.

69.   **INTERPRET**   This problem is about the electric field due to a 10-m-long straight wire, which is our source charge.

DEVELOP   For a uniformly charged wire of length L and charge Q, the line density is $\lambda = Q/L$. Approximating the wire as infinitely long, the electric field due to the line charge can be written as (see Example 20.7)

$$E = \frac{2k\lambda}{r}$$

EVALUATE   **(a)** The charge density is

$$\lambda = \frac{Q}{L} = \frac{25 \ \mu C}{10 \ m} = 2.5 \ \mu C/m$$

(b) Since $r = 15$ cm $\ll 10$ m $= L$ and the field point is far from either end, we may regard the wire as approximately infinite. Then Example 20.7 gives

$$E = \frac{2k\lambda}{r} = \frac{2\left(9.0 \times 10^9 \ N \cdot m^2/C^2\right)\left(2.5 \ \mu C/m\right)}{0.15 \ m} = 300 \ kN/C$$

(c) At $r = 350$ m and $L = 10$ m , the wire behaves approximately like a point charge, so the field strength is

$$E = \frac{kQ}{r^2} = \frac{\left(9.0 \times 10^9 \ N \cdot m^2/C^2\right)\left(25 \times 10^{-6} \ C\right)}{(350 \ m)^2} = 1.8 \ N/C$$

ASSESS   The finite-size, line charge distribution looks like a point charge at large distances.

71.   **INTERPRET**   In this problem we want to find the electric field due to a uniformly charged disk of radius R.

DEVELOP   We take the disk to consist of a large number of annuli. With uniform surface charge density σ, the amount of charge on an area element dA is $dq = \sigma dA$. Our strategy is to first calculate the electric field dE due to dq at a field point on the axis, simplify with symmetry argument, and then integrate over the entire disk to get E.

EVALUATE   **(a)** The area of an annulus of radii $R_1 < R_2$ is just $\pi(R_2^2 - R_1^2)$. For a thin ring, $R_1 = r$ and $R_2 = r + dr$, so the area is $\pi[(r + dr)^2 - r^2] = \pi(2rdr + dr^2)$. When dr is very small, the square term is negligible, and $dA = 2\pi rdr$. (This is equal to the circumference of the ring times its thickness.) **(b)** For surface charge density σ, $dq = \sigma dA = 2\pi\sigma rdr$. **(c)** From Example 20.6, the electric field due to a ring of radius r and charge dq is

$$dE_x = k\frac{xdq}{\left(x^2 + r^2\right)^{3/2}} = \frac{2\pi k\sigma xr}{\left(x^2 + r^2\right)^{3/2}}dr$$

which holds for x positive away from the ring's center. **(d)** Integrating from $r = 0$ to R, one finds

$$E_x = \int_0^R dE_x = 2\pi k\sigma x \int_0^R \frac{rdr}{\left(x^2 + r^2\right)^{3/2}} = 2\pi k\sigma x \frac{-1}{\sqrt{x^2 + r^2}}\bigg|_0^R = 2\pi k\sigma\left[\frac{x}{|x|} - \frac{x}{(x^2 + R^2)^{1/2}}\right]$$

For $x > 0$, $|x| = x$ and the field is

$$E_x = 2\pi k\sigma\left(1 - \frac{x}{\sqrt{x^2 + R^2}}\right)$$

On the other hand, for $x < 0$, $|x| = -x$, the electric field is

$$E_x = 2\pi k\sigma\left(-1 + \frac{|x|}{\sqrt{x^2 + R^2}}\right)$$

This is consistent with symmetry on the axis, since $E_x(x) = -E_x(-x)$.

ASSESS   One may readily verify that (see Problem 71), for $x \gg R$, $E_x \approx \frac{kQ}{x^2}$. In other words, the finite-size charge distribution looks like a point charge at large distances.

73. **INTERPRET** In this problem we want to show that at large distances, the electric field due to a uniformly charged disk of radius R reduces to that of a point charge.

DEVELOP The result of Problem 71 for the field on the axis of a uniformly charged disk, of radius R, at a distance $x > 0$ on the axis (away from the disk's center) is

$$E_x = 2\pi k\sigma\left(1 - \frac{x}{\sqrt{x^2 + R^2}}\right)$$

For $R^2/x^2 \ll 1$, we use the binomial expansion in Appendix A and write

$$\left(1 + \frac{R^2}{x^2}\right)^{-1/2} \approx 1 - \frac{1}{2}\frac{R^2}{x^2} +$$

EVALUATE Substituting the above expression into the first equation, we obtain

$$E_x = 2\pi k\sigma\left[1 - \left(1 + \frac{R^2}{x^2}\right)^{-1/2}\right] \approx 2\pi k\sigma\left[1 - \left(1 - \frac{1}{2}\frac{R^2}{x^2} + \cdots\right)\right] \approx \frac{2\pi k\sigma R^2}{2x^2} = \frac{kQ}{x^2}$$

which is the field from a point charge $Q = (\pi R^2)\sigma = \sigma A$ at a distance x.

ASSESS The result once again demonstrates that any finite-size charge distribution looks like a point charge at large distances.

75. **INTERPRET** We are to find the position of the charge Q in Example 20.2 for which the force is a maximum. At large distances, the force will be small because of the inverse-square nature of the force. At close distances, the net force will be small because the forces from the two charges tend to cancel. Somewhere between near and far will be a maximum.

DEVELOP The equation for force is found in the example, so we will differentiate to find the value of y where force is a maximum. We are given the equation for force:

$$\vec{F} = \frac{2kqQy}{(a^2 + y^2)^{3/2}}\hat{j}$$

We find the value of y at which this force is a maximum by setting $dF/dy = 0$.

EVALUATE

$$0 = \frac{dF}{dy} = \frac{d}{dy}\left[\frac{2kqQy}{(a^2+y^2)^{3/2}}\right] = 2kqQ\frac{d}{dy}\left[\frac{y}{(a^2+y^2)^{3/2}}\right] = 2kqQ\left[\frac{1}{(a^2+y^2)^{3/2}} - \frac{3}{2}\frac{2y^2}{(a^2+y^2)^{5/2}}\right]$$

$$= \frac{(a^2+y^2) - 3y^2}{(a^2+y^2)^{5/2}}$$

$$= (a^2+y^2) - 3y^2$$

$$a^2 = 2y^2$$

$$y = \frac{a}{\sqrt{2}}$$

ASSESS This is a bit less than one and a half times the distance from the center to one charge.

77. **INTERPRET** We are to find the electric field near a line of *non-uniform* charge density. This is an electric field calculation, and we will integrate to find the field.

DEVELOP The rod has charge density $\lambda = \lambda_0(\frac{x}{L})^2$, and extends from $x = 0$ to $x = L$. We want to find the electric field at $x = -L$. We will use $d\vec{E} = \frac{k\,dq}{r^2}\hat{r}$, with $dq = \lambda\,dx$ and $r = x + L$.

EVALUATE

$$dE = \frac{k\,dq}{r^2}\hat{r}$$

$$\vec{E} = k\hat{i}\int_0^L \frac{\lambda_0\left(\frac{x}{L}\right)^2}{(x+L)^2}\,dx = \frac{k\lambda_0}{L^2}\hat{i}\left[x - \frac{L^2}{x+L} - 2L\ln(x+L)\right]_0^L$$

$$= \frac{k\lambda_0\hat{i}}{L^2}\left[L - \frac{L^2}{2L} - 2L\ln\left(\frac{2L}{L}\right) - L\right] = \frac{k\lambda_0\hat{i}}{L^2}\left[-\frac{L}{2} - 2L\ln(2)\right] = -\frac{k\lambda_0\hat{i}}{L}\left[\frac{1}{2} + 2\ln(2)\right]$$

ASSESS   Since λ_0 is charge per length, the units are correct.

79. **INTERPRET**   An electric field is used to deflect ink drops in an ink-jet printer. You need to find the maximum electric field for which the ink drops still can exit the deflection device. This is a problem of projectile motion where the dynamics are controlled by the electric force, not the gravitational force.

DEVELOP   The time that it takes the ink drop to traverse the field region in the Figure 20.35 is: $t = L/v$. During that time it will undergo acceleration from the electric field: $a = qE/m$. Since the field is uniform, this acceleration is constant, so the amount of vertical deflection will be: $\Delta y = v_{y0}t + \frac{1}{2}at^2$. The drop initially has no vertical velocity, so $v_{y0} = 0$. The maximum field is that which deflects the drop by $|\Delta y| = d/2$, since the drop starts off in the middle between the two plates.

EVALUATE   Solving for the maximum field magnitude,

$$E_{max} = \frac{mdv^2}{qL^2}$$

ASSESS   You can check that this has the right units:

$$[E_{max}] = \left[\frac{mdv^2}{qL^2}\right] = \frac{kg\cdot m\cdot m/s^2}{C\cdot m^2} = N/C$$

Indeed, these are the right units for an electric field.

81. **INTERPRET**   We are considering the electric fields that operate in the heart muscle.

DEVELOP   The magnitude of the dipole field on a line that bisects the dipole axis is $E = kp/r^3$, from Equations 20.6a. Whereas, the magnitude of the dipole field along the dipole axis is $E = 2kp/r^3$, from Equations 20.6b. So the field is twice as large along the dipole axis.

EVALUATE   The extension of the line in Figure 20.36c bisects the dipole axes of all the dipoles in the heart muscle. A line perpendicular to this one will approximately correspond to the axes of all the dipoles. So the field on the extension should be weaker than the field on the perpendicular.

The answer is (a).

ASSESS   At the same distance from the heart, the field on the extension of the line in Figure 20.36c is the weakest compared to other directions. That's because the field contribution from the positive and negative charges are equal and approximately opposite.

83. **INTERPRET**   We are considering the electric fields that operate in the heart muscle.

DEVELOP   Equations 20.6b shows that the electric field on the dipole axis is parallel to the axis and points in the same direction as the dipole moment, i.e. in the direction from the negative charge to the positive charge.

EVALUATE   Inside the heart above and below the line in Figure 20.36c, the electric field should point in the same direction as the dipole moment.

The answer is (a).

ASSESS   If we had been asked about the internal field of the dipoles, the answer would have been opposite the direction of the dipole moment (see Figure 20.24). But this would only comprise a small sliver of the heart area. The majority of the electric field points in the dipole moment's direction.

GAUSS'S LAW

EXERCISES

Section 21.1 Electric Field Lines

17. **INTERPRET** This problem involves associating electric field lines to charges. Given the electric field lines and the magnitude of a single charge, we are to find the net charge.

DEVELOP The number of field lines emanating from (or terminating on) the positive (or negative) charges is the same (i.e., 14), the charges have the same magnitude. The field lines are pointing toward the middle charge, which means that a positive charge placed in this field will experience an force attracting it to the central charge. Thus, the central charge must be negative (i.e.−3 μC). The field lines point outward from the outer charges, so they are positive charges. Because the same number of field lines emanate from the outer charges as from the central charge, they must have the same magnitude, so the outer charges are +3 μC. Sum the charges to find the net charge.

EVALUATE The net charge is thus +3 μC + 3 μC − 3 μC = +3 μC.

ASSESS If the magnitude of the central charge were less (greater) than 3 μC then fewer (more) lines would terminate on this charge.

19. **INTERPRET** In this problem we are asked to identify the charges based on the pattern of the field lines and the given net charge.

DEVELOP From the direction of the lines of force (away from positive and toward negative charge) one sees that A and C are positive charges and B is a negative charge. Eight lines of force terminate on B, eight originate on C, but only four originate on A, so the magnitudes of B and C are equal, while the magnitude of A is half that value.

EVALUATE Based on the reasoning above, we may write $Q_C = -Q_B = 2Q_A$. The total charge is $Q = Q_A + Q_B + Q_C = Q_A$, so $Q_C = 2Q = -Q_B$.

ASSESS The magnitude of the charge is proportional to the number of field lines emerging from or terminating at the charge.

Section 21.2 Electric Flux and Field

21. **INTERPRET** This problem involves finding the electric field strength, given the flux through a given surface.

DEVELOP The flux is given by Equation 21.1,

$$\Phi = \vec{E} \cdot \vec{A} = EA \cos \theta$$

where $\vec{A}$ is a vector whose magnitude is the surface area and whose direction is normal (i.e., perpendicular) to the surface, and θ is the angle between $\vec{A}$ and $\vec{E}$. If the surface is perpendicular to the electric field, than $\theta = 0°$ or $180°$, so $\cos \theta = \pm 1$. The magnitude of the flux through such a surface is thus

$$|\Phi| = EA$$

EVALUATE From the equation above, we find the field strength to be

$$E = \frac{|\Phi|}{A} = \frac{65 \text{ N} \cdot \text{m}^2/\text{C}}{\left(1.0 \times 10^{-2} \text{ m}\right)^2} = 650 \text{ kN/C}$$

ASSESS We see that when $\theta = 0°$, $\Phi = E/A$ is a maximum, and when $\theta = 180°$, $\Phi = -E/A$ is a minimum.

23. **INTERPRET** This problem involves calculating the electric flux through the surface of a sphere.

DEVELOP   The general expression for the electric flux Φ is given by Equation 21.1: $\Phi = \vec{E} \cdot \vec{A} = EA \cos\theta$, where θ is the angle between the normal vector $\vec{A}$ and the electric field $\vec{E}$. The magnitude of the normal vector $\vec{A}$ is the surface area of the sphere: $A = 4\pi r^2$.

EVALUATE   For a sphere, with $\vec{E}$ parallel or antiparallel to $\vec{A}$, Equation 21.1 gives

$$\Phi = \pm 4\pi r^2 E = \pm 4\pi \left(\frac{0.10}{2}\ \text{m}\right)^2 (47\ \text{kN/C}) = \pm 1.5\ \text{kN} \cdot \text{m}^2/\text{C}$$

ASSESS   The flux Φ is positive if $\vec{A}$ points outward, and is negative if $\vec{A}$ points inward.

Section 21.3 Gauss's Law

25.   **INTERPRET**   This problem is about applying Gauss's law to find the electric flux through a closed surface.

DEVELOP   Gauss's law is given in Equation 21.3 and states that the flux through any closed surface is proportional to the charge enclosed:

$$\Phi = \oint \vec{E} \cdot d\vec{A} = \frac{q_{\text{enclosed}}}{\epsilon_0}$$

EVALUATE   For the surfaces shown, the results are as follows:

(a) $q_{\text{enclosed}} = q + (-2q) = -q \implies \Phi = -q/\epsilon_0$
(b) $q_{\text{enclosed}} = q + (-2q) + (-q) + 3q + (-3q) = -2q \implies \Phi = -2q/\epsilon_0$
(c) $q_{\text{enclosed}} = 0 \implies \Phi = 0$
(d) $q_{\text{enclosed}} = 3q + (-3q) = 0 \implies \Phi = 0$

ASSESS   The flux through the closed surface depends only on the charge enclosed, and is independent of the shape of the surface.

27.   **INTERPRET**   This problem involves applying Gauss's law to find the electric flux through the surface of a cube that encloses a given charge.

DEVELOP   Gauss's law, given in Equation 21.3, states that the flux through any closed surface is proportional to the charge enclosed:

$$\Phi = \oint \vec{E} \cdot d\vec{A} = \frac{q_{\text{enclosed}}}{\epsilon_0}$$

The symmetry of the situation guarantees that the flux through one face is $1/6$ the flux through the whole cubical surface.

EVALUATE   The flux through one face of a cube is

$$\Phi_{\text{face}} = \frac{1}{6} \oint_{\text{cube}} \vec{E} \cdot d\vec{A} = \frac{1}{6} \frac{q_{\text{enclosed}}}{\epsilon_0} = \frac{2.6\ \mu\text{C}}{6\left[8.85 \times 10^{-12}\ \text{C}^2/\left(\text{N} \cdot \text{m}^2\right)\right]} = 49\ \text{kN} \cdot \text{m}^2/\text{C}$$

ASSESS   Because the flux through each surface is the same, the total flux through the cube is simply $\Phi = 6\Phi_{\text{face}}$, which is proportional to q_{enclosed}.

Section 21.4 Using Gauss's Law

29.   **INTERPRET**   This problem involves applying Gauss's law to calculate the electric field. Our charge distribution has spherical symmetry.

DEVELOP   The charge distribution is exactly that considered in Example 21.1. This example derives Equations 21.4 and 21.5, which express the strength of the electric field inside and outside the sphere of radius $R = 25$ cm. The result is

$$E = \begin{cases} \dfrac{1}{4\pi\epsilon_0} \dfrac{Qr}{R^3}, & r \leq R \\[2ex] \dfrac{1}{4\pi\epsilon_0} \dfrac{Q}{r^2}, & r \geq R \end{cases}$$

which we can evaluate at the points given.

EVALUATE **(a)** At $r = 15$ cm, (i.e., inside the sphere), we can use the upper formula, which gives

$$E = \frac{1}{4\pi\epsilon_0}\frac{Qr}{R^3} = \frac{(14 \ \mu C)(0.15 \ m)}{4\pi\left[8.85\times10^{-12} \ C^2/(N\cdot m^2)\right](0.25 \ m)^3} = 1.2 \ MN/C$$

(b) At $r = R$, we can use either the upper or the lower formula. The result is

$$E = \frac{1}{4\pi\epsilon_0}\frac{Q}{R^2} = \frac{(14 \ \mu C)}{4\pi\left[8.85\times10^{-12} \ C^2/(N\cdot m^2)\right](0.25 \ m)^2} = 2.0 \ MN/C$$

(c) At $r = 2R = 50$ cm (i.e., outside the sphere), we use the lower formula, which gives

$$E = \frac{1}{4\pi\epsilon_0}\frac{Q}{(2R)^2} = \frac{14 \ \mu C}{4\pi\left[8.85\times10^{-12} \ C^2/(N\cdot m^2)\right](50 \ cm)^2} = 50\times10^4 \ N/C$$

ASSESS Inside the solid sphere where $r < R$, the electric field increases linearly with r. On the other hand, outside the sphere where $r > R$, the field strength decreases as $1/r^2$. Gauss's law can be applied in this problem because the charge configuration is spherically symmetric.

31. **INTERPRET** In this problem we are given the field strength at two different points outside the charge distribution and asked to determine the symmetry possessed by the configuration.

    DEVELOP The symmetry of the charge distribution can be determined by noting that the electric field strength decreases as $1/r^2$ for a spherically symmetric charge distribution, and as $1/r$ for a line charge (see Examples 21.4 and 21.1).

    EVALUATE We write $E = Cr^{-n}$, for some constant C. This gives

    $$\frac{E_2}{E_1} = \left(\frac{r_1}{r_2}\right)^n \quad \Rightarrow \quad \ln\left(\frac{E_2}{E_1}\right) = n\ln\left(\frac{r_1}{r_2}\right)$$

    Inserting the values given, we obtain

    $$n = \frac{\ln(E_2/E_1)}{\ln(r_2/r_1)} = \frac{\ln(55/43)}{\ln(23/18)} = 1.00$$

    Thus, we conclude that the charge distribution possesses line symmetry.

    ASSESS The $1/r$ dependence characteristic of line symmetry can be readily verified by taking the field strength to be of the form $E = 2k\lambda/r$.

33. **INTERPRET** We are to find the electric field produced by a uniformly charged sheet. This problem has planar symmetry, so we will use the result for the electric field near an infinite plane of charge (Example 21.6 and Equation 21.7).

    DEVELOP Example 21.6 derives the electric field near a uniform sheet of charge with charge density σ. The result is (Equation 21.7)

    $$E = \frac{\sigma}{2\epsilon_0}$$

    The charge density given in the problem is $\sigma = 87 \ pC/cm^2 = 87\times10^{-12} \ C/(0.01 \ m)^2$.

EVALUATE Inserting the given quantities into the expression above gives

$$E = \frac{\sigma}{2\epsilon_0} = \frac{\left(87\times10^{-12}\ \text{C}\right)/(0.01\ \text{m})^2}{2\left[8.85\times10^{-12}\ \text{C}^2/\left(\text{N}\cdot\text{m}^2\right)\right]} = 49\times10^3\ \text{N/C}$$

ASSESS Although this seems at first to be a small charge per area, the resulting field is quite large. Remember that a Coulomb is a very large charge.

Section 21.5 Field of Arbitrary Charge Distribution

35.   **INTERPRET** We are given a charge distribution with approximate line symmetry, so we can apply Gauss's law to compute the electric field. We are to consider two cases: the first at a distance that is small compared to the length of the rod and the second at a distance much, much greater than the length of the rod.

DEVELOP Close to the rod, but far from either end, the rod appears infinite, so the electric field strength is (see Example 21.4) $E = 2k\lambda/r$. For part (b), we are considering a position $r = 23$ m that is over an order of magnitude larger than the size of the rod (0.50 m). Therefore, the rod may be treated as a point charge, so the electric field will be given by Equation 20.3.

EVALUATE **(a)** Substituting the values given in the problem statement, we obtain

$$E = \frac{2k\lambda}{r} = \frac{2kQ/l}{r} = \frac{2\left(9.0\times10^9\ \text{N}\cdot\text{m}^2/\text{C}^2\right)(2.0\ \mu\text{C})/(0.50\ \text{m})}{0.014\ \text{m}} = 5.1\times10^6\ \text{N/C}$$

(b) Far away $(r \gg L)$, the rod appears like a point charge, so

$$E \approx \frac{kq}{r^2} = \frac{\left(9.0\times10^9\ \text{N}\cdot\text{m}^2/\text{C}^2\right)(2.0\ \mu\text{C})}{(23\ \text{m})^2} = 34\ \text{N/C}$$

ASSESS At distance much, much greater than the characteristic size of any charge distribution, (line charge in this case), the field always resembles that of a point charge.

37.   **INTERPRET** We are to approximate the electric field strength near the center of a flat, charged disk and far from the charged disk. In both cases, we will choose appropriate approximations.

DEVELOP The area of the disk is given as $A = 0.14\ \text{m}^2$, so the radius of the disk must be

$$\pi r^2 = A \quad\Rightarrow\quad r = \sqrt{\frac{A}{\pi}} = 21\ \text{cm}.$$

The point $r = 0.1$ cm from the center of the disk is 2 orders of magnitude smaller than the disk radius, so the disk will appear to be an infinite plane. Thus, we can use the result of Example 21.6,

$$E = \frac{\sigma}{2\epsilon_0} = \frac{Q}{2\epsilon_0 A}$$

The point $r = 250$ cm from the disk is one order of magnitude larger than the disk radius, so the disk will look more like a point charge and we can use Equation 20.3

$$E = \frac{kQ}{r^2}$$

The charge on the disk is $Q = 5.0\mu\text{C}$.

EVALUATE **(a)** Inserting the given quantities into the expression for a flat plane gives

$$E = \frac{Q}{2\epsilon_0 A} = \frac{5.0\ \mu\text{C}}{2\left[8.85\times10^{-12}\ \text{C}^2/\left(\text{N}\cdot\text{m}^2\right)\right](0.14\ \text{m})^2} = 2.0\times10^6\ \text{N/C}$$

(b) Inserting the given quantities into the expression for the point charge gives

$$E = \frac{kQ}{r^2} = \frac{\left(9.0\times10^9\ \text{N}\cdot\text{m}^2/\text{C}^2\right)(5.0\ \mu\text{C})}{(2.5\ \text{m})^2} = 7.2\times10^3\ \text{N/C}$$

ASSESS   If you are far enough from anything, it looks like a point charge; and if you are close enough, it looks like an infinite plane.

Section 21.6 Gauss's Law and Conductors

39.   **INTERPRET**   This problem involves finding the charge distribution of a conductor using Gauss's law. The charge distribution is spherically symmetric.

DEVELOP   As explained in Section 21.6, the electric field inside a conductor of arbitrary geometry is zero. In addition, the net charge must reside on the conductor surface.

EVALUATE   **(a)** Because the net charge inside the conductor is zero, the volume charge density inside the conductor is also zero.

(b) If the sphere is electrically isolated, the charge will be uniformly distributed (i.e., spherically symmetric), so the surface charge density is just the total charge divided by the surface area of the sphere. This gives

$$\sigma = \frac{Q}{4\pi R^2} = \frac{5.0\ \mu C}{4\pi (0.010\ \text{m})^2} = 4.0 \times 10^{-3}\ \text{C/m}^2$$

ASSESS   Since charges are mobile, the presence of other charges near the conductor will cause the charges on the surface to move, so the equilibrium charge distribution will not be spherically symmetric.

41.   **INTERPRET**   This problem is about finding the electric field near the surface of a conducting plate. The approximate plane symmetry of the system allows us to make use of Gauss's law.

DEVELOP   The net charge of $Q = 18\ \mu C$ must distribute itself over the outer surface of the plate, in accordance with Gauss's law for conductors (see Section 21.6). The outer surface of the plate consists of two plane square surfaces on each face, plus the edges and corners. Symmetry arguments imply that for an isolated plate, the charge density on each face is the same, but not necessarily uniform because the edges and corners also have charge. If the thickness of the plate is much, much less than its length and width, we can assume that the edges and corners have negligible charge and that the density on the faces is approximately uniform. With this assumption, the surface charge density is the total charge divided by the area of both faces,

$$\sigma = \frac{Q}{2A} = \frac{18\ \mu C}{2(0.75\ \text{m})^2} = 16.0\ \mu C$$

Given this surface charge density, we can apply Equation 21.8 to find the electric field near the surface (but not near an edge) of the conducting plate.

EVALUATE   Inserting the surface charge density into Equation 21.8, we find the field strength near the plate to be

$$E = \frac{\sigma}{2\epsilon_0} = \frac{16\ \mu C}{8.85 \times 10^{-12}\ \text{C}^2/\left(\text{N} \cdot \text{m}^2\right)} = 1.8\ \text{MN/C}$$

ASSESS   Note the distinction between a charged conducting plate and a uniformly charged plate. In the latter, charges are not free to move and the electric field is (see Example 21.6) $E = \sigma/(2\epsilon_0)$.

PROBLEMS

43.   **INTERPRET**   This problem is about finding the electric flux through a given surface.

DEVELOP   The electric flux through a surface is given by Equation 21.2:

$$\Phi = \int_{\text{surface}} \vec{E} \cdot d\vec{A}$$

Since the electric field depends only on y, we break up the square into strips of area $d\vec{A} = \pm a\ dy\ \hat{k}$ of length a parallel to the x axis and width dy (see figure below). The normal to the surface is $\pm \hat{k}$.

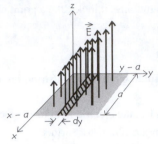

EVALUATE   The integral of Equation 21.2 gives

$$\Phi = \int_{\text{surface}} \vec{E} \cdot d\vec{A} = \int_0^a (E_0 y/a)\hat{k} \cdot (\pm a dy\, \hat{k}) = \pm E_0 \int_0^a y\, dy = \pm \frac{1}{2} E_0 a^2$$

ASSESS   Our result can be compared to the case where the field strength is constant. In that case, the flux through the surface would be $\Phi = \pm E_0 a^2$.

45.   INTERPRET   We're asked to estimate the electric field outside red blood cells of two different species, given the enclosed charges and the cell radii.

DEVELOP   We are told to approximate the red blood cells as perfect spheres. The charge from an excess of electrons $(Q = Ne)$ is spread uniformly around the surface. The symmetry implies Gauss's law can be used to find the electric field. Outside a charged sphere, the electric field is $E = kQ/r^2$, which is the same as for a point charge at the origin.

EVALUATE   For rabbit RBCs:

$$E = \frac{kQ}{r^2} = \frac{\left(9.0\times10^9 \;\frac{\text{N·m}^2}{\text{C}^2}\right)\left(4.4\times10^6\right)\left(1.6\times10^{-19}\,\text{C}\right)}{\left(30\times10^{-6}\,\text{m}\right)^2} = 7.0 \text{ MN/C}$$

For human RBCs:

$$E = \frac{kQ}{r^2} = \frac{\left(9.0\times10^9 \;\frac{\text{N·m}^2}{\text{C}^2}\right)\left(15\times10^6\right)\left(1.6\times10^{-19}\,\text{C}\right)}{\left(36\times10^{-6}\,\text{m}\right)^2} = 17 \text{ MN/C}$$

ASSESS   Because the charge is from electrons, the electric fields point in towards the center of the cells.

47.   INTERPRET   This problem involves a uniform, spherically symmetric charge distribution. We are to find the second location where the electric field has the given value, and find the net charge on the sphere. Doing so will involve Gauss's law.

DEVELOP   Example 21.1 uses Gauss's law to derive expressions for the electric fields inside and outside a spherical charge distribution. We are given the field strength inside the sphere (at $r_1 = R/2$), so we can equate it to the field strength outside the sphere at r_2 and solve for the position. Explicitly, we have

$$E_{\text{inside}} = \frac{Qr_1}{4\pi\epsilon_0 R^3} = E_{\text{outside}} = \frac{Q}{4\pi\epsilon_0 r_2^2}$$

$$\frac{1}{2R^2} = \frac{1}{r_2^2}$$

To find the net charge on the sphere, we can use the expression for E_{inside} and solve for Q.

EVALUATE   (a) Solving the expression above for r_2 gives $r_2 = R\sqrt{2} = (2.0 \text{ cm})\sqrt{2} = 2.8 \text{ cm}$

(b) Solving for the total charge Q gives

$$Q = \frac{4\pi\epsilon_0 R^3 E_{\text{inside}}}{r_1} = 8\pi\epsilon_0 R^2 E_{\text{inside}} = 8\pi\left[8.85\times10^{-12} \;\text{C}^2/\left(\text{N·m}^2\right)\right](0.020 \text{ m})^2 (39 \text{ kN/C}) = 3.5 \text{ nC}$$

ASSESS   For part (a), we assumed that the second location at which the field had the given value was outside the sphere. Were this not true, we would have found an unphysical answer (i.e., an imaginary field).

49. **INTERPRET**   You're helping your friend determine what charge is needed on a square plate to obtain the desired field at a point near the plate's center. The situation has planar symmetry.

DEVELOP   The field strength is specified at a distance from the plate, $y = 10$ cm, which is considerably less than the size of the square plate, $L = 4.5$ m. Therefore, we can assume that the field lines are essentially perpendicular to the plate, which means the electric field strength is given by Gauss's law: $E = \sigma / 2\epsilon_0$. The total charge on the plate will be $Q = \sigma L^2$.

EVALUATE   Solving for the total charge:

$$Q = \sigma L^2 = 2\epsilon_0 E L^2 = 2\left(8.85\times10^{-12}\ \tfrac{C^2}{N\cdot m^2}\right)(430\ \text{N/C})(4.5\ \text{m})^2 = 0.15\mu C$$

ASSESS   The answer doesn't depend on the distance y, but as one moves farther from the plate, the assumption that the field lines are all pointing in the direction perpendicular to the sheet will no longer be true. In fact, for $y \gg L$, the plate will seem like a point and the field will go as $1/y^2$.

51. **INTERPRET**   The charge distribution has spherical symmetry, so we can apply Gauss's law to find the electric field. In addition, it is composed of two distinct charge distributions, so we can consider each one separately and use the superposition principle to construct the net electric field. Note that this problem is exactly the same as Problem 21.50, so we will defer here to that solution.

DEVELOP   See the solution to Problem 21.50 for the derivation. The total electric field is the superposition of the fields due to the point charge and the spherical shell. The field is spherically symmetric about the center.

EVALUATE   **(a)** The field due to the shell is zero inside, so at $r = 5$ cm, the field is due to the point charge only. Thus,

$$E = \frac{kq}{r^2} = \frac{\left(9.0\times10^9\ \text{N}\cdot\text{m}^2/\text{C}^2\right)(1.0\ \mu C)}{(0.05\ \text{m})^2} = 3.6\times10^6\ \text{N/C}$$

The field points radially outward.

(b) Outside the shell, its field is like that of a point charge with a total charge equal to the sum of the charge on the sphere and the point charge, so at $r = 45$ cm, the field strength is

$$E = \frac{k(q+Q)}{r^2} = \frac{\left(9.0\times10^9\ \text{N}\cdot\text{m}^2/\text{C}^2\right)(86\ \mu C)}{(0.45\ \text{m})^2} = 3.8\times10^6\ \text{N/C}$$

The direction of the field is radially outward.

(c) If the charge on the shell were doubled, the charge enclosed inside the sphere would not change, so Gauss's law (Equation 21.3) tells us that the electric field inside the sphere would not change. However, the field outside would be

$$E = \frac{k(1.0\ \mu C + 2\times85\ \mu C)}{(45\ \text{cm})^2} = 7.6\ \text{MN/C}$$

which is essentially doubled (to two significant figures).

ASSESS   By Gauss's law, the shell produces no electric field in its interior. The field outside a spherically symmetric distribution is the same as if all the charges were concentrated at the center of the sphere.

53. **INTERPRET**   We are given a charge distribution with approximate line symmetry (provided we consider distances from the wire that are much, much smaller than the length of the wire, and provided we are not near the end of the wire). We can use this symmetry and Gauss's law to compute the electric field.

DEVELOP   With the assumption that the electric field is approximately that from an infinitely long, line symmetric charge distribution, we can use the result of Example 21.4 (i.e., Equation 21.6) to express the electric field near the wire:

$$E = \frac{\lambda_{\text{enclosed}}}{2\pi\epsilon_0 r}$$

where $\lambda_{\text{enclosed}}$ is the charge per unit length enclosed by a cylinder of radius r. We can evaluate this expression to find the electric field at different distances r from the wire.

EVALUATE   **(a)** For $r = 0.50$ cm $= 0.0050$ m, which is between the wire and the pipe, the enclosed charge per unit length is $\lambda_{\text{enclosed}} = \lambda_{\text{wire}}$, and

$$E = \frac{\lambda_{enclosed}}{2\pi\epsilon_0 r} = \frac{5.6 \text{ nC/m}}{2\pi\left[8.85\times10^{-12} \text{ C}^2/(\text{N}\cdot\text{m}^2)\right](0.0050 \text{ m})} = 20 \text{ kN/C}$$

The field is (positive) radially away from the axis of symmetry (i.e., from the wire).

(b) For r = 1.5 cm = 0.015 m, the enclosed charge per unit length is $\lambda'_{enclosed} = \lambda_{wire} + \lambda_{pipe}$, and

$$E' = \frac{\lambda'_{enclosed}}{2\pi\epsilon_0 r} = \frac{5.6 \text{ nC/m} - 4.2 \text{ nC/m}}{2\pi\left[8.85\times10^{-12} \text{ C}^2/(\text{N}\cdot\text{m}^2)\right](0.015 \text{ m})} = 1.7 \text{ kN/C}$$

The field is in the same direction as the field in part **(a)**.

ASSESS   Between the wire and the pipe, the enclosed charge per unit length is $\lambda_{enclosed} = \lambda_{wire}$, whereas outside the pipe, the enclosed charge is $\lambda'_{enclosed} = \lambda_{wire} + \lambda_{pipe}$. Since the pipe and the wire carry opposite charges, $\lambda'_{enclosed} < \lambda_{enclosed}$, so $E' < E$, as we have shown.

55.   **INTERPRET**   Because we are not concerned with the electric field near the end of the wire, but only with the field in the central part of the wire, we can consider that the given charge distribution has line symmetry (and this is true only for distances r that are much, much less than the length of the wire). Thus, we will apply Gauss's law to compute the electric field near the wire.

DEVELOP   We assume that the rod is long enough and approximate its field using line symmetry. This is the situation dealt with in Example 21.4, so we use that result:

$$E = \frac{\lambda_{enclosed}}{2\pi\epsilon_0 r}$$

where $\lambda_{enclosed}$ is the charge per unit length inside a cylindrical Gaussian surface of radius r about the symmetry axis. Since the charge is uniformly distributed throughout the solid rod, the line charge density is simply equal to the volume charge density times the cross-sectional area of the rod: $\lambda_{enclosed} = \rho A = \rho \pi r^2$. Combining the two equations yields

$$E = \frac{\lambda_{enclosed}}{2\pi\epsilon_0 r} = \frac{\rho\pi r^2}{2\pi\epsilon_0 r} = \frac{\rho r}{2\epsilon_0}$$

This equation (valid for $r \le R$, where R is the radius of the rod) allows us to solve for ρ.

EVALUATE   With the electric field at r = R given, we find the volume charge density to be

$$\rho = \frac{2\epsilon_0 E}{R} = \frac{2\epsilon_0 E}{R} = \frac{2\left[8.85\times10^{-12} \text{ C}^2/(\text{N}\cdot\text{m}^2)\right](16 \text{ kN/C})}{0.045 \text{ m}} = 6.3 \text{ μC/m}^3$$

This is the magnitude of ρ, since the direction of the field at the surface (i.e., radially inward or outward) was not specified.

ASSESS   If we were given the field strength at a point $r > R$, then

$$E = \frac{\lambda_{enclosed}}{2\pi\epsilon_0 r} = \frac{(\rho\pi R^2)}{2\pi\epsilon_0 r}$$

and the volume charge density would be

$$\rho = \frac{2\pi\epsilon_0 E r}{\pi R^2} = \frac{2\epsilon_0 E r}{R^2}$$

57.   **INTERPRET**   The infinitely large slab has plane symmetry, and we can apply Gauss's law to compute the electric field.

DEVELOP   When we take the slab to be infinitely large, the electric field is everywhere normal to the slab's surface and symmetrical about the center plane. We follow the approach outlined in Example 21.6 to compute the electric field. As the Gaussian surface, we choose a box that has area A on its top and bottom and that extends a distance x both up and down from the center of the slab. See figure below.

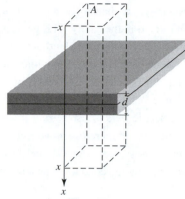

EVALUATE   (a) For points inside the slab $|x| \le d/2$, the charge enclosed by our Gaussian box is

$$q_{\text{enclosed}} = \rho V_{\text{enclosed}} = \rho A(2x)$$

Thus, Gauss's law gives

$$\Phi = \int \vec{E} \cdot d\vec{A} = E(2A) = \frac{q_{\text{enclosed}}}{\epsilon_0} \quad \rightarrow \quad E = \frac{\rho x}{\epsilon_0}$$

The direction of $\vec{E}$ is away from (toward) the central plane for positive (negative) charge density.
(b) For points outside the slab $|x| > d/2$, the enclosed charge is

$$q_{\text{enclosed}} = \rho V_{\text{enclosed}} = \rho A d$$

Applying Gauss's law again gives

$$E = \frac{\rho d}{2\epsilon_0}$$

ASSESS   Inside the slab, the charge distribution is equivalent to a sheet with $\sigma = 2\rho x$. On the other hand, outside the slab, it is equivalent to a sheet with $\sigma = \rho d$.

59.   **INTERPRET**   At distances much, much less than 75 cm from the plate, and not near the edges of the plate, it can be approximated by an infinite plane, so we can apply Gauss's law for plane symmetry. For distances much, much greater than 75 cm, the plate looks like a point charge. We can use these limiting cases to find the electric field strength 15 m from the plate.
DEVELOP   Close to the nonconducting plate $(x = 1 \text{ cm} \ll 75 \text{ cm} = a)$, the charge distribution has approximate plane symmetry, so the electric field is given by (see Equation 21.7)

$$E = \frac{\sigma}{2\epsilon_0}$$

Therefore, the charge on the plate is $q = \sigma A = 2\epsilon_0 E A = 2\epsilon_0 E a^2$, where a is the length of a side of the square plate. At a point sufficiently far from the plate $(r \gg a)$ the field strength will resemble that from a point charge, $E = kq/r^2$.
EVALUATE   Inserting the values given, we find the charge on the plate is

$$q = 2\epsilon_0 E a^2 = 2\left[8.85 \times 10^{-12} \text{ C}^2/\left(\text{N} \cdot \text{m}^2\right)\right](45 \text{ kN/C})(0.75)^2 = 450 \text{ nC}$$

The field strength at $r = 15 \text{ m} \gg 0.75 \text{ m}$ is like that from a point charge:

$$E = \frac{kq}{r^2} = \frac{\left(9.0 \times 10^9 \text{ N} \cdot \text{m}^2/\text{C}^2\right)(448 \text{ nC})}{(15 \text{ m})^2} = 18 \text{ N/C}$$

ASSESS   Far from the finite charge distribution (plane charge in this case), the field always resembles that of a point charge.

61.   INTERPRET   This problem involves a irregular-shaped conductor in electrostatic equilibrium. The conductor is hollow and we are to show there is zero field in the cavity. This geometry is nonsymmetric, so we cannot apply the formulas developed for circular, plane, or line charge distributions.

DEVELOP   Review Section 21.6 to see why the electric field within a conducting medium, in electrostatic equilibrium, is always zero. In addition, the net charge must reside on the conductor surface.

EVALUATE   (a) When there is no charge inside the cavity, the flux through any closed surface within the cavity (see S_1 in figure below) is zero, so the electric field is also zero. If it were not, there would be a nonzero electric field acting on the surface charges, which would therefore move and the conductor would not be in equilibrium.
(b) If the surface charge density on the outer surface (and also the electric field there) is to vanish, then the net charge inside a Gaussian surface S_2 containing the conductor must be zero. Thus, the point charge in the cavity must be $-Q$, so that $q_{enclosed} = Q + (-Q) = 0$.

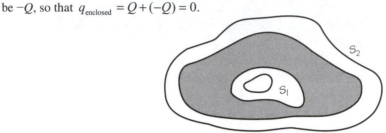

ASSESS   An alternative approach is to say that the total charge on the conductor is $q_c = q_{inner} + q_{outer} = Q$. Requiring $q_{outer} = 0$ means that $q_{inner} = Q$. Since electric field inside the cavity vanishes, the point charge placed inside must be $q = -q_{inner} = -Q$.

63.   INTERPRET   We are given a spherically symmetric charge distribution, so Gauss's law can be applied in this problem.

DEVELOP   The field from the given charges is spherically symmetric, so (from Gauss's law) is like that of a point charge located at the center with magnitude equal to the net charge enclosed by a sphere of radius equal to the distance to the field point.

EVALUATE   Thus, the electric field is $E = -kq/r^2$ inside the first shell (8 lines radially inward, see figure below), $E = +kq/r^2$ between the first and second shells (8 lines radially outward), and $E = -kq/(2r^2)$ outside the second shell (4 lines radially inward).

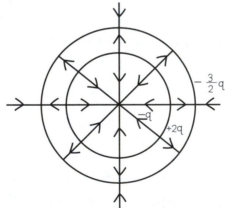

ASSESS   The direction of the electric field depends on the net charge enclosed. When $q_{enclosed} > 0$, $\vec{E}$ is radially outward. On the other hand, $\vec{E}$ is radially inward when $q_{enclosed} < 0$. The discontinuity in electric field across the shell is due to a net surface charge density on the shell.

65.   INTERPRET   The charge distribution has spherical symmetry, so we can apply Gauss's law. Also, since the charge distribution is non-uniform, integration is needed to find the electric field.

DEVELOP   The charge inside a sphere of radius $r \leq a$ is $q(r) = \int_0^r \rho \, dV$. For volume elements, take concentric shells of radius r and thickness dr, so $dV = 4\pi r^2 dr$.

EVALUATE   (a) Integrating over r, the charge enclosed by a Gaussian sphere of radius r is

$$q(r) = 4\pi \int_0^r \rho r^2 \, dr = \frac{4\pi\rho_0}{a} \int_0^r r^3 \, dr = \frac{\pi\rho_0 r^4}{a}$$

For $r = a$, the total charge is $Q = \pi\rho_0 a^3$.

(b) Inside the sphere, Gauss's law and Equation 21.5 give

$$\oint_{surface} \vec{E} \cdot d\vec{A} = \frac{q_{enclosed}}{\epsilon_0} 4\pi r^2$$

$$E(r) = \frac{q(r)}{\epsilon_0} = \frac{\pi\rho_0 r^4}{\epsilon_0 a} \quad \Rightarrow \quad E(r) = \frac{1}{4\pi\epsilon_0} \frac{\pi}{a} = \frac{\rho_0 r^2}{4\epsilon_0 a}$$

ASSESS   The r^2 dependence of E inside the sphere can be contrasted with the r dependence in the case (see Example 21.1) where the charge distribution is uniform.

67.   INTERPRET   We are given a charge distribution with spherical symmetry, which we can exploit with Gauss's law. Also, since the charge distribution is non-uniform, integration is needed to find the condition which gives zero electric field outside the sphere.

DEVELOP   Gauss's law tells us that the field outside the sphere will be zero if the total charge within the Gaussian surface is zero. Thus, using concentric shells of thickness dr and volume $dV = 4\pi r^2 dr$ as our charge elements, we require that

$$q_{enclosed} = \int_{sphere} \rho(r) \, dV = 0$$

EVALUATE   Inserting the expression for $\rho(r)$ into the integrand and integrating gives

$$0 = \int_0^R (\rho_0 - ar^2) 4\pi r^2 \, dr = 4\pi \left(\frac{1}{3}\rho_0 R^3 - \frac{1}{5}aR^5 \right)$$

$$a = \frac{5\rho_0}{3R^2}$$

ASSESS   The charge density starts out from the center of the sphere as ρ_0 and decreases as r^2. At $r = R$, the density is

$$\rho(R) = \rho_0 \left(1 - \frac{5}{3} \right) = -2\rho_0 / 3$$

Note that $\rho(r)$ must change sign (from positive to negative) in order for $q_{enclosed}$ to be zero at $r = R$.

69.   INTERPRET   Given a spherically symmetric charge distribution in a sphere, we are to find the electric field at the surface of the sphere. We will use Gauss's law, and integrate the charge density to find the charge enclosed.

DEVELOP   The charge density is given by $\rho = \rho_0 e^{r/R}$, where r is the distance from the center and R is the radius of the sphere. Gauss's law tells us that

$$\oint_{surface} \vec{E} \cdot d\vec{A} = \frac{q_{enclosed}}{\epsilon_0}$$

Because the charge density is spherically symmetric, we know that the electric field will not be angularly dependent. In other words, it will be constant over a spherical surface centered at the center of the charged sphere. Furthermore, the same reasoning tells us that the direction of the electric field must parallel to $d\vec{A}$ (i.e., radial) so the dot product gives a factor of unity. Finally, the charge enclosed can be calcualated by integrating the charge distribution in the sphere. Thus, we have

$$E \int_{\substack{spherical \\ surface}} dA = \frac{q_{enclosed}}{\epsilon_0} = \frac{1}{\epsilon_0} \int_V dq$$

EVALUATE   Performing the integration gives

$$E\left(4\pi R^2\right) = \frac{1}{\epsilon_0} \int_0^{2\pi} \int_0^{\pi} \int_0^{R} \rho_0 e^{r/R} r^2 \sin\theta \, dr \, d\theta \, d\phi$$

$$E = \frac{1}{4\pi\epsilon_0} 4\pi\rho_0 \int_0^{R} r^2 e^{r/R} \, dr = \frac{\rho_0}{\epsilon_0}\left[e^{r/R} R\left(r^2 - 2rR + 2R^2\right)\right]_0^{R}$$

$$E = \frac{\rho_0}{\epsilon_0}\left[eR\left(R^2\right) - R\left(2R^2\right)\right] = R^3 \frac{\rho_0}{\epsilon_0}(e - 2)$$

ASSESS   This answer is positive so the field points away from the sphere. It is also linear in ρ_0, as we would expect.

71.   **INTERPRET**   We consider the electric fields associated with a coaxial cable.

DEVELOP   The electric field has to be zero within the wire and the shield, since they are both conductors. Charges will line up on the surface of each to ensure that the net field cancels inside. To illustrate this, we draw the cross-section of the coaxial cable in the figure below, assuming for argument's sake that the inner conductor has negative charge and the outer has positive charge.

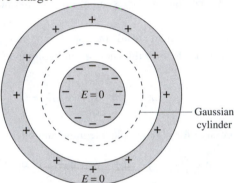

In the figure, we also include the cross-section of a Gaussian cylinder. If there is a field somewhere, it will be perpendicular to the curved surface of such a cylinder.

EVALUATE   In the region between the wire and shield, the enclosed charge is that of the wire. Since this is non-zero, the field in this region will be non-zero. As for the region outside the shield, the enclosed charge is both that of the wire and the shield, which are equal and opposite. The net charge will be zero, so the field will be zero in this region. To sum up, the only place where the field is not zero is between the wire and shield.

The answer is (a).

ASSESS   Unlike the conductors, the insulation can have a non-zero electric field. It's charges are not free to move around until the internal field is zero. However, the insulation may be dielectric, in which case, the alignment of electric dipoles in the material will partially reduce the field (see Figure 20.24).

73.   **INTERPRET**   We consider the electric fields associated with a coaxial cable.

DEVELOP   To find the field between the wire and the shield, we can use a Gaussian cylinder with length L and radius r that lies between the two conductors (see figure in Problem 21.71).

EVALUATE   The cylinder will enclose the charge on the wire: $-Q = \lambda L$, where λ is the charge per unit length. The electric flux is limited to just the outer curved surface: $\Phi = 2\pi rL \cdot E$ (see Example 21.4). So the electric field in between the wire and shield will be proportional to $1/r$.

The answer is (b).

ASSESS   Because of the symmetry, the wire's electric field is the same as that of an infinite line of charge. The shield only contributes to the field outside its inner radius.

ELECTRIC POTENTIAL

EXERCISES

Section 22.1 Electric Potential Difference

15. **INTERPRET** For this problem, we are to find the work required to move 50 μC of charge through a potential difference of 12 V. Recall that the SI units of potential difference in volts is in J/C, so it represents the potential energy difference per unit charge.

    DEVELOP The potential difference in volts is the negative of the work required per unit charge in moving a positive charge from point A to point B. Mathematically, this may be expressed as

$$W_{AB} = -qV_{AB}$$

    (see also Equation 22.1a). For this problem, the potential difference is -12 V because we are moving against the potential difference (or against the electric field, so from high potential to low potential), so we can solve for W_{AB}.

    EVALUATE Inserting the given values gives

$$W = -q\Delta V = (50~\mu\text{C})(-12~\text{V}) = 600~\mu\text{J}$$

    ASSESS Work is required to increase the potential energy of a charge.

17. **INTERPRET** This problem involves calculating a potential difference per unit charge between two points given the work required to move a given charge between the two points.

    DEVELOP The work done by an external agent against the electric field is the potential energy change,

    $\Delta U_{AB} = 45~\text{J} = q~\Delta V_{AB}$.

    EVALUATE Solving for ΔV_{AB} and inserting the given values gives

$$\Delta V_{AB} = (45~\text{J})/(15~\text{mC}) = 3.0~\text{kV}.$$

    ASSESS Note that the work done *by* the electric field is the negative of the potential difference between two points.

19. **INTERPRET** This problem is an exercise in calculating the potential difference between two points, given their separation and the magnitude of the uniform electric field between the two points.

    DEVELOP Apply Equation 22.1b, which applies for a uniform electric field.

$$\Delta V = -\vec{E} \cdot \vec{r}$$

    Because we are moving parallel to the electric field, the dot product gives $\cos(0°) = 1$. Also, because we are only interested in the magnitude of the potential difference, we can omit the negative sign.

    EVALUATE The magnitude of the potential difference is

$$|\Delta V| = Er = (650~\text{N/C})(1.4~\text{m}) = 910~\text{V}$$

    ASSESS If this involves a charge moving with (against) the electric field, the potential will decrease (increase).

21. **INTERPRET**   We are to find the energy gained by the three given charged particles as they move through a 100-V potential difference.

DEVELOP   The energy gained is $q\Delta V$ (see Example 22.1). For the proton, alpha particle, and singly ionized He atom, $q = e, 2e, e$, respectively.

EVALUATE   For proton and the ionized He atom, the energy gained is

$$\Delta K = q\Delta V = e(100 \text{ V}) = 100 \text{ eV} = (1.60 \times 10^{-19} \text{ C})(100 \text{ V}) = 1.6 \times 10^{-17} \text{ J}$$

For the alpha particle, the charge is twice that of the other particles, so the energy gained is twice: 3.2×10^{-17} J.

ASSESS   Note that the velocity of each particle is different at the output because each has a different mass.

Section 22.2 Calculating Potential Difference

23. **INTERPRET**   In this problem, we are given a uniform electric field and asked to calculate the potential difference between two points.

DEVELOP   For a uniform field, the potential difference between two points a and b is given by Equation 22.1b:

$$\Delta V_{AB} = V_B - V_A = -\vec{E} \cdot \Delta \vec{r}$$

where $\Delta \vec{r}$ is a vector from a to b.

EVALUATE   With $\Delta \vec{r} = \vec{r}_B - \vec{r}_A = y\hat{j}$, we obtain

$$V(y) - V(0) = V(y) = -\vec{E} \cdot \Delta \vec{r} = -\left(E_0 \hat{j}\right) \cdot \left(y\hat{j}\right) = -E_0 y$$

$$V(y) = -E_0 y$$

ASSESS   The electric potential decreases in the direction of the electric field. In other words, electric field lines always point in the direction of decreasing potential.

25. **INTERPRET**   We are asked to find the charge on a sphere, given the potential at its surface. Because the charge distribution is spherically symmetric, we will use the equation for the potential of a point charge.

DEVELOP   Equation 22.3 gives the potential for a point charge as

$$V(r) = \frac{kq}{r}$$

Since any spherically symmetric charge distribution looks like a point charge from outside the distribution, we can solve this for q. The potential at the surface of the sphere is $V = 4.8$ kV and the radius is $r = 0.10$ m.

EVALUATE   Inserting the given quantities yields

$$q = \frac{Vr}{k} = \frac{(4.8 \text{ N} \cdot \text{m/C})(0.10 \text{ m})}{(9.0 \times 10^9 \text{ N} \cdot \text{m}^2/\text{C}^2)} = 53 \text{ nC}$$

ASSESS   The key is to recognize that spherically symmetric charge distributions look like point charges from the outside. This is the same as for gravitational potentials. Note also that the units in the result cancel to give coulombs.

27. **INTERPRET**   This problem is about the electric potential of a spherically symmetric charge distribution. We are to find the potential at the surface of a charged conducting sphere and the kinetic energy (or speed) of a proton accelerated from the surface to infinity by the sphere's potential.

DEVELOP   Since the electric field outside the spherical charge distribution is the same as that of a point charge, the electric potential outside the metal sphere $(r \geq R)$ is given by Equation 22.3:

$$V(r) = \frac{kQ}{r}$$

Note that we have taken the zero of the potential to be at infinity.

EVALUATE   **(a)** An isolated metal sphere has a uniform surface charge density, so the potential at its surface is

$$V(R) = \frac{kQ}{R} = \frac{(9.0 \times 10^9 \text{ N} \cdot \text{m}^2/\text{C}^2)(0.86 \text{ }\mu\text{C})}{0.035 \text{ m}/2} = 440 \text{ kV}$$

(b) The work done by the repulsive electrostatic field (the negative of the change in the proton's potential energy) equals the proton's kinetic energy at infinity:

$$W_{AB} = -qV_{AB} = -e\left[V_\infty - V(R)\right] = eV(R) = \frac{1}{2}mv^2$$

Thus, the speed of the proton far from the sphere is

$$v = \sqrt{\frac{2eV(R)}{m}} = \sqrt{\frac{2(1.60 \times 10^{-19}\ \text{C})(442\ \text{kV})}{1.67 \times 10^{-27}\ \text{kg}}} = 9.2 \times 10^6\ \text{m/s}$$

ASSESS   As the proton moves away from the metal sphere, its potential energy decreases. However, by energy conservation, its kinetic energy increases. Also, notice that the result of part (a) is given to two significant figures, but that three significant figures are used when we insert that result into part (b) because it serves as an intermediate result for part (b).

Section 22.3 Potential Difference and the Electric Field

29.   **INTERPRET**   This problem involves calculating the electric field from the electric potential (or voltage).
DEVELOP   Given electric potential $V(x)$, the x component of the electric field may be obtained as $E_x = -dV/dx$ (see Equation 22.9). Use this equation to estimate E_x for the seven straight-line segments shown in Fig. 22.20.
EVALUATE   Using the equation above, we find $E_x = 0$ for $x = 0$ to 2 m. Similarly, for x = 2 to 4 m, $E_x = -(-2\ \text{V} - 2\ \text{V})/(4\ \text{m} - 2\ \text{m}) = 2\ \text{V/m}$. The field strength in other regions can be calculated in a similar manner. The result is sketched below.

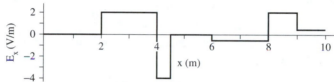

ASSESS   The field component $E_x = -dV/dx$ is the negative of the rate of change of V with respect to x. The negative sign means that if we move in the direction of increasing potential, then we're moving against the electric field.

31.   **INTERPRET**   This problem is about calculating electric field given the electric potential.
DEVELOP   Given the electric potential V, the corresponding electric field is (see Equation 22.9)

$$\vec{E} = E_x\hat{i} + E_y\hat{j} + E_z\hat{k} = -\left(\frac{\partial V}{\partial x}\hat{i} + \frac{\partial V}{\partial y}\hat{j} + \frac{\partial V}{\partial z}\hat{k}\right)$$

Thus, taking the partial derivatives of V allows us to get the field components.
EVALUATE   **(a)** Direct substitution gives the voltage at $(x, y, z) = (1\ \text{m}, 1\ \text{m}, 1\ \text{m})$:

$$V(x,y,z) = 2xy - 3zx + 5y^2 = (2\ \text{Vm}^{-2})(1\ \text{m})(1\ \text{m}) - (3\ \text{Vm}^{-2})(1\ \text{m})(1\ \text{m}) + (5\ \text{Vm}^{-2})(1\ \text{m})^2 = 4\ \text{V}$$

(b) Use of Equation 22.9 gives the components of the electric field:

$$E_x = -\frac{\partial V}{\partial x} = -2y + 3z$$

$$E_y = -\frac{\partial V}{\partial y} = -2x - 10y$$

$$E_z = -\frac{\partial V}{\partial z} = 3x$$

At $(x, y, z) = (1\ \text{m}, 1\ \text{m}, 1\ \text{m})$, we obtain $E_x = 1\ \text{V/m}$, $E_y = -12\ \text{V/m}$ and $E_z = 3\ \text{V/m}$.
ASSESS   Electric field is strong in the region where the potential changes rapidly. At (1 m, 1 m, 1 m), the potential changes most rapidly in the direction of the electric field

$$\vec{E} = (\hat{i} - 12\hat{j} + 3\hat{k})\text{V/m}$$

Section 22.4 Charged Conductors

33. **INTERPRET** This problem is about finding the minimum potential that leads to a dielectric breakdown in air.

DEVELOP We shall treat the field from the central electrode as if it were from an isolated sphere, for which Equation 20.3 gives the electric field to be $E = kq/R^2$ and Equation 22.3 gives the potential to be $V = kq/R$. Combining these two expressions gives $V = RE$.

EVALUATE Breakdown of air occurs at a field strength of $E = 3 \times 10^6$ V/m. Therefore, dielectric breakdown in air would occur for potentials exceeding

$$V = RE = (1.0 \times 10^{-3} \text{ m})(3 \times 10^6 \text{ V/m}) = 3 \text{ kV}$$

ASSESS The result means that if we attempt to raise the potential of the electrode in air above 3 kV, then the surrounding air would become ionized and conductive; the extra added charge would leak into the air, resulting in plug sparks.

PROBLEMS

35. **INTERPRET** This problem is about finding the electric field strength, given the potential difference between two points a given distance apart. We are also given the orientation of the electric field with respect to the line joining the two points.

DEVELOP Since the field $\vec{E}$ is uniform, Equation 22.1b, $\Delta V_{AB} = -\vec{E} \cdot \Delta \vec{r}$, can be used to relate $\vec{E}$ to the potential difference ΔV_{AB}. Since the path AB is parallel to $\vec{E}$, the angle between $\vec{E}$ and $\Delta \vec{r}$ is 0°. Because $\cos(0°) = 1$ the dot product reduces to

$$\Delta V_{AB} = E \Delta r$$

where E is the field strength, and Δr is the separation between points A and B.

EVALUATE The field strength is

$$E = \frac{\Delta V_{AB}}{\Delta r} = \frac{840 \text{ V}}{0.15 \text{ m}} = 5.6 \text{ kV/m}$$

ASSESS Since $dV = -\vec{E} \cdot d\vec{r}$ the potential always decreases in the direction of the electric field. Note that the angle between $\vec{E}$ and $\Delta \vec{r}$ is 180° if the two are antiparallel.

37. **INTERPRET** This problem involves finding the potential difference between two points (i.e., the terminals of the battery) given the work done on each elementary charge that moves between these points.

DEVELOP From the discussion in Section 22.1, we know that the work done by the electric field on each charge between two points is the potential difference between the same two points. Thus, $|W/q| = \Delta V$.

EVALUATE Inserting the given quantities gives

$$\frac{W}{q} = \frac{7.2 \times 10^{-19} \text{ J}}{1.6 \times 10^{-19} \text{ C}} = 4.5 \text{ V}$$

ASSESS The energy imparted per electron is 4.5 eV.

39. **INTERPRET** This problem involves finding the potential difference between two conducting plates separated by a distance d and having opposite charge densities.

DEVELOP We first calculate the electric field between the plates. Using the result obtained in Example 21.6 for one sheet of charge, and applying the superposition principle, the electric field strength between the plates is

$$\vec{E} = \vec{E}_1 + \vec{E}_2 = \sigma/(2\epsilon_0)\hat{i} + [-\sigma/(2\epsilon_0)](-\hat{i}) = (\sigma/\epsilon_0)\hat{i}$$

where $\hat{i}$ is directed from the positive plate to the negative plate. Once E is known, we can use Equation 22.1b to calculate V.

EVALUATE Equation 22.1b gives

$$V = V_+ - V_- = -\vec{E} \cdot \Delta \vec{r} = -\left(\frac{\sigma}{\epsilon_0}\right)(-d) = \frac{\sigma d}{\epsilon_0}$$

ASSESS   The displacement from the negative to the positive plate is opposite to the field direction. In other words, the potential always decreases in the direction of the electric field.

41.   **INTERPRET**   The problem asks for the charge of the particle which has been accelerated through a potential difference. We can find the magnitude of this potential difference given the information of the speed acquired upon traversing the potential difference by the mass with the given charge.

DEVELOP   The speed acquired by a charge q, starting from rest at point A and moving through a potential difference of V can be found using the work-energy theorem (see Problem 22.40). The result is

$$\Delta K_{AB} = q \Delta V_{AB} \quad \Rightarrow \quad \frac{1}{2} m v^2 = q \, \Delta V_{AB} \quad \Rightarrow \quad v = \sqrt{\frac{2 q \Delta V_{AB}}{m}}$$

This is the work-energy theorem for the electric force. A positive charge is accelerated in the direction of decreasing potential (i.e., increasing electric field). If we have two masses moving through the same potential difference, the ratio of their speeds would be

$$\frac{v_2}{v_1} = \sqrt{\frac{2 q_2 V / m_2}{2 q_1 V / m_1}} = \sqrt{\frac{q_2}{q_1} \frac{m_1}{m_2}}$$

EVALUATE   If the second object acquires twice the speed of the first object $(v_2/v_1 = 2)$, moving through the same potential difference we find its charge from the equation above to be

$$q_2 = \left(\frac{m_2}{m_1}\right)\left(\frac{v_2}{v_1}\right)^2 q_1 = \left(\frac{2g}{5g}\right)(2)^2 (3.8 \,\mu\text{C}) = 6.1 \,\mu\text{C}$$

ASSESS   The speed of the particle moving through a potential difference is proportional to the square root of its charge, and inversely proportional to the square root of its mass.

43.   **INTERPRET**   The positively charged proton is attracted to the negatively charged sphere via Coulomb interaction. Work must be done to pull the proton away from the sphere.

DEVELOP   From the work-energy theorem (Equation 6.14), the work done by the electric field when a proton escapes from the surface to an infinite distance, equals the change in kinetic energy, or

$$W_{\text{surf},\infty} = -e\left(\overbrace{V_\infty}^{=0} - V_{\text{surf}}\right) = e V_{\text{surf}} = \overbrace{K_\infty}^{=0} - K_{\text{surf}} = -\frac{1}{2} m v_{\text{surf}}^2$$

where we have assumed zero kinetic energy for the proton at infinity and that the sphere is stationary.

EVALUATE   For a uniformly charged sphere with a total charge $-Q$, $V_{\text{surf}} = -kQ/R$ (see Equation 22.3). Inserting this into the expression above and solving for v_{surf} gives

$$v_{\text{surf}} = \sqrt{\frac{-2 e V_{\text{surf}}}{m}} = \sqrt{\frac{2 k e Q}{mR}}$$

ASSESS   The escape speed of a proton from the electric field of the charged sphere in this problem is analogous to the escape speed of a rocket from the Earth's gravitational field.

45.   **INTERPRET**   For this problem, we are to find the potential at the center of a hollow spherical shell that carries a uniform charge density on its surface.

DEVELOP   From Gauss's law, we know that the electric field inside the shell is zero, because there is no charge in the shell. Because the electric field is constant, we can apply Equation 22.1b, which shows that the change in the potential between any two points within the sphere must be zero. Consider the boundary condition for the electric potential (at the surface of the sphere) to find the electric potential everywhere in the sphere.

EVALUATE   At the surface of the sphere, the electric potential must be (see Equation 22.3)

$$V = \frac{kQ}{R}$$

Thus, this must be the potential everywhere inside the sphere, since we have argued above that the potential inside the sphere must be constant.

ASSESS   The potential at the surface of the sphere is measured with respect to infinity, which is to say a distance $r \gg R$ from the sphere.

47.   **INTERPRET**   For this problem, we are given the electric field as a function of position, and we are to find the electric potential as a function of position. We are also given the electric potential at a given point, so we will define our electric potential with respect to this point.

DEVELOP   Apply Equation 22.1a,

$$\Delta V_{AB} = -\int_A^B \vec{E} \cdot d\vec{r}$$

where $\vec{E} = ax(\hat{i})$ and $d\vec{r} = dx(\hat{i})$. Furthermore, we take point A to be x = 0, so $V_A = V(x = 0) = 0$, and point B to be an arbitrary point x.

EVALUATE   Evaluating the integral gives

$$\Delta V_{AB}(x) = -\int_A^B ax' dx' (\hat{i} \cdot \hat{i}) = -\int_0^x ax' dx' = -\frac{a}{2} x^2$$

so

$$V(x) = V(0) + \Delta V_{AB}(x) = -\frac{a}{2} x^2$$

ASSESS   This potential increases quadratically with position, whereas the electric field is linear in position.

49.   **INTERPRET**   This problem involves finding the charge density on a power line given the potential difference over a given distance away. To use Gauss's law for geometries with line symmetry, we will assume that the power line is much, much longer than 1.0 m.

DEVELOP   The electric potential around an object with line symmetry (such as our power line) is derived in Example 22.4. The result is

$$\Delta V_{AB} = \frac{\lambda}{2\pi\epsilon_0} \ln\left(\frac{r_A}{r_B}\right)$$

From the problem statement, make a sketch showing the location of the given voltages (see figure below). From this sketch, we see that r_A = 3.0 cm, r_B = 1.0 m, and $\Delta V_{AB} = V_B - V_A$ = +3.9 kV for this problem.

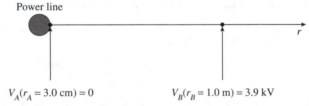

EVALUATE   Solving the expression above for the line charge density λ gives

$$\lambda = \frac{2\pi\epsilon_0 \Delta V_{AB}}{\ln(r_A / r_B)} = \frac{2\pi (9.0 \times 10^9 \text{ N} \cdot \text{m}^2/\text{C}^2)(3.9 \text{ kV})}{\ln(0.030 \text{ m}/1.0 \text{ m})} = -52 \text{ nC/m}$$

ASSESS   Thus, the wire carries excess negative charge.

51.   **INTERPRET**   This problem involves using the superposition principle to find two points on a line joining the two given charges where the electric potential is zero.

DEVELOP   Using the superposition principle in the form of Equation 22.5, the potential at the x-axis is

$$V(x) = \sum_i \frac{kq_i}{x_i} = \frac{kQ}{|x|} + \frac{k(-3Q)}{|x-a|}$$

Set this expression equal to zero and solve for the position x.

EVALUATE   The potential is zero when $3|x| = |x - a|$. For $x < 0$, this implies $-3x = a - x$, or $x = -a/2$. For $0 < x < a$, the condition is $3x = a - x$, or $x = a/4$. For $x > a$, there are no solutions.

ASSESS   The same results follow from the quadratic $8x^2 + 2ax - a^2 = 0$, which results from the square of the above condition.

53.   **INTERPRET**   We're asked to find the electric potential around an electric dipole.

DEVELOP   Equation 22.6 gives the potential from a dipole at a distance, r, much larger than the charge separation: $V(r, \theta) = kp \cos \theta / r^2$, where θ is the angle from the dipole axis.

EVALUATE   (a) For $\theta = 0°$,

$$V(r, \theta) = \frac{\left(9.0 \times 10^9 \ \frac{\text{N·m}^2}{\text{C}^2}\right)(2.9 \ \text{nC·m})\cos 0°}{(10 \ \text{cm})^2} = 2.6 \ \text{kV}$$

(b) For $\theta = 45°$,

$$V(r, \theta) = \frac{\left(9.0 \times 10^9 \ \frac{\text{N·m}^2}{\text{C}^2}\right)(2.9 \ \text{nC·m})\cos 45°}{(10 \ \text{cm})^2} = 1.8 \ \text{kV}$$

(c) For $\theta = 90°$,

$$V(r, \theta) = \frac{\left(9.0 \times 10^9 \ \frac{\text{N·m}^2}{\text{C}^2}\right)(2.9 \ \text{nC·m})\cos 90°}{(10 \ \text{cm})^2} = 0$$

ASSESS   The results seem reasonable. It should be made clear that these values assume the potential is zero at infinity. This is an arbitrary choice, since the only physical quantity is the potential difference between two points.

55.   **INTERPRET**   This problem involves finding the electric potential at the center of a non-uniform circular charge distribution. Because the charge is non-uniform, we will need to integrate over it to find the potential.

DEVELOP   Note that the result of Example 22.6 does not depend on the ring being uniformly charged. For a point on the axis of the ring, the geometrical factors are the same, and $\int_{\text{ring}} dq = Q_{\text{tot}}$ for any arbitrary charge distribution, so

$$V = \frac{kQ_{\text{tot}}}{\sqrt{x^2 + a^2}}$$

still holds.

EVALUATE   Thus, at the center (i.e., $x = 0$) of a ring of total charge $Q_{\text{tot}} = 3Q - Q = 2Q$, and radius $a = R$, the potential is $V = 2kQ/R$.

ASSESS   This integration was simple, because the charge was all located the same distance from the point of interest (i.e., from the center of the circle).

57.   **INTERPRET**   This problem involves a circularly symmetric, uniform charge distribution for which we are to find an expression for the electric potential at arbitrary points along its axis.

DEVELOP   The annulus can be considered to be composed of thin rings of radius r ($a \leq r \leq b$) and charge $dq = 2\pi\sigma \, r dr$ (see Example 22.7 and Figs. 22.12 and 12.13). The contribution from a ring to the electric potential on the axis, a distance x from the center, is $dV = kdq/\sqrt{x^2 + r^2}$ (see Example 22.6), which we can integrate from $r = a$ to $r = b$ to find the potential V.

EVALUATE   The potential from the whole annulus is:

$$V(x) = \int dV = 2\pi\sigma k \int_a^b \frac{r \, dr}{\sqrt{x^2 + r^2}} = 2\pi k\sigma \left| \sqrt{x^2 + r^2} \right|_a^b = 2\pi k\sigma \left(\sqrt{x^2 + b^2} - \sqrt{x^2 + a^2} \right)$$

ASSESS   This reduces to the potential on the axis of a uniformly charged disk if $a \to 0$.

59.   **INTERPRET**   In this problem we are to use the expression for the electric dipole potential to find the electric field at a point on the perpendicular bisector of the dipole.

DEVELOP   The dipole potential is given by Equation 22.6:

$$V(r, \theta) = \frac{kp\cos\theta}{r^2}$$

Using Equation 22.9, the general expressions for the r and θ components of the electric fields are

$$E_r = -\frac{\partial V}{\partial r} = \frac{2kp\cos\theta}{r^3}$$

$$E_\theta = -\frac{1}{r}\frac{\partial V}{\partial\theta} = \frac{kp\sin\theta}{r^3}$$

EVALUATE   On the bisecting plane, $\theta = 90°$, which yields $E_r = 0$ and $E_\theta = kp/r^3$, or $\vec{E} = E_\theta\hat{\theta} = (kp/r^3)\hat{\theta}$. To compare with Equation 20.6a, we take the origin at the center of the dipole, the dipole moment along the x axis $(\vec{p} = p\hat{i})$, and the y axis up in Fig. 22.10, so $\hat{\theta} = -\hat{i}\sin(90°) + \hat{j}\cos(90°) = -\hat{i}$ and $r = \sqrt{0^2 + y^2} = y$ on the bisecting plane. This leads to $\vec{E} = -(kp/y^3)\hat{i}$.

ASSESS   Instead of using polar coordinates, one could first express V in terms of x and y (using $x = r\cos\theta$ and $y = r\sin\theta$):

$$V(x, y) = \frac{kpx}{(x^2 + y^2)^{3/2}}$$

and then differentiate, $E_x = -\partial V/\partial x$ and $E_y = -\partial V/\partial y$. The result is the same.

61.   **INTERPRET**   We are given the electric potential and asked to find the corresponding electric field.

DEVELOP   We first note that the potential $V(r) = -V_0 r/R$ depends only on r. This implies that the electric field is spherically symmetric and points in the radial direction. The field can be calculated using Equation 22.9,

$$\vec{E} = -\left(\frac{\partial V}{\partial x}\hat{i} + \frac{\partial V}{\partial y}\hat{j} + \frac{\partial V}{\partial z}\hat{k}\right)$$

In spherical coordinates, this is

$$\vec{E} = -\left[\left(\frac{\partial V}{\partial r}\right)\hat{r} + \frac{1}{r\sin\theta}\left(\frac{\partial V}{\partial\phi}\right)\hat{\phi} + \frac{1}{r}\left(\frac{\partial V}{\partial\theta}\right)\hat{\theta}\right]$$

Because the potential depends only on r, the second two terms in this expression will give zero.

EVALUATE   The electric field is

$$\vec{E} = -\frac{dV}{dr}\hat{r} = \frac{V_0}{R}\hat{r}$$

where $\hat{r}$ is a unit vector that points radially outward.

ASSESS   The electric field is uniform, but the potential is linear in r. The difference of one power in r is because the potential is an integral of the field over distance.

63.   **INTERPRET**   We are given two charge-carrying conducting spheres, and we want to find the electric potential and electric field at various points.

DEVELOP   Since the spheres are separated by a distance that is over an order of magnitude greater than the radii of the spheres, we can consider them to be isolated spheres. Thus, their charge distributions are essentially spherical and we can apply Equation 22.3 to find the potential.

EVALUATE   (a) Using the result obtained in Example 22.3, at the surface of either sphere, the potential is

$$V(R) = \frac{kq}{R} = \frac{(9.0\times10^9 \text{ N}\cdot\text{m}^2/\text{C}^2)(1.2\times10^{-7} \text{ C})}{0.025 \text{ m}} = 43 \text{ kV}$$

(b) From Equation 22.9 (or Equation 21.3), the electric field at the surface of each sphere is

$$E(R) = \frac{kq}{R^2} = \frac{V(R)}{R} = \frac{43.2 \text{ kV}}{0.025 \text{ m}} = 1.7\times10^6 \text{ N/C}$$

(c) Midway between the spheres, the potential from each one is the same, so we apply the principle of superposition and sum the two potentials to find

$$V_{\text{mid-pt.}} = \frac{2kq}{r} = \frac{2(9.0\times10^9\ \text{N}\cdot\text{m}^2/\text{C}^2)(1.2\times10^{-7}\ \text{C})}{4.0\ \text{m}} = 540\ \text{V}$$

(d) Since the spheres are at the same potential, the difference is zero.

ASSESS   In this problem, the two conducting spheres can be treated as being isolated because they are far apart $(r \gg R)$ and the superposition principle applies. If they were brought close to each other, then the charge distribution would no longer be spherical.

65.   **INTERPRET**   This problem gives the electric field of a spherically symmetric charge distribution (i.e., it only depends on r, not on θ or ϕ), and we are to find the difference in electric potential between the sphere center and its outer edge.

DEVELOP   Apply Equation 22.1a,

$$\Delta V_{AB} = -\int \vec{E}(r)\cdot d\vec{r}$$

where $\vec{E}(r) = E_0(r/R)^2\,\hat{r}$, A is the sphere center, and B is the outer surface of the sphere.

EVALUATE   Evaluating the integral gives

$$\Delta V_{AB} = V_B - V_A = -\int_0^R \vec{E}(r)\hat{r}\cdot d\vec{r} = -E_0\int_0^R (r/R)^2\,dr = -\frac{E_0 R}{3}$$

ASSESS   The outer surface of the sphere (point B) is thus at a lower potential than the inner surface, which is normal because the potential decreases in the direction of the electric field, which in this case points radially outward.

67.   **INTERPRET**   This problem concerns a conducting sphere surrounded by a concentric conducting shell. We are given the charge on each and are to find the electric potential at the sphere's surface.

DEVELOP   Recall from Example 21.1 that the electric field outside a spherically symmetric charge distribution is that of a point charge with all the charge Q at that point (i.e., $\vec{E}(r) = kQ/r^2$). Use this result in Equation 22.1a to find the potential of the inner sphere with respect to the potential at infinity. This gives

$$V_{\text{sphere}} = \Delta V_{AB} = -\int_{A=R_1}^{B=\infty} \vec{E}(r)\cdot d\vec{r} = -\int_{R_1}^{R_2} \frac{kQ_1}{r^2}dr - \int_{R_2}^{\infty} \frac{k(Q_1+Q_2)}{r^2}dr$$

$$= -kQ_1\left(\frac{1}{R_2}-\frac{1}{R_1}\right) + \frac{k(Q_1+Q_2)}{R_2}$$

EVALUATE   (a) Inserting $Q_1 = 60$ nC and $Q_2 = -60$ nC into the expression above gives

$$V_{\text{sphere}} = -kQ_1\left(\frac{1}{R_2}-\frac{1}{R_1}\right) = -(9.0\times10^9\ \text{N}\cdot\text{m}^2/\text{C}^2)(60\ \text{nC})\left(\frac{1}{0.15\ \text{m}}-\frac{1}{0.050\ \text{m}}\right) = 7.2\ \text{kV}$$

(b) Inserting $Q_1 = 60$ nC and $Q_2 = 60$ nC into the expression above gives

$$V_{\text{sphere}} = -kQ_1\left(\frac{1}{R_2}-\frac{1}{R_1}\right) + \frac{k(Q_1+Q_2)}{R_2} = -(9.0\times10^9\ \text{N}\cdot\text{m}^2/\text{C}^2)(60\ \text{nC})\left(\frac{-1}{15\ \text{cm}}-\frac{1}{5.0\ \text{cm}}\right) = 14\ \text{kV}$$

ASSESS   As discussed in Problem 22.64, the potential inside a conducting shell is constant and is equal to the potential at the surface of the shell (with respect to the potential at infinity). Thus, the potential of any object inside the shell "floats" on the potential of the shell.

69.   **INTERPRET**   We are given the potential of a disk on its axis at two distances from the disk and are to find the disk radius and its total charge.

DEVELOP   Combining the given data with the potential in Example 22.7, we find

$$150\ \text{V} = \frac{2kQ}{a^2}\left(\sqrt{(5.0\ \text{cm})^2 + a^2} - 5.0\ \text{cm}\right)$$

$$110\ \text{V} = \frac{2kQ}{a^2}\left(\sqrt{(10\ \text{cm})^2 + a^2} - 10\ \text{cm}\right)$$

Taking the ratio of these two equations eliminates the charge, so we can solve for the radius a, following which we can solve for the charge Q.

EVALUATE   The ratio of these two expressions gives

$$\left(\frac{150}{110}\right) = \frac{\sqrt{1+\left(a/5.0\text{ cm}\right)^2}-1}{\sqrt{4+\left(a/5.0\text{ cm}\right)^2}-2}$$

Several lines of algebra to remove the square roots finally yields

$$a = \left(5.0\text{ cm}\right)\frac{\sqrt{105\times209}}{52} = 14\text{ cm}$$

We can now solve for Q from either of the first two equations, which gives

$$Q = \frac{\left(110\text{ V}\right)a^2}{2k\left(\sqrt{\left(10\text{ cm}\right)^2+a^2}-10\text{ cm}\right)}$$

$$= \frac{\left(110\text{ V}\right)\left(0.142\text{ m}\right)^2}{2\left(9.0\times10^9\text{ N}\cdot\text{m}^2/\text{C}^2\right)\left(\sqrt{\left(0.10\text{ m}\right)^2+\left(0.142\text{ m}\right)^2}-0.10\text{ m}\right)} = 1.7\text{ nC}$$

ASSESS   The units for the expression for charge is $V\cdot C^2/(N\cdot m) = (N/C)\cdot m\cdot C^2/(N\cdot m) = C$.

71.   **INTERPRET**   This problem involves a disk with a circularly symmetric charge distribution. We are to find an expression for the potential on the disk axis, the electric field on the disk axis, and show that the electric field decays as $1/x^2$ for $x \gg a$ (a is the disk radius).

DEVELOP   For part (a), use the integral expression for voltage of Example 22.7,

$$V(x) = \int_0^a \frac{kdq}{\sqrt{x^2+r^2}}$$

For this problem, $dq = 2\pi\sigma\,rdr$, with $\sigma = \sigma_0 r/a$, which gives

$$V(x) = \frac{2\pi k\sigma_0}{a}\int_0^a \frac{r^2 dr}{\sqrt{x^2+r^2}}$$

The electric field is the spatial derivative of the potential (see Equation 22.9).

EVALUATE   **(a)** Reference to standard integral tables gives

$$V(x) = \frac{2\pi k\sigma_0}{a}\int_0^a \frac{r^2\,dr}{\sqrt{x^2+r^2}} = \frac{2\pi k\sigma_0}{a}\left[\frac{a}{2}\sqrt{x^2+a^2}-\frac{x^2}{2}\ln\left(\frac{a+\sqrt{x^2+a^2}}{x}\right)\right]$$

$$= \pi k\sigma_0 a\left[\sqrt{1+\left(x/a\right)^2}-\left(x/a\right)^2\ln\left(a/x+\sqrt{1+\left(a/x\right)^2}\right)\right]$$

(b) As in Example 22.8, $E_x = -dV/dx$ results in

$$E_x = \pi k\sigma_0\left[\frac{2x}{a}\ln\left(\frac{a+\sqrt{x^2+a^2}}{x}\right)-\frac{x}{\sqrt{x^2+a^2}}+\frac{x^2}{a}\left(\frac{x}{a+\sqrt{x^2+a^2}}\right)\left(\frac{1}{\sqrt{x^2+a^2}}-\frac{a+\sqrt{x^2+a^2}}{x^2}\right)\right]$$

$$= \frac{2\pi k\sigma_0 x}{a}\left[\ln\left(\frac{a}{x}+\sqrt{1+\left(\frac{a}{x}\right)^2}\right)-\frac{a}{x\sqrt{1+\left(a/x\right)^2}}\right]$$

(c) The logarithm has to be expanded carefully, up to order $\left(a/x\right)^3$ to evaluate E_x for $x \gg a$. Thus,

$$\ln\left(\frac{a}{x}+\sqrt{1+\left(\frac{a}{x}\right)^2}\right) = \ln\left(1+\frac{a}{x}+\frac{a^2}{2x^2}+\cdots\right) \approx \left(\frac{a}{x}+\frac{a^2}{2x^2}\right)-\frac{1}{2}\left(\frac{a}{x}+\frac{a^2}{2x^2}\right)^2+\frac{1}{3}\left(\frac{a}{x}+\frac{a^2}{2x^2}\right)^3+\cdots \approx \frac{a}{x}-\frac{a^3}{6x^3}$$

Also,

$$\frac{a}{x}\left(1+\frac{a^2}{x^2}\right)^{-1/2} \approx \frac{a}{x}\left(1-\frac{a^2}{2x^2}\right) = \frac{a}{x} - \frac{a^3}{2x^3}$$

Then,

$$E_x \approx \frac{2\pi k\sigma_0 x}{a}\left[\frac{a}{x} - \frac{a^3}{6x^3} - \frac{a}{x} + \frac{a^3}{2x^3}\right] = \frac{2\pi k\sigma_0 a^2}{3x^2}$$

which is the field for a point charge with

$$Q = \frac{2\pi\sigma_0}{a}\int_0^a r^2 dr = 2\pi\sigma_0 a^2/3$$

ASSESS Checking the limits is a good manner to test the validity of expressions. If we were to have found that our expression for the electric field did not reduce at large distances to that of a point charge, then we could be sure that the expression is wrong. Of course, the correct asymptotic behavior does not guarantee that the expression is valid!

73. **INTERPRET** We need to find the potential due to a line charge with a non-constant charge density. The problem is one-dimensional: the line and the point of interest are all on the x axis. We will use the integral expression for potential. We must also show that the result reduces to that of a point charge for distances much, much larger than the length of the line charge.

DEVELOP Consider the line charge to be a superposition of many point charges. The potential for each point charge is given by Equation 22.3, $dV = kdq/r$. Integrate this expression over the length of the line charge to find the total potential:

$$V(x) = \int_{-L/2}^{L/2} \frac{kdq}{r}$$

with $dq = \lambda\, dx' = \lambda_0(x'/L)^2\, dx'$ and $r = x - x'$.

EVALUATE Evaluating the integral gives

$$V = \int_{-L/2}^{L/2} \frac{kdq}{r} = \int_{-L/2}^{L/2} \frac{k\lambda_0}{x-x'}\left(\frac{x'}{L}\right)^2 dx' = \frac{k\lambda_0}{L^2}\int_{-L/2}^{L/2} \frac{x'^2}{x-x'}dx'$$

$$= \frac{k\lambda_0}{L^2}\left[-\frac{x'^2}{2} + x'x - x^2\ln(x-x')\right]_{-L/2}^{L/2}$$

$$= -\frac{k\lambda_0}{L^2}\left[Lx + x^2\ln\left(\frac{2x-L}{2x+L}\right)\right]$$

For $x \gg L$, we rearrange slightly and use the approximation $\ln(1+\xi) \approx \xi$ for small ξ

$$V = -\frac{k\lambda_0}{L^2}x^2\left[\frac{L}{x} + \ln\left(1-\frac{L}{2x}\right) - \ln\left(1+\frac{L}{2x}\right)\right]$$

$$\approx -\frac{k\lambda_0}{L^2}x^2\left[\frac{L}{x} + \left(-\frac{L}{2x} - \frac{1}{2}\left(-\frac{L}{2x}\right)^2 + \frac{1}{3}\left(-\frac{L}{2x}\right)^3\right) - \left(\frac{L}{2x} - \frac{1}{2}\left(\frac{L}{2x}\right)^2 + \frac{1}{3}\left(\frac{L}{2x}\right)^3\right)\right]$$

$$\approx -\frac{k\lambda_0}{L^2}x^2\left[\frac{L}{x} - \frac{L}{x} - \frac{L^2}{8x^2} + \frac{L^2}{8x^2} - \frac{L^3}{24x^3} - \frac{L^3}{24x^3}\right] = \frac{k\lambda_0}{L^2}x^2\frac{L^3}{12x^3} = \frac{k\lambda_0 L}{12x}$$

The total charge on the rod is

$$q = \int dq = \int_{-L/2}^{L/2} \lambda_0\left(\frac{x'}{L}\right)^2 dx' = \frac{\lambda_0}{L^2}\left[\frac{x'^3}{3}\right]_{-L/2}^{L/2} = \frac{\lambda_0}{3L^2}\left[\frac{2L^3}{8}\right] = \frac{\lambda_0 L}{12}$$

so in the limit of $x \gg L$,

$$V = \frac{k\lambda_0 L}{12x} = \frac{kq}{x}$$

ASSESS At large distances, the potential looks like that of a point charge, as expected.

75. **INTERPRET** You need to find the minimum possible wire diameter that won't be susceptible to breakdown, which is when the air gets ionized by the electric field near the wire.

DEVELOP You're told to neglect any charge distributions on the ground. Therefore, the electric field around the wire can be found with Gauss's law: $E = \lambda/2\pi\epsilon_0 r$ (Equation 21.6). The maximum field, which will be at the outer surface of the wire at radius r_A, needs to be at most 25% of the breakdown field in air, 3MV/m. The potential difference for a long wire was given in Equation 22.4: $\Delta V_{AB} = \lambda/2\pi\epsilon_0 \ln(r_A/r_B)$. You know that there will be 115 kV between the wire's outer surface and the ground below $(r_B = 60 \text{ m})$. You can combine the field and potential equations to solve for the radius r_A.

EVALUATE Combining the above information gives

$$r_A \ln(r_A/r_B) = \frac{\Delta V_{AB}}{E} \quad \to \quad r_A \ln(r_A/60\text{m}) = \frac{-115 \text{ kV}}{3 \text{ MV/m}} = -0.0383 \text{ m}$$

where a negative sign has been added to the potential difference to be consistent with the fact that $r_A < r_B$. One way to solve this equation is with Newton's method. Let $y = r_A \ln(r_A/60) + 0.0383$ and let $r_{A,0}$ be your best guess for the root of y. Then you use the derivative of y, which in this case is $y' = \ln(r_A/60) + 1$, to find a better guess:

$$r_A = r_{A,0} - \frac{y(r_{A,0})}{y'(r_{A,0})} = r_{A,0} - \frac{r_{A,0} \ln(r_{A,0}/60) + 0.0383}{\ln(r_{A,0}/60) + 1}$$

This process can be repeated several times with the r_A of one iteration becoming the $r_{A,0}$ of the next iteration. A good first guess for the radius might be 1 cm. The table below shows how quickly Newton's method converges on the root.

$r_{A,0}$	$y(r_A)$	$y'(r_A)$	r_A
0.01	−0.048695147	−7.699514748	0.003675556
0.003675556	0.002645651	−8.700395344	0.00397964
0.00397964	1.22454E-05	−8.62090841	0.003981061
0.003981061	2.53465E-10	−8.620551548	0.003981061
0.003981061	0	−8.620551541	0.003981061

The minimum radius is 4.0 mm, so the minimum diameter is 8.0 mm.

ASSESS An 8-mm wire would likely be too fragile for a transmission line, but this at least gives you the lower limit on what you could use.

77. **INTERPRET** We are asked to analyze an electrocardiograph showing equipotentials in a human torso.

DEVELOP From the $V = 0$ and the $V = 0.5$ mV equipotentials, we can surmise that the potential rises to a peak in the upper left corner of the heart, and drops to a valley in the lower right corner of the heart.

EVALUATE Since electric field lines point downhill, the electric field in the heart must point from the upper left to the lower right.
The answer is (a).

ASSESS Between the two charges of the dipole, the electric field is parallel to the dipole moment (see Figure 22.16).

79. **INTERPRET** We are asked to analyze an electrocardiograph showing equipotentials in a human torso.

DEVELOP The equipotential line at point A has $V = 0.2$ mV. This line is approximately parallel and midway between the two surrounding equipotentials at 0.1 mV and 0.3 mV. Therefore, we can assume that the electric potential is approximately linear in this region with regard to the distance x from the $V = 0.2$ mV equipotential, i.e.,

$$V = 0.2 \text{ mV} + 0.1 \text{ mV}\left(\frac{x}{x_0}\right)$$

where x_0 is the distance between the equipotential lines. Assuming the torso is about 30 cm across, we estimate that $x_0 \approx 3$ cm.

EVALUATE The electric field is the derivative of the electric potential (Equation 22.9), so at the point A the field should be roughly:

$$\left|\vec{E}\right| = \frac{\partial}{\partial x}V \approx \frac{0.1 \text{ mV}}{3 \text{ cm}} \approx 3 \text{ mV/m} = 3 \text{ mN/C}$$

The closest answer is (b).

ASSESS The units work out because $1 \text{ V} = 1$ J/C. You can arrive at a similar answer by taking the derivative in radial components of the dipole potential in Equation 22.6:

$$\left|\vec{E}\right| = \left|\frac{\partial}{\partial r}V + \frac{1}{r}\frac{\partial}{\partial \theta}V\right| = V(r,\theta)\left(\frac{2 + \tan\theta}{r}\right)$$

If we assume point A is located about 15 cm from the dipole center and at an angle of about 60° from the dipole axis, then the magnitude of the electric field at point A is about $E \approx (0.2 \text{ mV})(25 \text{ m}^{-1}) = 5 \text{ mN/C}$.

ELECTROSTATIC ENERGY AND CAPACITORS

EXERCISES

Section 23.1 Electrostatic Energy

13. **INTERPRET** We are to find the work required to assemble a linear sequence of charges.

    DEVELOP We use the technique described for assembling the three charges in Figure 23.1. For this problem, we have 4 charges, to be arranged as shown in the figure below. Number the charges $q_i = 50 \ \mu C$, $i = 1, 2, 3, 4$, as they are spaced along the line at $a = 2$ cm intervals. There are six pairs, so

$$W = \sum_{\text{pairs}} \frac{kq_i q_j}{r_{ij}}$$

    which we can evaluate to find the work W.

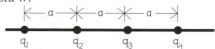

    EVALUATE Evaluating the expression above gives

$$W = k\left(\frac{q_1 q_2}{a} + \frac{q_1 q_3}{2a} + \frac{q_1 q_4}{3a} + \frac{q_2 q_3}{a} + \frac{q_2 q_4}{2a} + \frac{q_3 q_4}{a}\right)$$

$$= \frac{kq^2}{a}\left(1 + \tfrac{1}{2} + \tfrac{1}{3} + 1 + \tfrac{1}{2} + 1\right) = \frac{13 kq^2}{3a} = \frac{13\left(9.0 \times 10^9 \ \text{N} \cdot \text{m}^2/\text{C}^2\right)\left(50 \ \mu C\right)^2}{3.0 \times 0.20 \ \text{m}} = 4.9 \ \text{kJ}$$

    ASSESS The work required does not depend on how the charge configuration is assembled, only on its final state.

15. **INTERPRET** We are to repeat the preceding problem, with the final charge changed from $q_4 = q/2$ to $q_4 = -q$.

    DEVELOP Use the same strategy as for Problem 23.14, but with $q_4 = -q$

    EVALUATE Evaluating the expression for work required to assemble a collection of point charges gives

$$W = k\left(\frac{q_1 q_2}{a} + \frac{q_1 q_3}{a} + \frac{q_2 q_3}{a\sqrt{2}} + \frac{q_1 q_4}{a\sqrt{2}} + \frac{q_2 q_4}{a} + \frac{q_3 q_4}{a}\right)$$

$$= \frac{kq^2}{a}\left(1 + 1 + \frac{1}{\sqrt{2}} - \frac{1}{\sqrt{2}} - 1 - 1\right) = 0$$

    ASSESS Although it takes work to bring the second and third charges into place, that energy is regained by the negative work done in bringing in the fourth charge.

17. **INTERPRET** For this problem, we are to find the work required to assemble a crude model of a water molecule. Note that if the work is negative, then energy is released in forming the molecule.

    DEVELOP In this approximation, electrostatic potential energy of the water molecule (i.e., the work required to assemble the molecule) is

$$U = W = \sum_{\text{pairs}} \frac{kq_i q_j}{r_{ij}}$$

    The two oxygen-hydrogen pairs have separation $a = 10^{-10}$ m, while the hydrogen-hydrogen pair has separation $2a\cos\left(37.5°\right) = 1.59a$.

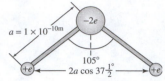

EVALUATE   Evaluating this expression gives

$$U = 2ke\left(\frac{-2e}{a}\right) + ke\left(\frac{e}{1.59a}\right) = -\frac{3.37ke^2}{a}$$

$$= -3.37\left(9.0\times10^9 \text{ N}\cdot\text{m}^2/\text{C}^2\right)\left(1.60\times10^{-19} \text{ C}\right)^2 = -7.76\times10^{-18} \text{ J} = -48.5 \text{ eV}.$$

ASSESS   Because the potential energy is negative, assembling this molecules releases energy (or does work). Note that the electrostatic potential energy of the assembled molecule is with respect to the constituents being infinitely far apart, so the work done equates to the change in potential energy caused by bringing the charges together from infinity.

Section 23.2 Capacitors

19.   **INTERPRET**   We are to find the work required to charge a capacitor with the given charge, then find the additional work required to double the charge.

DEVELOP   The separation between capacitor plates is much smaller than the linear dimensions of the plates, so the discussion in Section 23.2 applies. From Equation 23.3, we see that the work is

$$W = \frac{1}{2}CV^2$$

where V is the final voltage and may be expressed using Equation 23.1, $C = Q/V$. This gives

$$W = \frac{1}{2}C\left(\frac{Q}{C}\right)^2 = \frac{1}{2}\frac{Q^2}{C}$$

The capacitance can be expressed in terms of the geometry of the capacity (Equation 23.2, $C = \epsilon_0 A/d$ which leads to

$$W = \frac{1}{2}\frac{Q^2}{C} = \frac{1}{2}\frac{Q^2 d}{\epsilon_0 A}$$

EVALUATE   (a) The work required to transfer Q =7.2 µC is

$$W = \frac{1}{2}\frac{Q^2 d}{\epsilon_0 A} = \frac{1}{2}\frac{(7.2 \text{ µC})(0.0012 \text{ m})}{\left[8.85\times10^{-12} \text{ C}^2/(\text{N}\cdot\text{m}^2)\right](0.050 \text{ m})^2} = 1.4 \text{ J}$$

(b) The additional work required to double the charge on each plate is

$$\Delta W = \frac{(2Q)^2 d}{2\epsilon_0 A} - W = 3W = 4.2 \text{ J}$$

ASSESS   This energy is stored in the capacitor and can be released by electrically connecting the two capacitor faces.

21.   **INTERPRET**   We are given the charge and voltage of a capacitor and are to find the capacitance.

DEVELOP   Apply Equation 2.31, $C = Q/V$.

EVALUATE   From Equation 23.1, $C = Q/V = (1.3 \text{ µC})/(60 \text{ V}) = 22 \text{ nF}$.

ASSESS   The capacitance is the charge per unit voltage.

23.   **INTERPRET**   We are given the capacitance and charge of a capacitor and are to find the voltage.

DEVELOP   Solve Equation 23.1, $C = Q/V$, for the voltage V.

EVALUATE   Equation 23.1 gives $V = Q/C = (1.6 \text{ mC})/(100 \text{ µF}) = 16 \text{ V}$

ASSESS   This is a typical-sized capacitor.

25.   **INTERPRET**   We are given the separation of a parallel-plate capacitor, its charge, and its voltage and are to find its capacitance.

DEVELOP   Because the plate separation $d \ll r$, the radius of the capacitor, and apply Equation 23.2, $C = \varepsilon_0 A / d$, with $A = \pi r^2$.

EVALUATE   Inserting the given quantities gives

$$C = \frac{\varepsilon_0 \pi r^2}{d} = \frac{(8.85 \text{ pF/m}) \pi (0.20 \text{ m})^2}{0.0015 \text{ m}} = 740 \text{ pF}$$

to two significant figures.

ASSESS   This is a typical value for a capacitance.

27.   **INTERPRET**   We are given the capacitance and the voltage of a capacitor and are to find the stored energy.

DEVELOP   Apply Equation 23.3, $U = CV^2/2$.

EVALUATE   From Equation 23.3,

$$U_C = \tfrac{1}{2} CV^2 = \tfrac{1}{2}(2500 \text{ } \mu\text{F})(35 \text{ V})^2 = 1.5 \text{ J}$$

ASSESS   This is the energy it would take to lift 1.0 liter of water through a height of

$$U = mgh$$

$$h = \frac{U}{mg} = \frac{1.5 \text{ J}}{(1.0 \text{ kg})(9.8 \text{ m/s}^2)} = 15 \text{ cm}$$

Section 23.3 Using Capacitors

29.   **INTERPRET**   This problem involves calculating the equivalent capacitance for the two given capacitors connected in series or in parallel.

DEVELOP   Apply Equations 23.5 to find the equivalent series capacitance and Equation 23.23.6b for the equivalent parallel capacitance.

EVALUATE   In parallel, the capacitance is

$$C = C_1 + C_2 = 1.0 \text{ } \mu\text{F} + 2.0 \text{ } \mu\text{F} = 3.0 \text{ } \mu\text{F}$$

In series, the capacitance is

$$C = \frac{C_1 C_2}{C_1 + C_2} = \frac{(1.0 \text{ } \mu\text{F})(2.0 \text{ } \mu\text{F})}{1.0 \text{ } \mu\text{F} + 2.0 \text{ } \mu\text{F}} = 2/3 \text{ } \mu\text{F}$$

ASSESS   Connecting the capacitors in parallel results in a higher equivalent capacitance than connecting them in series.

31.   **INTERPRET**   This problem requires us to find the equivalent capacitance of the given arrangement of individual capacitors, which combines series and parallel connections.

DEVELOP   For part (a), compute the equivalent capacitance of C_2 and C_3 (which are in parallel), then combine this in series with C_1 to find the overall capacitance. For part (b), note that the charge on C_2 and C_3 must be the same as on C_1, which must be the same as the total charge (see discussion accompanying Figure 23.8), so

$$Q_T = Q_1 = Q_2 + Q_3$$

In addition, the total charge and capacitance must satisfy Equation 23.1,

$$C_T = \frac{Q_T}{V_T}$$

where $V_T = 100$ V. Finally, the voltage drop across C_2 must be the same as across C_3, so

$$V_2 = Q_2/C_2 = V_3 = Q_3/C_3$$

$$Q_2 = Q_3 \frac{C_2}{C_3}$$

EVALUATE   (a) Capacitors C_2 and C_3 combined in parallel give an equivalent capacitance of

$$C_{2,3} = C_1 + C_2 = 0.03\,\mu\text{F}$$

Combining this in series with C_1 gives

$$C_T = \frac{C_1 C_{2,3}}{C_1 + C_{2,3}} = \frac{(0.02\,\mu\text{F})(0.03\,\mu\text{F})}{0.02\,\mu\text{F} + 0.03\,\mu\text{F}} = 0.012\,\mu\text{F}$$

(b) Knowing C_T and V_T, we can find Q_T, which must be the same as Q_1 (see Example 23.3)

$$C_T = \frac{Q_T}{V_T} = \frac{Q_1}{V_T}$$

$$Q_1 = C_T V_T = (0.0012\,\mu\text{F})(100\text{ V}) = 0.12\,\mu\text{C}$$

Knowing Q_1, we can solve for Q_2 and Q_3 using the expressions above. The result is

$$Q_1 = Q_2 + Q_3 = Q_3\frac{C_2}{C_3} + Q_3 = Q_3\left(1 + \frac{C_2}{C_3}\right)$$

$$Q_3 = \frac{Q_1}{1 + C_2/C_3} = \frac{0.12\,\mu\text{C}}{1 + 1/2} = 0.080\,\mu\text{C}$$

Finally, Q_2 is

$$Q_2 = Q_3\frac{C_2}{C_3} = (0.080\,\mu\text{C})\frac{1}{2} = 0.040\,\mu\text{C}$$

(c) The voltage on C_1 can be found using Equation 23.1. The result is

$$V_1 = \frac{Q_1}{C_1} = \frac{0.12\,\mu\text{C}}{0.02\,\mu\text{F}} = 60\text{ V}$$

$$V_2 = V_3 = V_T - V_1 = 100\text{ V} - 60\text{ V} = 40\text{ V}$$

ASSESS   The voltage across the parallel capacitors may also be found using Equation 23.1:

$$V_2 = V_3 = \frac{Q_2}{C_2} = \frac{Q_3}{C_3} = 40\text{ V}$$

Section 23.4 Energy in the Electric Field

33.   **INTERPRET**   This problem involves finding the uniform electric field that carries the given energy density.

DEVELOP   Apply Equation 23.7, ($E = \sqrt{2u/\epsilon_0}$) which relates the field strength and the electric energy density.

EVALUATE   Inserting the given energy density into Equation 23.7 gives

$$E = \sqrt{2u/\epsilon_0} = \sqrt{\frac{2(3.0\text{ J/m}^3)}{8.85 \times 10^{-12}\text{ F/m}}} = 8.2 \times 10^5\text{ V/m}$$

ASSESS   The manipulation of units is facilitated by the relations V = J/C and F = C/V. Thus,

$$(\text{J/m}^3)/(\text{F/m}) = \frac{\text{VC/m}^3}{\text{C}/(\text{V}\cdot\text{m})} = \text{V/m}^2$$

35.   **INTERPRET**   This problem involves finding the maximum electrical energy density possible in air.

DEVELOP   From Appendix C, we find that the energy content of gasoline is 44×10^6 J/kg , and the density of gasoline is 670 kg/m^3, so the equivalent energy density is

$$u_{\text{gas}} = (44 \times 10^6\text{ J/kg})(670\text{ kg/m}^3) = 2.95 \times 10^{10}\text{ J/m}^3$$

EVALUATE From Equation 23.7, the field strength giving the same electrostatic energy density is

$$E = \sqrt{2u/\epsilon_0} = \sqrt{\frac{2\left(2.95\times10^{10}\ \text{J/m}^3\right)}{8.85\times10^{-12}\ \text{F/m}}} = 8.16\times10^{10}\ \text{V/m}$$

which greatly exceeds the breakdown field in air.

ASSESS Gasoline is actually a very dense form of energy storage, which is one reason it is hard to replace!

PROBLEMS

37. **INTERPRET** This problem involves finding an expression of the work required to assemble the given charge configuration, and using this expression to find the relative charge on one of the charges with respect to the initial charge.

DEVELOP The work necessary to position Q_x is

$$W_x = \frac{kQ_0Q_x}{a} = \frac{2kQ_0^2}{a}$$

while the work necessary to bring up Q_y is

$$W_y = \frac{kQ_0Q_y}{a} + \frac{kQ_xQ_y}{\sqrt{2}a} = \frac{kQ_0Q_y\left(1+\sqrt{2}\right)}{a}$$

Given that $W_y = 2W_x$, we can solve for Q_y in terms of Q_0.

EVALUATE $W_y = 2W_x$ gives

$$\frac{kQ_0Q_y\left(1+\sqrt{2}\right)}{a} = \frac{4kQ_0^2}{a}$$

$$Q_y = \frac{4Q_0}{\sqrt{2}+1}$$

ASSESS More explicitly, we can write $Q_y = 1.66Q_0$, so Q_y is a little less than twice Q_0.

39. **INTERPRET** This problem requires us to find the charge on a pair of parallel, square conducting plates (i.e., a parallel-plate capacitor) given the energy density in the electric field between the plates.

DEVELOP Combine Equation 23.7, which relates the electric field to the energy density,

$$E = \pm\sqrt{2u/\epsilon_0}$$

with Equation 21.8, which gives the electric field near the surface of a charged conducting plate:

$$E = \sigma/\epsilon_0$$

EVALUATE Eliminating the unknown electric field E and solving for the surface charge density σ gives

$$\sigma = \pm\epsilon_0\sqrt{2u/\epsilon_0}$$

Using the given surface area, the total charge on a plate is

$$q = \sigma A = \pm A\sqrt{2u/\epsilon_0} = \pm(0.10\ \text{m})^2\sqrt{2\left(4.5\ \text{kJ/m}^3\right)\left[8.85\times10^{-12}\ \text{C}^2/\left(\text{N}\cdot\text{m}^2\right)\right]} = \pm2.8\ \mu\text{C}$$

ASSESS Notice that we do not know the sign on the charge because of the charge symmetry involved.

41. **INTERPRET** We are to find the capacitance of a sphere surrounded by a concentric shell.

DEVELOP This geometry is the same as for Problem 23.38, for which we found the potential between the spheres to be

$$V = kQ\left(a^{-1} - b^{-1}\right)$$

from which we can find the capacitance using Equation 23.1, $C = Q/V$.

EVALUATE   The capacitance is

$$V = kQ\left(a^{-1} - b^{-1}\right) = \frac{Q(b-a)}{4\pi\epsilon_0 ab} = \frac{Q}{C}$$

$$C = \frac{4\pi\epsilon_0 ab}{(b-a)}$$

ASSESS   The capacitance depends only on the geometry of the capacitor, as expected.

43.   **INTERPRET**   This problem involves finding the energy stored in a capacitor for a given voltage and finding the capacitance of the capacitor.

   DEVELOP   Apply Equation 23.3 to express the energy stored as a function of voltage. Take the ratio of the two expressions to find the energy stored at 25 V. The same equation may be used to find the capacitance.

   EVALUATE   (a) Equation 23.3, expressed as a ratio for the same capacitor charged to two different voltages, gives $U_2/U_1 = (V_2/V_1)^2$. Therefore, $U_2 = (25/100)^2(0.04 \text{ J}) = 2.5$ mJ.
   (b) Solving Equation 23.3 for the capacitance gives $C = 2U_1/V_1^2 = 2(0.04 \text{ J})/(100 \text{ V})^2 = 8.0\,\mu\text{F}$.

   ASSESS   The energy stored scales as the voltage squared so, for example, twice the voltage gives four times the energy.

45.   **INTERPRET**   This problem requires us to find the voltage across a capacitor given its capacitance and stored energy.

   DEVELOP   Solve Equation 23.3 for the voltage.

   EVALUATE   The capacitor must withstand a potential difference of $V = \sqrt{2U_c/C} = \sqrt{2(12 \text{ mJ})/(10\,\mu\text{F})} = 49$ V, so one rated at 50 V would just suffice.

   ASSESS   If more than 50 V were to put across the capacitor, dielectric breakdown would occur.

47.   **INTERPRET**   We are to find the voltage across the given capacitor and the power it discharges if all its energy is discharged in the given time.

   DEVELOP   Solve Equation 23.3 for voltage to find the voltage across the capacitor. Use the definition of average power, $\bar{P} = E/\Delta t$ to find the average power.

   EVALUATE   (a) From Equation 23.3, $V = \sqrt{2\,U/C} = \sqrt{2(950 \text{ J})/(100\,\mu\text{F})} = 4.4$ kV.
   (b) $\bar{P} = \Delta U/\Delta t = (300 \text{ J})/(2.5 \text{ ms}) = 120$ kW.

   ASSESS   This is like connecting yourself to 2000 60-W light bulbs for 2.5 ms. Care to try?

49.   **INTERPRET**   This problem involves calculating the equivalent capacitance of a group of capacitors connected as shown in Figure 23.15.

   DEVELOP   Consider the same circuit, but drawn as shown in the figure below. This circuit has two parallel components that consist of (1) C_1 and (2) $C_{2,3,4}$. The second component may be further divided into two components in series: (3) C_2 and (4) $C_{3,4}$. Finally, the fourth component consists of C_3 and C_4 in parallel. Thus, the entire circuit can be described as

$$C_{\text{tot}} = C_1 \| C_{2,3,4} = C_1 \| \left(C_2 \leftrightarrow C_{3,4}\right) = C_1 \| \left[C_2 \leftrightarrow \left(C_3 \| C_4\right)\right]$$

where $\|$ means "in parallel with" and $\leftrightarrow$ means "in series with". Apply Equations 23.5 for the parallel capacitances and Equation 23.6b for the series capacitances to find the total capacitance C_{tot}.

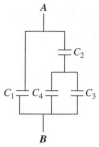

A

C_2

$C_1 \quad C_4 \quad C_3$

B

EVALUATE Because C_1 and $C_{2,3,4}$ are in parallel, the total capacitance is

$$C_{\text{tot}} = C_1 + C_{234}.$$

Because C_2 is in series with $C_{3,4}$, $C_{2,3,4}$ is

$$C_{234} = \frac{C_2 C_{34}}{C_2 + C_{34}}$$

Because C_3 and C_4 are in parallel, $C_{3,4}$ is

$$C_{34} = C_3 + C_4$$

Therefore, the total capacitance is

$$C_{\text{tot}} = C_1 + \frac{C_2 C_{34}}{C_2 + C_{34}} = C_1 + \frac{C_2(C_3 + C_4)}{C_2 + C_3 + C_4}$$

Because the capacitors are all identical ($C_1 = C_2 = C_3 \equiv C$), this reduces to

$$C_{\text{tot}} = C_1 + \frac{C_2(C_3 + C_4)}{C_2 + C_3 + C_4} = C + \frac{2C^2}{3C} = \frac{5}{3}C$$

ASSESS Note that redrawing the circuit made it easier to understand (hopefully).

51. **INTERPRET** For this problem, we are to find the equivalent capacitance for the given capacitor.

    DEVELOP Number the capacitors as shown in the figure below. Relative to points A and B, C_1, C_4, and the combination of C_2 and C_3 are in series, so the capacitance is given by Equation 23.6a:

    $$C_{AB}^{-1} = C_1^{-1} + C_4^{-1} + C_{23}^{-1}.$$

    C_{23} is a parallel combination, so $C_{23} = C_2 + C_3$, and we can find C_{AB}.

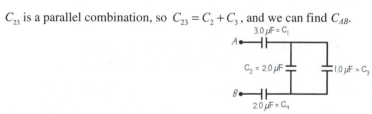

    EVALUATE Inserting the expression for C_{23} into that for C_{AB} gives

    $$C_{AB}^{-1} = (3.0\,\mu\text{F})^{-1} + (2.0\,\mu\text{F})^{-1} + (2.0\,\mu\text{F} + 1.0\,\mu\text{F})^{-1}$$
    $$C_{AB} = \tfrac{6.0}{7.0}\,\mu\text{F} = 0.86\,\mu\text{F}$$

    ASSESS The result is given to two significant figures, as justified by the data.

53. **INTERPRET** This problem involves two capacitors in series. We are to find an expression for the voltage across each capacitor.

    DEVELOP From Equation 23.5, we have

    $$\frac{1}{C} = \frac{1}{C_1} + \frac{1}{C_2}$$

Furthermore, the charge on each capacitor must be the same, as argued in the text accompanying Figure 23.8 so, from Equation 23.1, $Q = CV_1 = CV_2 = CV$. In addition, the voltages across each capacitor must sum to the voltage across the equivalent capacitor, so $V = V_1 + V_2$.

EVALUATE   Combining the expressions above to solve for V_1 in terms of V, C_1, and C_2 gives

$$V_1 = V - V_2 = V - Q/C_2 = V - \frac{CV}{C_2} = V - \frac{V}{\left(C_1^{-1} + C_2^{-1}\right)C_2} = V\left(1 - \frac{C_1}{C_2 + C_1}\right) = \frac{VC_2}{C_2 + C_1}$$

Solving for V2 likewise gives

$$V_2 = V - V_1 = V - Q/C_1 = V - \frac{CV}{C_1} = V - \frac{V}{\left(C_1^{-1} + C_2^{-1}\right)C_1} = V\left(1 - \frac{C_2}{C_2 + C_1}\right) = \frac{VC_1}{C_2 + C_1}$$

ASSESS   These expressions agree with that given in the problem statement.

55.   **INTERPRET**   For this problem, we are to find the capacitance and working voltage of a capacitor given its geometry and dielectric material.

DEVELOP   Apply Equations 22.4 and use Table 23.1 for the dielectric constant of polyethylene.

EVALUATE   **(a)** From Equation 23.4, and Table 23.1, one obtains

$$C = \kappa \frac{\varepsilon_0 A}{d} = (2.3)\frac{(8.85 \text{ pF/m})(50 \times 10^{-4} \text{ m}^2)}{25 \times 10^{-6} \text{ m}} = 4.1 \text{ nF}$$

(b) Dielectric breakdown in polyethylene occurs at a field strength of 50 kV/mm, corresponding to a maximum voltage for this capacitor of

$$V = Ed = \left(50 \times 10^6 \text{ V/m}\right)(25 \times 10^{-6} \text{ m}) = 1.3 \text{ kV}$$

ASSESS   The results are given to two significant figures, as warranted by the data.

57.   **INTERPRET**   In this problem, we are given the capacitance per unit area and the dielectric strength for a capacitor and are asked to find the plate separation.

DEVELOP   If we assume that the inner and outer surfaces of the membrane act like a parallel plate capacitor, with the space between the plates filled with material of dielectric constant $\kappa = 3$, then we can use Equation 23.4 to find the separation d.

EVALUATE   The capacitance per unit area is $C/A = \kappa \epsilon_0/d$. Thus, $d = 3(8.85 \text{ pF/m})/\left(1 \text{ } \mu\text{F/cm}^2\right) = 2.7 \text{ nm}$.

ASSESS   This result is about an order of magnitude larger than the Bohr radius, which gives an idea of the thickness of the membrane in terms of atoms (biological molecules being largely hydrogen and carbon).

59.   **INTERPRET**   We're asked to find the total electric-field energy in a cubical region with a variable field.

DEVELOP   The electric-field energy for a variable field is given by Equation 23.8: $U = \frac{1}{2}\epsilon_0 \int E^2 dV$. For the cubical region in this case, $dV = dx\,dy\,dz$, and the integration for each variable is from 0 to 1 m.

EVALUATE   Performing the integration, the energy stored in the field is

$$U = \frac{1}{2}\epsilon_0 \int_0^{1\text{m}} dz \int_0^{1\text{m}} dy \int_0^{1\text{m}} \left[E_0\left(\frac{x}{x_0}\right)\right]^2 dx$$

$$= \frac{1}{2}\left(8.85 \times 10^{-12} \tfrac{C^2}{N \cdot m^2}\right)(1 \text{ m})^2 (24 \text{ kV/m})^2 \left[\frac{\frac{1}{3}(1\text{ m})^3}{(6 \text{ m})^2}\right] = 24 \text{ } \mu\text{J}$$

ASSESS   The value seems reasonable. One can check that the units work out by using the fact that $1 \text{ V} \cdot \text{C} = 1 \text{ J} = 1 \text{ N} \cdot \text{m}$.

61.   **INTERPRET**   This problem involves a spherical charge distribution outside of which we are to find the total energy contained in the electric field.

DEVELOP   This problem is the same Example 23.5 if we let $R_2 = R$ and $R_1 \rightarrow \infty$ (i.e., we can consider that we are compressing an infinitely separated charge distribution to one that is on the surface of the sphere of radius R_2).
EVALUATE   The energy is thus

$$U = \frac{kQ^2}{2R}$$

ASSESS   The energy is quadratic in charge so, for example, it would take 4 times the energy to assemble twice the charge.

63.   **INTERPRET**   This problem involves finding the change in electrostatic potential energy between two separated, charged, water drops and a single drop with the same charge.

DEVELOP   The initial electrostatic energy of two isolated spherical drops, with charge Q on their surfaces and radii R, $U_i = 2(\frac{1}{2}kQ^2/R)$ (see Problem 61 and Example 23.5). Together, a drop of charge $2Q$, radius $2^{1/3}R$, and energy $U_f = \frac{1}{2}k(2Q)^2/(2^{1/3}R) = 2^{2/3}kQ^2/R$ is created. The difference in potential energy is $W = U_f - U_i$

EVALUATE   Inserting the given quantities gives

$$\Delta U = \frac{(2^{2/3} - 1)kQ^2}{R} = \frac{(0.587)(9.0 \times 10^9 \text{ N} \cdot \text{m}^2/\text{C}^2)(1.5 \times 10^{-8} \text{ C})^2}{2.0 \times 10^{-3} \text{ m}} = 6.0 \times 10^{-4} \text{ J}$$

ASSESS   In eV, this is

$$\Delta U = \frac{6.0 \times 10^{-4} \text{ J}}{1.6 \times 10^{-19} \text{ J/eV}} = 3.8 \times 10^{15} \text{ eV}$$

65.   **INTERPRET**   For this problem, we are asked to find the time required for lightening strikes to deplete a reservoir of energy, given the charge transferred, the electric potential energy difference (per charge), and the frequency of the lightening strikes.

DEVELOP   The energy in the thunderstorm of Example 23.4 is about 1.4×10^{11} J, while the energy in a lightning flash is $qV = (30 \text{ C})(30 \text{ MV}) = 9.0 \times 10^8$ J. Thus, there is energy for about $(1.4 \times 10^{11})/(9.0 \times 10^8) = 156$ flashes.
EVALUATE   At a rate of one flash every 5 s, there is enough energy to last $156 \times (5 \text{ s}) = 13$ min.

ASSESS   This seems like a reasonable time frame for a summer thunderstorm.

67.   **INTERPRET**   This problem involves a spherical, uniform charge density. We are to find the fraction of the total energy contained within the sphere.

DEVELOP   In Problem 23.60, we found that the energy within just such a charged sphere is

$$U_{\text{inside}} = \frac{kQ^2}{10R}$$

In Problem 23.61, we found that the energy outside such a sphere is

$$U_{\text{outside}} = \frac{kQ^2}{2R}$$

Divide U inside by the sum to find the fraction of energy inside the sphere.
EVALUATE   The fraction of energy inside the sphere is

$$f_{\text{inside}} = \frac{U_{\text{inside}}}{U_{\text{inside}} + U_{\text{outside}}} = \frac{kQ^2/(10R)}{kQ^2/(10R) + kQ^2/(2R)} = \frac{1}{1+5} = \frac{1}{6}$$

ASSESS   Ignoring gravity, this result is independent of the size of the sphere.

69.   **INTERPRET**   This problem involves a charged capacitor into which we insert a dielectric material so that it occupies half the volume of the capacitor. We are to find the new capacitance, the stored energy, and the force on the dielectric in this configuration.

DEVELOP   Use the coordinate system defined in the figure below. In so far as fringing fields can be neglected, the electric field between the plates is a uniform $E = V/d$ (but when the dielectric is inserted, $V \neq V_0$ and E depends on x). In fact, on the left side, where the slab has penetrated, $E = (1/\kappa)(\sigma_L/\epsilon_0)$ and on the right, $E = \sigma_R/\epsilon_0$, where σ_L and σ_R are the charge densities on the left and right sides, respectively. Thus, $\sigma_L = \kappa\epsilon_0 E$ and $\sigma_R = \epsilon_0 E$, and the charge can be written (in terms of geometrical values taken from Fig. 23.19) as

$$Q = \sigma_L wx + \sigma_R w(L - x) = \epsilon_0 Ew(\kappa x + L - x) = \epsilon_0 (V/d)w(\kappa x + L - x).$$

When the battery is disconnected, the capacitor is isolated and the charge on it is a constant, $Q = Q_0$, and we can use Equation 23.3 $U = CV^2/2$ to find the energy stored in the capacitor. The force on a part of an isolated system is related to the potential energy of the system by Equation 8.9. The force on the slab is therefore

$$F_x = -\frac{dU}{dx} = -\frac{d}{dx}\left(\frac{U_0 L}{\kappa x + L - x}\right) = \frac{U_0 L(\kappa - 1)}{(\kappa x + L - x)^2}$$

in the direction of increasing x (so as to pull the slab into the capacitor).

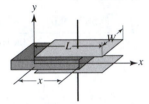

EVALUATE   (a) From Equation 23.1,

$$C = \frac{Q}{V} = C_0 \frac{(\kappa x + L - x)}{L}$$

where $C_0 = \epsilon_0 A/d$ and $A = Lw$. Inserting $x = L/2$, we find

$$C = C_0 \frac{\kappa + 1}{2}$$

(b) The stored energy is

$$U = \frac{Q^2}{2C} = \frac{q_0^2 L}{2C_0(\kappa x + L - x)} = \frac{U_0 L}{\kappa x + L - x}$$

where $U_0 = Q_0^2/(2C_0) = C_0 V_0^2/2$. For $x = L/2$, the energy is $C_0 V_0^2/(\kappa + 1)$.
(c) For $x = L/2$, the magnitude of the force is

$$F_x = \frac{2C_0 V_0^2(\kappa - 1)}{L(\kappa + 1)^2}$$

ASSESS   Notice that the results depend only on geometrical factors and the dielectric strength of the material. It turns out that if we rewrite the force, for any value of x, in terms of the voltage for that x, using $q_0 = C_0 V_0 = CV = C_0 V(\kappa x + L - x)/L$, the expression can be used in the succeeding problem. Thus,

$$F_x = \frac{C_0 V_0^2 L(\kappa - 1)}{2(\kappa x + L - x)^2} = \frac{C_0}{2}\left(\frac{V}{L}\right)^2 L(\kappa - 1) = \frac{C_0 V^2(\kappa - 1)}{2L}$$

71. **INTERPRET** We are to find the capacitance per unit length of two parallel wires whose separation is much, much greater than their radii, so we can assume line symmetry (i.e., infinite wires). The wires carry opposite charge density.

    DEVELOP Use the coordinate system shown in the figure below. Apply Equation 22.1a:

    $$\Delta V_{AB} = -\int_A^B \vec{E} \cdot d\vec{r}$$

    The total electric field is the superposition of the electric fields due to the two wires, each of which is given by Equation 21.6. Summing the two contributions gives

    $$\vec{E} = \frac{\lambda}{2\pi\epsilon_0 r_-}(-\hat{i}) + \frac{\lambda}{2\pi\epsilon_0 r_+}(-\hat{i}) = -\frac{\lambda}{2\pi\epsilon_0}\left(\frac{1}{x} + \frac{1}{b-x}\right)(\hat{i})$$

    Insert this into the integral and perform the integration to find the voltage difference between the wires. The wire capacitance per unit length can then be found using Equation 23.1, $Q = CV$, which we can transform into per unit length by dividing each side by an arbitrary length L:

    $$\frac{Q}{L} = \frac{C}{L}V \quad \Rightarrow \quad \lambda = C_L V$$

    where C_L is the capacitance per unit length.

    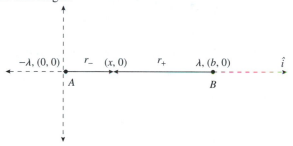

    EVALUATE Evaluating the integral gives

    $$\Delta V_{AB} = -\int_A^B \vec{E} \cdot d\vec{r} = \frac{\lambda}{2\pi\epsilon_0}\int_a^{b-a}\left(\frac{1}{x} + \frac{1}{b-x}\right)dx = \frac{\lambda}{2\pi\epsilon_0}\left[\ln(x) - \ln(b-x)\right]_a^{b-a} = \frac{\lambda}{\pi\epsilon_0}\ln\left(\frac{b-a}{a}\right)$$

    Inserting this into the expression for capacitance gives

    $$C_L = \pi\epsilon_0 \ln\left(\frac{b-a}{a}\right)^{-1}$$

    ASSESS The capacitance per unit length depends only on geometrical parameters, and is positive.

73. **INTERPRET** We are to find the electrostatic energy stored between two parallel plates of a parallel-plate capacitor, and then differentiate to find the force between the plates.

    DEVELOP Using Equation 21.8, find the electric field between the plates:

    $$E = \frac{\sigma}{\epsilon_0} = \frac{Q}{A\epsilon_0}$$

    This is constant, so total energy stored in this field is then $U = uV$, where u is the energy density (energy per unit volume). We can find the force by using $F_x = -dU/dx$.

    EVALUATE (a) The electrostatic potential energy is

    $$U = uV = \left(\frac{1}{2}\epsilon_0 E^2\right)(Ax) = \frac{Q^2}{2A\epsilon_0}x$$

    (b) The force between the plates is

    $$F_x = -\frac{dU}{dx} = -\frac{Q^2}{2A\epsilon_0}$$

This is half the value you would obtain by multiplying the charge on one plate by the field between the plates.

ASSESS   The answer we get for **(b)** is half the field times the charge on one plate: but we must remember that the field between the plates is created by *both* charged plates. A charge is not affected by the field it creates. Only the field created by the *other* plate causes a force on each plate, and the other plate creates half the field.

75.   **INTERPRET**   We're considering the energy used by the National Ignition Facility.

DEVELOP   The energy stored in a capacitor is given by $U = \frac{1}{2}CV^2$ (Equation 23.3).

EVALUATE   We're told that the NIF capacitor system stores 400 MJ at 20 kV, so the capacitance is

$$C = \frac{2U}{V^2} = \frac{2(400 \text{ MJ})}{(20 \text{ kV})^2} = 2 \text{ F}$$

The answer is (d).

ASSESS   This is an impressively large capacitance. The advantage of using capacitors in this application is that they can discharge rapidly and thus supply a large amount of power over a short time.

77.   **INTERPRET**   We're considering the energy used by the National Ignition Facility.

DEVELOP   The power is energy divided by time.

EVALUATE   We're told that the lasers deliver 2 MJ of energy in 1 ns. So the power is

$$P = \frac{\Delta E}{\Delta t} = \frac{2 \times 10^6 \text{ J}}{1 \times 10^{-9} \text{ s}} = 2 \times 10^{15} \text{ W} = 2 \text{ PW}$$

The answer is (d).

ASSESS   This is over 100 times the world's average power consumption, but it only lasts for a fraction of a second.

24

ELECTRIC CURRENT

EXERCISES

Section 24.1 Electric Current

13. **INTERPRET** This problem involves converting current in amperes to the number of electrons moving past a given point in a wire per unit time.

    DEVELOP Current in amperes is the amount of charge in Coulombs passing a given point in the wire, per second. Mathematically, this is given by Equation 24.1a:

    $$I = \frac{\Delta q}{\Delta t}$$

    To find the number of electrons, divide the charge in Coulombs by the charge of a single electron in Coulombs ($|e|$ = 1.60×10^{-19} C).

    EVALUATE Solving the expression above for the charge and dividing it by the charge of an electron gives the number n of electrons as

    $$n = \frac{\Delta q}{|e|} = \frac{I\Delta t}{|e|} = \frac{(15\text{ A})(1.0\text{ s})}{1.6 \times 10^{-19}\text{ C}} = 9.4 \times 10^{18}$$

    ASSESS Because electrons carry a negative charge, they actually move in the direction opposite that of the given current.

15. **INTERPRET** This problem involves finding the charge that moves through a membrane over a given time intervale given the current.

    DEVELOP From the definition of an ampere, we know that the charge moving through the membrane each second is 30 nC (see Equation 24.1a; $q = I\Delta t$). Since singly-charged ions carry one elementary charge (about q_{ion} = 160 zC), we can find the total charge that passes through the membrane in one second.

    EVALUATE Letting the units guide us, we find the number n of ions that pass the membrane in one second is

    $$n = \frac{I\Delta t}{q_{ion}} = \frac{(30\text{ nC/s})(1\text{ s})}{160\text{ zC/ion}} = 1.9 \times 10^{11}\text{ ion}$$

    ASSESS This may also be expressed in terms of the number n_{mol} of moles:

    $$n_{mol} = \frac{1.88 \times 10^{11}\text{ ion}}{6.02 \times 10^{23}\text{ ion/mol}} = 0.31\text{ pmol}$$

    Chemical drug-testing instrumentation can detect amounts of substances this low.

Section 24.2 Conduction Mechanisms

17. **INTERPRET** The problem involves the microscopic version of Ohm's law.

    DEVELOP Aluminum obeys Ohm's law, so from Equation 24.4b: $J = E/\rho$, where the resistivity of aluminum is given in Table 24.1: $\rho = 2.65 \times 10^{-8}\ \Omega \cdot \text{m}$.

EVALUATE   The current density in the wire is:

$$J = \frac{E}{\rho} = \frac{85 \text{ mV/m}}{2.65 \times 10^{-8} \Omega \cdot \text{m}} = 3.2 \times 10^6 \text{ A/m}^2$$

ASSESS   The units work out, since $1 \Omega = 1$ V/A.

19.   INTERPRET   We are asked to find the diameter of a current-carrying cylinder, given the current and the electric field. The resistivity of the cylinder is also known via Table 24.1.

DEVELOP   Assuming a uniform current density obeying Ohm's law, apply Equation 24.4b, $J = E/\rho$ to find the current density J. Knowing J, we can find the cylinder diameter using $J = I/A$ (see Example 24.2), where $A = \pi(d/2)^2$.

EVALUATE   Combining the expressions above to solve for the diameter d gives

$$J = E/\rho = \frac{I}{\pi(d/2)^2}$$

$$d = \sqrt{\frac{4\rho I}{\pi E}} = 2\sqrt{\frac{(0.22 \ \Omega \cdot \text{m})(350 \text{ mA})}{\pi(21 \text{ V/m})}} = 6.8 \text{ cm}$$

ASSESS   This seems like a reasonable inner diameter for the tube.

21.   INTERPRET   Given the resistivity of two materials, we are to find their conductivity.

DEVELOP   Equation 24.4a and 24.4b show that the conductivity and the resistivity are reciprocals of one another.

EVALUATE   (a) For copper, $\rho^{-1} = \sigma = (1.68 \times 10^{-8} \ \Omega \cdot \text{m})^{-1} = 5.95 \times 10^7 (\Omega \cdot \text{m})^{-1}$.
(b) For seawater, $\sigma = (0.22 \ \Omega \cdot \text{m})^{-1} = 4.55 \ (\Omega \cdot \text{m})^{-1}$.

ASSESS   The salinity of open-ocean water varies between 33 and 37 parts per thousand, but can vary from 1 to 80 parts per thousand in shallow coastal waters. This variation has a proportional effect on the conductivity.

Section 24.3 Resistance and Ohm's Law

23.   INTERPRET   This problem involves using the macroscopic form of Ohm's law to find the voltage needed to produce a desired current.

DEVELOP   For an Ohmic resistor, the macroscopic version of Ohm's law (Equation 24.5) is V = IR.
EVALUATE   Inserting the given quantities gives $V = IR = (300 \text{ mA})(1.2 \text{ k}\Omega) = 360$ V.

ASSESS   This is three times the peak voltage provided by standard power outlets in the USA, so it would not be possible to generate this current without employing some method to increase the voltage from the power outlet.

25.   INTERPRET   We are to find the resistance of an iron rail with the given dimensions. This involves using both macroscopic (resistantce) and microscopic (resitivity) concepts.

DEVELOP   The microscopic resistivity ρ is connected to the macroscopic resistance R by Equation 24.6, $R = \rho L/A$.
EVALUATE   Inserting the given quantities gives $R = \rho L/A = (9.71 \times 10^{-8} \ \Omega \cdot \text{m})(5.0 \text{ km})/(10 \times 15 \text{ cm}^2) = 32 \text{ m}\Omega$..

ASSESS   This resistance is quite low, which explains in part why this material is used for this purpose.

27.   INTERPRET   We are to find the resistance of a wire whose length is doubled while its density remains constant.
DEVELOP   If the density (and mass) of the wire remains constant, then its volume is constant, so $V = LA =$ constant. Using the fact that the length is doubled ($2L_1 = L_2$), we can find the new resistance by using Equations 24.6, $R = \rho L/A$.
EVALUATE   Taking the ratio of the Equation 24.6 applied to the wire before and after stretching gives

$$\frac{R_1}{R_2} = \frac{\rho L_1/A_1}{\rho L_2/A_2} = \frac{L_1^2/(L_1 A_1)}{L_2^2/(L_2 A_2)} = \frac{L_1^2/(V)}{L_2^2/(V)} = \left(\frac{L_1}{L_2}\right)^2$$

$$R_2 = R_1 \left(\frac{L_2}{L_1}\right)^2 = 4R_1$$

ASSESS   The resistance is quadratic in L, so doubling L increases the resistance by a factor of 4.

Section 24.4 Electric Power

29. **INTERPRET** We are given the power required to operate a device and the current it draws. From this information, we are to find its voltage and resistance.

    DEVELOP Apply Equation 24.7, $V = P/I$, to find the voltage and Equation 24.5, $R = V/I$ to find the resistance.

    EVALUATE (a) From Equation 24.7, $V = P/I = (4.5 \text{ W})/(750 \text{ mA}) = 6.0 \text{ V}$.

    (b) From Equation 24.5, $R = V/I = (6.0 \text{ V})/(0.75 \text{ A}) = 8.0 \,\Omega$.

    ASSESS This is a small voltage and a small resistance, with the ratio giving rise to a significant current.

31. **INTERPRET** We are to find the voltage at which a device operates, given its resistance and the power it consumes.

    DEVELOP Solve Equation 24.8b, $P = V^2/R$, for the voltage V.

    EVALUATE Equation 24.8b gives $V = \sqrt{PR} = \sqrt{(1.5 \text{ kW})(35 \,\Omega)} = 230 \text{ V}$.

    ASSESS This is greater than the peak power provided by standard power outlets in the USA, which is not surprising for an electric stove.

Section 24.5 Electrical Safety

33. **INTERPRET** We use the macroscopic version of Ohm's law to find the resistance necessary for the given voltage to drive the given current.

    DEVELOP The macroscopic version of Ohm's law is $V = IR$ (Equation 24.5). We are given $V = 30$ V and $I = 100$ mA, so we can solve for R.

    EVALUATE Inserting the given quantities yields

    $$R = \frac{V}{I} = \frac{30 \text{ V}}{100 \text{ mA}} = 300 \,\Omega$$

    ASSESS The resistance of dry skin is much higher than this, so in dry conditions this voltage does not pose a threat.

35. **INTERPRET** Given a resistance and voltage, what current would flow? We will use the macroscopic version of Ohm's law, and also estimate whether the resulting current could be felt.

    DEVELOP The macroscopic version of Ohm's law is $V = IR$. The resistance is $R = 100$ kΩ, and the voltage is $V = 12$ V. The threshold for sensation is listed in Table 24.3 as 0.5–2 mA.

    EVALUATE (a) Inserting the given quantities gives a current of

    $$I = \frac{V}{R} = \frac{12 \text{ V}}{100 \times 10^3 \,\Omega} = 0.12 \text{ mA}$$

    (b) This current is below the threshold for sensation, and would not be felt.

    ASSESS The resistance of human skin varies considerably with moisture. If your hand was wet, the resistance would be lower and the current would be high enough to deliver a noticeable shock.

PROBLEMS

37. **INTERPRET** The problem asks us to compare the current density in a light bulb filament and the wire that supplies electricity to the bulb.

    DEVELOP The current density is just current divided by area. The same amount of current flows through the filament and the wire; only the area changes.

    EVALUATE (a) For the filament, the current density is:

    $$J = \frac{I}{A} = \frac{0.833 \text{ A}}{\pi \left(\frac{1}{2} 0.050 \text{ mm}\right)^2} = 420 \text{ A/mm}^2$$

    (b) For the 12-gauge wire, the current density is:

    $$J = \frac{I}{A} = \frac{0.833 \text{ A}}{\pi \left(\frac{1}{2} 2.1 \text{ mm}\right)^2} = 0.24 \text{ A/mm}^2$$

    ASSESS The thinner filament has the higher current density, as we would expect.

39. **INTERPRET** This problem involves two wires of different radii, each carrying the same current I. Given the density of charge carriers in each, we are to find the drift speed and the current density between wires.

DEVELOP Apply Equation 24.3 to each wire, and take the ratio to find the ratio of the drift speeds, using the fact that I = JA and we are given $A_{Al} = 4A_{Cu}$. The same strategy may be used to find the ratio of the current densities.

EVALUATE Equation 24.3 gives

$$\frac{I_{Cu}}{I_{Al}} = 1 = \frac{J_{Cu}A_{Cu}}{J_{Al}A_{Al}} = \frac{J_{Cu}A_{Cu}}{J_{Al}(4A_{Cu})} = \frac{n_{Cu}ev_{d,Cu}}{4n_{Al}ev_{d,Al}}$$

$$\frac{v_{d,Cu}}{v_{d,Al}} = 4\left(\frac{n_{Al}}{n_{Cu}}\right) = 4\left(\frac{2.1\times10^{29}}{1.1\times10^{29}}\right) = 7.6$$

(b) The same equation also gives $J_{Cu}A_{Cu}/J_{Al}A_{Al} = 1$, or $J_{Cu}/J_{Al} = A_{Al}/A_{Cu} = 4$.

ASSESS Thus, the current density is 4 times greater in copper than in aluminum. Which wire would you expect to heat up faster for the same current? This is the fundamental concept of a fuse.

41. **INTERPRET** For this problem, we are given the current as a function of time and are asked to find the current of a 5.0-s interval.

DEVELOP Integrate Equation 24.1b from t = 0.0 to t = 5.0 s, with I(t) as given in the problem statement.

EVALUATE Performing the integration gives

$$q = \int_{0.0}^{5.0} I(t)\,dt = \int_{0.0}^{5.0}\left(60t + 200t^2 + 4.0t^3\right)dt = \left|30t^2 + \frac{200}{3}t^3 + 1.0t^4\right|_{0.0}^{5.0} = \left(30 + \frac{1000}{3} + 25\right)(25)\text{ nC} = 9.7\ \mu\text{C}$$

ASSESS Because the formula for I is in nA, we get nC when we multiply by seconds.

43. **INTERPRET** Given the current and the cross-sectional area of a copper wire, we are to find the current density and the electric field.

DEVELOP The current density is $J = I/A$. From Table 24.1, the resistivity is $\rho = 1.68 \times 10^{-8}$ $\Omega\cdot$m, which we can use in Equation 24.4b to find the electric field.

EVALUATE **(a)** For a wire carrying uniform current density,

$$J = \frac{I}{A} = \frac{I}{\pi d^2/4} = \frac{20\text{ A}}{\frac{1}{4}\pi\left(2.1\times10^{-3}\text{ m}\right)^2} = 5.8\text{ MA/m}^2.$$

(b) The electric field is $E = \rho J = \left(1.68\times10^{-8}\ \Omega\cdot\text{m}\right)\left(5.77\text{ MA/m}^2\right) = 97\text{ mV/m}$.

ASSESS This is a rather small electric field, which is not surprising because copper is a good conductor so the charges react quickly in an attempt to cancel out any electric field in the material.

45. **INTERPRET** We are given the dimensions and electrical characteristics of a material (i.e., voltage and current) and are asked to identify the material by matching its resitivity to one from Table 24.1.

DEVELOP For a uniform piece of material, Equations 24.5 and 24.6 imply $\rho = RA/L = VA/IL = V\left(\frac{1}{4}\pi d^2\right)/IL$.

EVALUATE Inserting the given quantities gives

$$\rho = \frac{(9\text{ V})\left(\frac{1}{4}\pi\right)(2.0\text{ mm})^2}{(2.6\text{ mA})(2.4\text{ cm})} = 0.453\ \Omega\cdot\text{m}$$

which is closest to the resistivity of germanium (= 0.47 $\Omega\cdot$m) in Table 24.1.

ASSESS This is a useful concept for identifying materials.

47. **INTERPRET** You need to specify the maximum length of an extension cord for a power saw. The cord has resistance that will reduce the voltage difference across the power saw's motor.

DEVELOP Let's define 4 points along the current's path, see figure below. You're told that the voltage across the outlet, ΔV_{14}, is 120 V, and the voltage across the motor, ΔV_{23}, is 115 V.

$$\Delta V_{12} = \Delta V_{34} = IR$$

The copper wires in the extension cord will produce a voltage drop due to their resistance: $\Delta V_{12} = \Delta V_{34} = IR$. The resistance is given by Equation 24.6: $R = \rho L / A$, where $\rho = 1.68 \times 10^{-8} \Omega \cdot m$ for copper, A is the area of the 1-mm-wide wire, and L is the length of the cord. For the voltage differences to be consistent with each other:

$$\Delta V_{14} = \Delta V_{12} + \Delta V_{23} + \Delta V_{34} \quad \rightarrow \quad 2IR = 5 \text{ V}$$

Since the saw draws 7 A, the resistance in one direction of the extension cord is 0.357 Ω.

EVALUATE Solving for the length of the cord, we get:

$$L = \frac{RA}{\rho} = \frac{(0.357 \ \Omega)\pi(\frac{1}{2} \cdot 1 \text{ mm})^2}{1.68 \times 10^{-8} \Omega \cdot m} = 16.7 \text{ m}\left(\frac{1 \text{ ft}}{0.3048 \text{ m}}\right) = 55 \text{ ft}$$

Because the cords come in increments of 25 ft, the maximum extension cord one can use with the saw is 50 ft.

ASSESS We generally tend to ignore the resistance in the wires that carry current to an electric device. But for an extension cord, the long length of the wires means their resistance will have to be accounted for.

49. **INTERPRET** We are to find the resistance between opposing faces of a rectangular block of iron, given that the opposing faces are all equipotentials (i.e., the electric potential is the same across any given pair of faces).

DEVELOP If opposite faces are equipotentials, the current density is uniform over any parallel cross-section. In this case, Equation 24.6 gives $R_1 = \rho L_1 = (L_2 \times L_3)$, where L_1 is the length in the direction of the potential drop and $L_2 \times L_3$ is the cross-sectional area of an equipotential face.

EVALUATE For $L_1 = 20$ cm, $L_2 = 1.0$ cm, and $L_3 = 0.50$ cm, and permutations thereof, we find $R_1 = (9.71 \times 10^{-8} \ \Omega \cdot m)(20 \text{ cm})/(1.0 \times 0.50 \text{ cm}^2) = 388 \ \mu\Omega$, $R_2 = 0.971 \ \mu\Omega$, and $R_3 = 0.243 \ \mu\Omega$.

ASSESS The greatest resistance is greatest for current traversing the greatest length per unit area.

51. **INTERPRET** This problem involves the dependence of resistance and power dissipation on the geometry of the material. We are to find the diameters of resistors given their length is the same and the same voltage is applied across both.

DEVELOP Equation 24.6, $R = \rho L/A$, relates the resistance of a material to its geometry (length and cross-sectional area). With a fixed voltage, the power dissipated is given by Equation 24.8b, $P = V^2/R$. Thus, we see that at the same voltage, the ratio of the power dissipated is the inverse of the ratio of the resistances, which in turn, goes as the inverse of the square of the ratio of the diameters:

$$\frac{P_1}{P_2} = \frac{V^2/R_1}{V^2/R_2} = \frac{R_2}{R_1} = \frac{\rho L/A_2}{\rho L/A_2} = \frac{A_1}{A_1} = \frac{\pi d_1^2/4}{\pi d_2^2/4} = \left(\frac{d_1}{d_2}\right)^2$$

EVALUATE From the equation above, we see that if $P_1 = 2P_2$, then $d_1 = \sqrt{2}d_2$.

ASSESS Our result shows that power dissipated increases with the area, or the square of the diameter, $P \propto A \propto d^2$.

53. **INTERPRET** This problem is about using electric power to do mechanical work.

DEVELOP If there are no losses, the electrical power supplied to the motor, $P_{in} = IV$, must equal the mechanical power expended lifting the weight, $P_{out} = Fv$ (see Equation 6.19).

EVALUATE With $P_{in} = P_{out}$, the current is

$$I = \frac{Fv}{V} = \frac{(15 \text{ N})(0.25 \text{ m/s})}{6.0 \text{ V}} = 0.63 \text{ A}$$

ASSESS In reality, the motor will not be 100% efficient. So the current drawn will be somewhat higher than 0.63 A.

55. **INTERPRET**   You are looking to save money on utility costs by comparing copper to aluminum wires.

DEVELOP   You're given the specifications for the resistance per unit length along the power line. From Equation 24.6: $R/L = \rho/A$, where the resistivity of copper and aluminum are $\rho_{\text{Cu}} = 1.68 \times 10^{-8} \, \Omega \cdot \text{m}$ and $\rho_{\text{Al}} = 2.65 \times 10^{-8} \, \Omega \cdot \text{m}$, respectively. This allows you to specify the cross-sectional area, A, of the wire. Since the mass per length is $m/L = \eta A$, where η is the mass density for either copper or aluminum, the cost per unit length will be

$$C/L = (C/m)(m/L) = \frac{(C/m)\eta\rho}{(R/L)}$$

where C/m is cost per kg of the material.

EVALUATE   The cost per length of copper wire is

$$C/L = \frac{(\$4.65/\text{kg})(8.9 \text{ g/cm}^3)(1.68 \times 10^{-8} \, \Omega \cdot \text{m})}{(50 \text{ m}\Omega/\text{km})} = \$14/\text{m}$$

The cost per length of aluminum wire is

$$C/L = \frac{(\$2.30/\text{kg})(2.7 \text{ g/cm}^3)(2.65 \times 10^{-8} \, \Omega \cdot \text{m})}{(50 \text{ m}\Omega/\text{km})} = \$3.30/\text{m}$$

Clearly, the aluminum is more economical.

ASSESS   Copper is a better conductor, so the required diameter of the wire $(d = 2.1 \text{ cm})$ is smaller than that for aluminum $(d = 2.6 \text{ cm})$. But aluminum is lighter and costs less per kilo, so it would save money to use aluminum wires.

57. **INTERPRET**   This problem involves a nonuniform current density in a metal bar. We are given the dimensions of the bar and are to find the total current in the bar.

DEVELOP   Use the coordinate system given below in the figure. The current density in the bar is $\vec{J} = (0.10 \text{ A/cm}^2)(x/10 \text{ cm})\hat{k}$. The cross section can be divided into strips of area $d\vec{A} = (5.0 \text{ cm}) \, dx\hat{k}$ (over which $\vec{J}$ is constant), so we can integrate over x to find the total current:

$$I = \int_x \vec{J} \cdot d\vec{A}$$

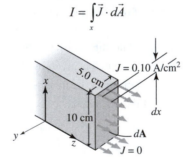

EVALUATE   Performing the integration gives

$$I = \int_x \vec{J} \cdot d\vec{A} = \int_0^{10 \text{ cm}} (0.010 \text{ A/cm}^3)(5.0 \text{ cm}) x dx$$

$$= (0.050 \text{ A/cm}^2)\frac{1}{2}(10 \text{ cm})^2 = 2.5 \text{ A}$$

ASSESS   This is quite a significant current!

59. **INTERPRET**   This problem gives the resistivity of copper as a function of temperature, from which we are to find the temperature at which the resistivity is twice its room-temperature value.

DEVELOP   We are to find the resistivity $\rho(T) = 2\rho(T_0)$ (where we let $T_0 = 20 \text{ °C}$ be room temperature). Take the ratio of this equation for these two temperatures and solve for T.

EVALUATE   The temperature at which the resistivity doubles is

$$\frac{\rho(T_0)}{\rho(T)}=\frac{1}{2}=\frac{\rho_0}{\rho_0\left[1+\alpha(T-T_0)\right]}$$

$$T=T_0+1/\alpha=20\,°C+(1°C)/(4.3\times10^{-3})=250\,°C$$

to two significant figures.

ASSESS   This ambient temperature is much higher than is encountered in normal circumstances. At a very hot ambient 50 °C, the resistivity is 13% higher than at room temperature.

61.   **INTERPRET**   For this problem, we are to find an expression for the resistivity through the solution with resistivity ρ from the cylinder side to the center disk.

DEVELOP   Consider a concentric cylindrical surface S, of radius r and height h, between the two metal electrodes. S completely surrounds the inner disk, so the current flowing (assumed from the center to the sides) is

$$I=\int_S \vec{J}\cdot d\vec{A}=\int_S (\vec{E}/\rho)\cdot d\vec{A}$$

for an ohmic solution. If the electrodes behave like perfect conductors $(\rho_{\text{metal}}=0)$ they are essentially equipotentials, and the electric field in the solution has cylindrical (line) symmetry. (There are no fringing fields because outside the solution, the current density is zero.) Thus, without changing the integral, flat circular surfaces above and below may be added to close the surface S, so that

$$\rho I=\int_S \vec{E}\cdot d\vec{A}$$

This integral is the same as the one appearing in Gauss's law for two conductors in an identical configuration, as in Example 21.5. Thus,

$$\rho I=\int_S \vec{E}\cdot d\vec{A}=\frac{2\pi h V}{\ln(b/a)}$$

where $V=V_a-V_b$ is the potential difference and we used h instead of L for the length. Comparing this with the macroscopic version of Ohm's law (Equation 24.5) $V=IR$, we can solve for the resistivity ρ.

EVALUATE   A comparison of this with Ohm's law, $V=IR$, shows that the resistance between the electrodes is

$$R=\frac{\rho\ln(b/a)}{2\pi h}$$

ASSESS   In terms of the capacitance of the same configuration of electrodes with air between the electrodes,

$$q/\epsilon_0=CV/\epsilon_0=\rho I=\rho(V/R),\text{ or }R=\rho\epsilon_0/C$$

This relation holds for electrodes of arbitrary shape.

63.   **INTERPRET**   We are to find the current in a particle beam, given its current density as a function of the beam radius. The beam is circular, has a current density J_0 at the center, and the current density falls to half that value at the edge.

DEVELOP   Integrate the current density over the beam to find the total current. From the problem statement, we can express the current density as a function of radius as

$$J(r)=J_0-\frac{J_0 r}{2a}$$

To find the total current, integrate over circular rings of radius r, each of which has area $dA=2\pi r dr$.

EVALUATE   Performing the integration gives

$$dI = JdA$$

$$I = \int_0^a J(r)\, dA = 2\pi J_0 \int_0^a \left(1 - \frac{r}{2a}\right) r\, dr$$

$$= 2\pi J_0 \left(\frac{r^2}{2} - \frac{r^3}{6a}\right)_0^a = \frac{2}{3} J_0 \pi a^2$$

ASSESS   The beam current is 2/3 as much as it would be if the current density was constant.

65.   **INTERPRET**   You want to find the steepest hill that a hybrid car can climb using only its battery.

DEVELOP   The maximum power that the battery can supply is given by $P_{max} = I_{max} V$. If all this power is used to propel the car forward, then the force will be equal to the power divided by the velocity: $F = P_{max}/v$ (recall Equation 6.19). If we neglect wind resistance, this force must be equal and opposite to the component of the gravitational force that is parallel to the slope of the incline, $mg\sin\theta$.

EVALUATE   Solving for the angle of the incline gives:

$$\theta = \sin^{-1}\left(\frac{I_{max} V}{mgv}\right) = \sin^{-1}\left(\frac{(180\ \text{A})(360\ \text{V})}{(1200\ \text{kg})(9.8\ \text{m/s}^2)(60\ \text{km/h})}\right) = 19°$$

ASSESS   It's rare that one encounters a slope this steep. But in a real situation, you would have to account for not all of the battery's power being used to turn the wheels. Moreover, the drag from wind resistance will reduce the car's speed for a given battery output.

67.   **INTERPRET**   We explore the effects of a brownout on a electrical network.

DEVELOP   We will use the fact that the current density $(J = I/A)$ decreases during a brownout.

EVALUATE   The current density is related to the electric field by Equation 24.4b: $E = \rho J$, so the electric field strength should decrease with the current density. The current density is also related to the number of charge carriers and the drift speed through Equation 24.3: $J = nqv_d$. The number of free electrons does not vary in conductors under normal conditions. Therefore, the drift speed must fall when the current density decreases. Finally, the number of electron collisions depends primarily on the temperature.

The answer is (a).

ASSESS   Another way to think of this is that the electric field is directly related to the voltage: $E = dV/dx$ in the one-dimensional case. So a drop in voltage corresponds to a drop in the electric field. Moreover, the electric field provides a force on the charges that results in the drift velocity.

69.   **INTERPRET**   We explore the effects of a brownout on a electrical network.

DEVELOP   The temperature of the resistor in a light bulb or a stove top should be proportional to the power being dissipated. Since we've just argued in the previous problem that the power decreases during a brownout, the temperature in these resistors should be lower than normal periods of electricity supply.

EVALUATE   As the problem states, the resistance scales with the temperature, so these devices should have slightly higher current going through them than if their resistance were constant. In other words, their current decreases by less than 10%.

The answer is (c).

ASSESS   This temperature dependence reflects the fact that at higher temperatures there will be more electron collisions. This reduces the net flow of charge in the direction of the applied electric potential.

25 ELECTRIC CIRCUITS

EXERCISES

Section 25.1 Circuits, Symbols, and Electromotive Force

13. **INTERPRET** This problem is an exercise in drawing a circuit diagram, given a written description of the circuit in terms of its capacitors, resistors, and battery.

DEVELOP A literal reading of the circuit specifications results in connections like those in sketch **(a)**, below.

EVALUATE See sketch below. Because the connecting wires are assumed to have no resistance (a real wire is represented by a separate resistor), a topologically equivalent circuit diagram is shown in sketch **(b)**.

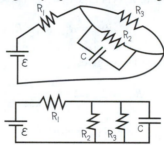

ASSESS There are three paths to ground (or to the negative battery terminal), two of which go through resistors and the third of which goes through the capacitor.

15. **INTERPRET** This problem involves drawing a circuit diagram from the description given in the problem statement.

DEVELOP The circuit has three parallel branches: one with R_1 and R_2 in series; one with just R_3; and one with the battery.

EVALUATE See figure below. R_{int} is the internal resistance of the battery.

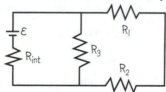

ASSESS This circuit has no capacitors, so we could replace R_1, R_2, and R_3 by an equivalent resistance

$$\frac{1}{R_{eq}} = \frac{1}{R_3} + \frac{1}{R_1 + R_2}$$

$$R_{eq} = \frac{R_3(R_1 + R_2)}{R_3 + R_2 + R_1}$$

17. **INTERPRET** This problem involves finding for how long a battery can supply the given current while maintaining its rated voltage. We are given the total energy stored in the battery, so if we can find the power the battery must deliver, we can find the time needed to deplete this energy reservoir.

DEVELOP Delivering the given current at the rated voltage results in a power expenditure of

$$\overline{P} = IV$$

(see Equation 24.7). Because the average power is defined as $\overline{P} = \Delta W / \Delta t$, we can find Δt, given that the energy we have to spend is $\Delta W = 4.5$ kJ.

EVALUATE Solving for the time interval Δt and inserting the known quantities gives

$$\Delta t = \frac{\Delta W}{\overline{P}} = \frac{\Delta W}{IV} = \frac{4.5 \text{ kJ}}{(1.5 \text{ V})(0.60 \text{ A})} = 5.0 \times 10^3 \text{ s} = 1.4 \text{ h}$$

ASSESS This result assumes that the battery voltage does not decrease as it depletes its store of energy, which is not a realistic assumption (although for most batteries, the departure from the ideal situation assumed in this problem is not huge).

Section 25.2 Series and Parallel Circuits

19. **INTERPRET** This problem involves calculating the equivalent resistance of the given resister combination.

DEVELOP Apply Equations 25.1 and 25.3b. For the parallel pair of resistors R_1 and R_2, Equation 25.3b gives

$$R_{1,2} = \frac{R_1 R_2}{R_1 + R_2}$$

Combining this in series with resistor R_3 (via Equation 25.1) gives

$$R_{1,2,3} = R_3 + R_{1,2} = R_3 + \frac{R_1 R_2}{R_1 + R_2}$$

EVALUATE Inserting the given resistances gives

$$R_{1,2,3} = 22 \text{ k}\Omega + \frac{(47 \text{ k}\Omega)(39 \text{ k}\Omega)}{47 \text{ k}\Omega + 39 \text{ k}\Omega} = 43 \text{ k}\Omega$$

ASSESS The final resistance is greater than the resistance of R_3 alone, as expected, but it is less than the resistance of R_1 alone. This is because R_2 is parallel to R_1 and allows some current to flow through the circuit without traversing R_1 (i.e., it adds another "traffic lane").

21. **INTERPRET** This problem involves analyzing a real battery circuit. We are given the rated voltage of the battery and its real voltage when it powers the defective starter, and are asked to find its real voltage for a proper starter.

DEVELOP This problem is similar to Example 25.2, which shows that the battery is connected in series with the load (i.e., the starter motor, which we treat as a resistor). To find the internal resistance of the battery, we use the macroscopic version of Ohm's law, $V = IR$:

$$V_{\text{terminals}} = 6 \text{ V} = \varepsilon - IR_{\text{int}} = 12 \text{ V} - (300 \text{ A})R_{\text{int}}$$
$$R_{\text{int}} = 0.020 \text{ }\Omega$$

where $V_{\text{terminals}}$ is the actual voltage across the battery's terminals. Knowing the internal resistance of the battery, we can repeat this calculation to find $V_{\text{terminals}}$ when a proper starter is used.

EVALUATE The voltage across the terminals when a proper starter is used is

$$V_{\text{terminals}} = 12 \text{ V} - (100 \text{ A})(0.020 \text{ }\Omega) = 10 \text{ V}$$

ASSESS Because the battery has an internal resistance, the voltage it can deliver is reduced by the voltage drop across the internal battery.

23. **INTERPRET** We are to find the internal resistance of a battery, given its short-circuit current and its rated voltage.

DEVELOP The battery contains an internal resistance (see section "Real Batteries"), which we can find using the macroscopic version of Ohm's law, $V = IR$, where $V = \varepsilon = 9$ V .

EVALUATE   Inserting the given quantities into Ohm's law gives

$$R_{int} = \varepsilon/I = (9 \text{ V})/(0.2 \text{ A}) = 50 \text{ } \Omega$$

to a single significant figure.

ASSESS   This is a rather large value for an internal resistance of a 9-V battery.

Section 25.3 Kirchhoff's Laws and Multiloop Circuits

25.   **INTERPRET**   This problem requires us to find the currents in all parts of a multi-loop circuit.

DEVELOP   The general solution of the two loop equations and one node equation given in Example 25.4 can be found using determinants (or I_1 and I_2 can be found in terms of I_3, as in Example 25.4). The equations and the solution are:

$$I_1 R_1 + 0 + I_3 R_3 = \varepsilon_1 \quad \text{(loop 1)}$$
$$0 - I_2 R_2 + I_3 R_3 = \varepsilon_2 \quad \text{(loop 2)}$$
$$I_1 - I_2 - I_3 = 0 \quad \text{(node A)}$$

$$\Delta \equiv \begin{vmatrix} R_1 & 0 & R_3 \\ 0 & -R_2 & R_3 \\ 1 & -1 & -1 \end{vmatrix} = R_1 R_2 + R_2 R_3 + R_3 R_1, \quad I_1 = \frac{1}{\Delta}\begin{vmatrix} \varepsilon_1 & 0 & R_3 \\ \varepsilon_2 & -R_2 & R_3 \\ 0 & -1 & -1 \end{vmatrix} = \frac{\varepsilon_1(R_2 + R_3) - \varepsilon_2 R_3}{\Delta}$$

$$I_2 = \frac{1}{\Delta}\begin{vmatrix} R_1 & \varepsilon_1 & R_3 \\ 0 & \varepsilon_2 & R_3 \\ 1 & 0 & -1 \end{vmatrix} = \frac{\varepsilon_1 R_3 - \varepsilon_2 (R_1 + R_3)}{\Delta}, \quad I_3 = \frac{1}{\Delta}\begin{vmatrix} R_1 & 0 & \varepsilon_1 \\ 0 & -R_2 & \varepsilon_2 \\ 1 & -1 & 0 \end{vmatrix} = \frac{\varepsilon_2 R_1 + \varepsilon_1 R_2}{\Delta}$$

EVALUATE   With the particular values of emfs and resistors in this problem, we have

$$\Delta = R_1 R_2 + R_2 R_3 + R_3 R_1 = (2 \text{ }\Omega)(4 \text{ }\Omega) + (4 \text{ }\Omega)(1 R_1) + (1 \text{ }\Omega)(2 \text{ }\Omega) = 14 \text{ }\Omega^2$$

and the currents are

$$I_1 = \left[(R_2 + R_3)\varepsilon_1 - R_3 \varepsilon_2\right]\Delta^{-1} = \frac{(4 \text{ }\Omega + 1 \text{ }\Omega)(6 \text{ V}) - (1 \text{ }\Omega)(1 \text{ V})}{14 \text{ }\Omega^2} = 2.07 \text{ A} = 2 \text{ A}$$

$$I_2 = \left[R_3 \varepsilon_1 - (R_1 + R_3)\varepsilon_2\right]\Delta^{-1} = \frac{(1 \text{ }\Omega)(6 \text{ V}) - (2 \text{ }\Omega + 1 \text{ }\Omega)(1 \text{ V})}{14 \text{ }\Omega^2} = 0.214 \text{ A} = 0.2 \text{ A}$$

$$I_3 = (R_2 \varepsilon_1 + R_1 \varepsilon_2)\Delta^{-1} = \frac{(4 \text{ }\Omega)(6 \text{ V}) + (2 \text{ }\Omega)(1 \text{ V})}{14 \text{ }\Omega^2} = 1.86 \text{ A} = 2 \text{ A}$$

to a single significant figure.

ASSESS   The same results could be obtained by retracing the reasoning of Example 25.4, with $\varepsilon_2 = 1.0$ V replacing the original value in loop 2. Then, everything is the same until the equation for loop 2: $1.0 + 4I_2 - I_3 = 0$.

27.   **INTERPRET**   We are to find the current through a resistor in a given circuit, for which we can use Kirchhoff's laws. We will use the loops and nodes drawn in Example 25.4.

DEVELOP   The circuit is given to us in Figure 25.14, with one change: $\varepsilon_2 = 2.0$V. We will use node A and loops 1 and 2. These will give us three equations, which we will use to solve the three unknown currents. At node A, $-I_1 + I_2 + I_3 = 0$. For loop 1, $\varepsilon_1 - I_1 R_1 - I_3 R_3 = 0$. For loop 2, $\varepsilon_2 + I_2 R_2 - I_3 R_3 = 0$.

EVALUATE   Because we are interested in the current I_2, we eliminate the other two currents. The node equation gives us $I_1 = I_2 + I_3$. Substitute this into the equation for loop 1 and solve for I3

$$\varepsilon_1 - (I_2 + I_3)R_1 - I_3 R_3 = 0$$

$$I_3 = \frac{\varepsilon_1 - I_2 R_1}{R_1 + R_3}$$

Now substitute this value into the equation for loop 2 and solve for I_2:

$$\mathcal{E}_2 + I_2 R_2 - \frac{\mathcal{E}_1 - I_2 R_1}{R_1 + R_3} R_3 = 0$$

$$\mathcal{E}_2(R_1 + R_3) + I_2 R_2(R_1 + R_3) - \mathcal{E}_1 R_3 + I_2 R_1 R_3 = 0$$

$$\mathcal{E}_2(R_1 + R_3) + I_2(R_1 R_2 + R_2 R_3 + R_1 R_3) - \mathcal{E}_1 R_3 = 0$$

$$I_2 = \frac{\mathcal{E}_1 R_3 - \mathcal{E}_2(R_1 + R_3)}{R_1 R_2 + R_2 R_3 + R_1 R_3} = \frac{(6\text{V})(1\Omega) - (2\text{V})(3\Omega)}{8\Omega^2 + 4\Omega^2 + 2\Omega^2} = 0 \text{ V}/\Omega = 0 \text{ A}$$

ASSESS   The current through resistor R_2 is zero! Looking back at the original diagram, we can see that this would mean that battery 2 is supplying no current and the voltage drops through resistors 1 and 3 equal the voltage supplied by battery 1. This is a somewhat unexpected solution, but it is consistent.

Section 25.4 Electrical Measurements

29.   **INTERPRET**   This problem involves finding the measurement error caused by the nonzero resistance of the ammeter used to measure the current.

DEVELOP   The current in the circuit of Fig. 25.27 is

$$I = \frac{V}{R_{\text{tot}}} = \frac{V}{R_1 + R_2} = \frac{150 \text{ V}}{5 \text{ k}\Omega + 10 \text{ k}\Omega} = 10 \text{ mA}$$

With the ammeter inserted (in series with the resistors), the resistance R_{tot} is increased by $R_A = 100 \ \Omega$.

EVALUATE   The resulting current after including R_A is

$$I' = \frac{V}{R_1 + R_2 + R_A} = \frac{150 \text{ V}}{5 \text{ k}\Omega + 10 \text{ k}\Omega + 0.10 \text{ k}\Omega} = 9.93 \text{ mA}$$

which is about 0.66% lower than I.

ASSESS   The current reading by the ammeter is lower due to its internal resistance.

Section 25.5 Capacitors in Circuits

31.   **INTERPRET**   In this problem we are asked to show that the quantity RC, the product of resistance and capacitance, has units of time.

DEVELOP   The SI units for R and C are Ω and F, respectively. The units can be rewritten as

$$1\,\Omega = 1\frac{V}{A} = 1\frac{V}{C/s} = 1\frac{V \cdot s}{C}, \quad 1\,F = 1\frac{C}{V}$$

EVALUATE   From the expressions above, the SI units for the time constant, RC, are

$$\Omega \cdot F = \left(\frac{V \cdot s}{C}\right)\left(\frac{C}{V}\right) = s$$

as stated.

ASSESS   The quantity RC is the characteristic time for changes to occur in an RC circuit.

33.   **INTERPRET**   This problem involves the time dependence of the capacitor voltage in a charging RC circuit. We are to find the charging ratio of a capacitor after 5 RC time constants.

DEVELOP   The capacitor voltage as a function of time is given by Equation 25.6:

$$V_{\text{cap}} = \mathcal{E}(1 - e^{-t/RC})$$

EVALUATE   When $t = 5RC$, the equation above gives a voltage of

$$\frac{V_{\text{cap}}}{\mathcal{E}} = 1 - e^{-5} = 1 - 6.74 \times 10^{-3} \approx 99.3\%$$

of the applied voltage.

ASSESS   As time goes on and after many more time constants, we find essentially no current flowing to the capacitor, and the capacitor could be considered as being fully charged for all practical purposes.

35.   **INTERPRET**   We are to find the voltage across the capacitor in Figure 25.24a when it is fully charged, which implies that the current through the capacitor is zero.

DEVELOP   Use the results of Example 25.7b and Ohm's law to find the voltage required. If the capacitor is fully charged, then no current flows through it and the circuit is equivalent to the circuit shown in 25.24c. So we find the current through resistor R_2 in Figure 25.24c and then determine the voltage across resistor R_2, which will be the same as the voltage across the capacitor.

EVALUATE   The current through resistor R_2 is given in Example 25.7 as $I = \varepsilon/(R_1 + R_2)$. The voltage is given by Ohm's law as

$$V = IR = \left(\frac{\varepsilon}{R_1 + R_2}\right) R_2 = \varepsilon \left(\frac{R_2}{R_1 + R_2}\right)$$

ASSESS   In the limit of long charging times, this circuit behaves like a voltage divider.

PROBLEMS

37.   **INTERPRET**   This problem asks for the current in a resistor which is part of a more complex multiloop circuit. We will find the voltage drop over this resistor, which is part of a parallel combination of resistors, to find the current passing through it.

DEVELOP   The circuit in Fig. 25.28, with a battery connected across points A and B, is similar to the circuit analyzed in Example 25.3. In this case, we have one 1.0-Ω resistor in parallel with two 1.0-Ω resistors in series. Thus, combining Equations 25.1 and 25.3b, we find

$$\frac{1}{R_\parallel} = \frac{1}{1.0\,\Omega} + \frac{1}{1.0\,\Omega + 1.0\,\Omega} = \frac{3}{2\,\Omega} \;\rightarrow\; R_\parallel = \frac{2}{3}\,\Omega$$

and the total resistance is $R_\parallel$ in series with two 1.0-Ω resistors: $R_{\text{tot}} = 1.0\,\Omega + 1.0\,\Omega + \tfrac{2}{3}\Omega = \tfrac{8}{3}\Omega$. The total current through the battery is

$$I_{\text{tot}} = \frac{\varepsilon}{R_{\text{tot}}} = \frac{6.0\text{ V}}{8/3\,\Omega} = \frac{9}{4}\text{A} = 2.25\text{ A}.$$

EVALUATE   Using the macroscopic version of Ohm's law, the voltage across the parallel combination is

$$V_\parallel = I_{\text{tot}} R_\parallel = \left(\frac{9}{4}\text{A}\right)\left(\frac{2}{3}\Omega\right) = \frac{3}{2}\text{V}$$

which is the voltage across the vertical $R_v = 1\,\Omega$ resistor. Thus, the current through this resistor is then

$$I_v = \frac{V_\parallel}{R_v} = \frac{3/2\text{ V}}{1.0\,\Omega} = 1.5\text{ A}$$

ASSESS   We have a total of 2.25 A of current flowing around the circuit. At the vertex of the triangular loop, it is split into $I_v = 1.5$ A and $I' = I_{\text{tot}} - I_v = 0.75$A. The voltage drop across the vertical resistor ($V_\parallel = 1.5$ V) is the same as that going through point C and the two 1.0-Ω resistors: $V' = (0.75\text{ A})(1.0\,\Omega + 1.0\,\Omega) = 1.5$V. Thus, the result is consistent.

39.   **INTERPRET**   The circuit has two batteries connected in series. We will apply Kirchhoff's law to find the current that flows through the discharged battery.

DEVELOP   Terminals of like polarity are connected with jumpers of negligible resistance, giving a circuit as shown below. Kirchhoff's voltage law gives

$$\varepsilon_1 - \varepsilon_2 - IR_1 - IR_2 = 0$$

EVALUATE   Solving the equation above for I, we obtain

$$I = \frac{\varepsilon_1 - \varepsilon_2}{R_1 + R_2} = \frac{12\text{ V} - 9\text{V}}{0.02\,\Omega + 0.08\,\Omega} = 30\text{ A}$$

ASSESS   When you try to jump start a car, you connect positive to positive and negative to negative terminals. The current is quite significant, which is why you want to have the charged car running to prevent the battery from being drained.

41.   **INTERPRET**   This problem involves finding the rate of energy dissipation in the internal resistor of a battery if the terminals are shorted (i.e., connected together with a zero-resistance connection).

DEVELOP   For a short-circuited battery, the macroscopic version of Ohm's law (see Table 24.2) gives $I = \varepsilon/R_{int}$, so the dissipated power is (from Equation 24.8a)

$$P = I^2 R_{int} = \frac{\varepsilon^2}{R_{int}}$$

EVALUATE   Inserting the quantities given in the problem, the rate of energy dissipation is

$$P = \frac{\varepsilon^2}{R_{int}} = \frac{(6.0 \text{ V})^2}{2.5 \text{ }\Omega} = 14 \text{ W}$$

ASSESS   With ε held fixed at 6.0 V, we see that the power dissipated is inversely proportional to the internal resistance R_{int}.

43.   **INTERPRET**   To check the safety of a battery, you must determine if a lethal dose of current could potentially flow through a person who is damp or sweaty.

DEVELOP   The battery is not ideal. It has an internal resistance that will reduce the terminal voltage when current is flowing out of the battery. This internal resistance will be in series with the human body's resistance.

EVALUATE   The total resistance will be the sum of the internal resistance and the human body's resistance. Therefore, the current that could potentially flow through a person with wet skin touching the battery terminals is

$$I = \frac{V}{R_{int} + R_{human}} = \frac{72 \text{ V}}{100 \text{ }\Omega + 500 \text{ }\Omega} = 120 \text{ mA}$$

Yes, this current could be fatal.

ASSESS   You'll likely need to introduce a safety feature, such as a fuse, that can prevent such a high current from flowing out of the battery.

45.   **INTERPRET**   The circuit in this problem contains a battery—the emf source, and three resistors. We want to analyze the voltage across the one which is a variable resistor.

DEVELOP   The resistors in parallel have an equivalent resistance of $R_{\parallel} = RR_1/(R + R_1)$ from Equation 25.3b. The other R, and $R_{\parallel}$, is a voltage divider in series with voltage $\mathcal{E}$.

EVALUATE   (a) Using Equation 25.2, we find the voltage across R_1 to be

$$V_{\parallel} = \frac{R_{\parallel}}{R + R_{\parallel}} \mathcal{E} = \frac{R_1}{R + 2R_1} \mathcal{E}$$

(b) The voltage across R_1 is sketched below.

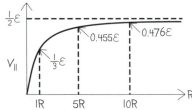

ASSESS   As $R_1 \to \infty$, the voltage across R_1 goes to $\mathcal{E}/2$, which is what the voltage would be if there were only two equal resistors in series.

47.   **INTERPRET**   This problem asks for the power dissipated in a resistor that is part of a multiloop circuit.

DEVELOP   The three resistors in parallel have an effective resistance of

$$\frac{1}{R_{\parallel}} = \frac{1}{2 \text{ }\Omega} + \frac{1}{4 \text{ }\Omega} + \frac{1}{6 \text{ }\Omega} = \frac{11}{12 \text{ }\Omega} \quad \Rightarrow \quad R_{\parallel} = \frac{12}{11} \text{ }\Omega$$

The equivalent resistance of the circuit is $R_{tot} = R_1 + R_{\parallel} = 1\Omega + \frac{12}{11}\Omega = \frac{23}{11}\Omega$. Equation 25.2 gives the voltage across them as

$$V_\parallel = \frac{\mathcal{E}R_\parallel}{R_{tot}} = \frac{(6\ \text{V})(12/11\ \Omega)}{23/11\ \Omega} = \frac{72}{23}\ \text{V}$$

EVALUATE   Using Equation 24.8b, the power dissipated in the 4-Ω resistor is

$$P_4 = \frac{V_\parallel^2}{R_4} = \frac{(72\ \text{V}/23)^2}{4\ \Omega} = 2.4\ \text{W}$$

which rounds to 2 W when retaining only a single significant figure.

ASSESS   With $\mathcal{E}$ held fixed at 6 V, we see that the power dissipated is inversely proportional to the resistance.

49.   **INTERPRET**   The problem asks for the equivalent resistance between two points in a multiloop circuit. Make a circuit diagram and label the nodes and currents as shown below.

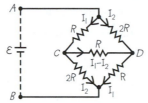

DEVELOP   The equivalent resistance is determined by the current which would flow through a pure emf if it were connected between A and B which, by Ohm's law, is $R_{AB} = \mathcal{E}/I$. Since I is but one of six branch currents, the direct solution of Kirchhoff's circuit laws is tedious (6 × 6 determinants). However, because of the values of the resistors in Fig. 25.32, a symmetry argument greatly simplifies the calculation. The equality of the resistors on opposite sides of the square implies that the potential difference between A and C equals that between D and B:

$$V_A - V_C = V_D - V_B$$

Equivalently, $V_A - V_D = V_C - V_B$. The symmetry argument requires that both R resistors on the perimeter carry the same current I_1 and both $2R$ resistors carry current I_2. Kirchhoff's current law then implies that the current through B is $I_1 + I_2$ and the current through the central resistor is $I_1 - I_2$ (as added to Figure 25.34). Now there are only two independent branch currents, which can be found from Kirchhoff's voltage law:

$$\mathcal{E} - I_1 R - I_2(2R) = 0 \quad \text{(loop ACBA)}$$
$$-I_1 R - (I_1 - I_2)R + I_2(2R) = 0 \quad \text{(loop ACDA)}$$

These equations may be rewritten as

$$I_1 + 2I_2 = \frac{\mathcal{E}}{R}$$
$$-2I_1 + 3I_2 = 0$$

with solution $I_1 = 3\mathcal{E}/(7R)$ and $I_2 = 2\mathcal{E}/(7R)$.

EVALUATE   The sum of the two currents gives $I = I_1 + I_2 = 5\mathcal{E}/(7R)$ which leads to

$$R_{AB} = \frac{\mathcal{E}}{I} = \frac{7R}{5}$$

ASSESS   The configuration of resistors considered here is called a Wheatstone bridge.

51.   **INTERPRET**   In this problem, we are to find the voltage across a given resistor as measured using a voltmeter with the given internal resistances. Because the voltmeter is connected in parallel with the 30-kΩ resistor, the voltmeter's resistance adds in parallel to the resistor's resistance.

DEVELOP   With a meter of resistance R_m connected as indicated in the figure below, the circuit reduces to two pairs of parallel resistors in series. The total resistance is the sum of these parallel resistances (Equations 25.1 and 25.3b):

$$R_{tot} = \frac{(30\ \text{k}\Omega)R_m}{30\ \text{k}\Omega + R_m} + \frac{40\ \text{k}\Omega}{2}$$

Using Ohm's law (Chapter 24), the voltage reading is

$$V_m = R_m I_m = \frac{R_m(30\ \text{k}\Omega)I_{tot}}{30\ \text{k}\Omega + R_m}$$

where $I_{tot} = (100 \text{ V})/R_{tot}$ (the expression for V_m follows from Equation 25.2, with R_1 and R_2 as the above pairs, or from I_m as a fraction of I_{tot}).

EVALUATE For the three voltmeter resistances specified, $I_{tot} = 2.58 \text{ mA}, \ 2.14 \text{ mA, and } 2.00 \text{ mA, and}$ $V_m = 48 \text{ V}, \ 57 \text{ V, and } 60 \text{ V, respectively.}$

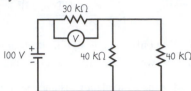

ASSESS Of course, 60 V is the ideal voltmeter reading. This reading corresponds to an ideal voltmeter that has infinite resistance. Thus, to two significant figures, the 10-MΩ voltmeter is an ideal voltmeter.

53. **INTERPRET** In this problem an ammeter is used to measure the current in a circuit. The ammeter is connected in series with the resistor.

DEVELOP The internal resistance of an ideal battery is zero, so the resistor has a value of $R = \mathcal{E}/I = 12.0 \ \Omega$. With the ammeter in place, the total resistance increases, and the current through the ammeter will be

$$I_a = \frac{\mathcal{E}}{R + R_a} = I\left(\frac{1}{1 + \frac{R_a}{R}}\right)$$

EVALUATE (a) The ammeter will read whatever current goes through it:

$$I_a = I\left(\frac{1}{1 + \frac{R_a}{R}}\right) = (1.00 \text{ A})\left(\frac{1}{1 + (0.10\Omega/12.0\Omega)}\right) = 0.992 \text{ A}$$

(b) If this current measurement were used to measure the resistance in the resistor, one would arrive at $R' = \mathcal{E}/I_a = 12.1 \text{V}$, which is an error of

$$\frac{R' - R}{R} = \frac{I}{I_a} - 1 = \frac{IR_a}{\mathcal{E}} = 0.83\%$$

ASSESS This is a small error. If one needed better accuracy, one could calculate the resistance in the resistor by accounting for the resistance in the ammeter.

55. **INTERPRET** You need to design a defibrillator that meets the desired discharge time. This is essentially an RC circuit, where the resistor is the human chest.

DEVELOP The defibrillator specs call capacitor to discharge to half its initial voltage in 10 ms. In terms of Equation 25.8, this implies: $e^{-t/RC} = \frac{1}{2}$. You can figure out the initial voltage using Equation 23.3: $U = \frac{1}{2}CV^2$.

EVALUATE Using $R = 40 \ \Omega$ for the transthoracic resistance, the needed capacitance is to the nearest 10 μF:

$$C = \frac{-t}{R \ln\left(\frac{1}{2}\right)} = \frac{(10 \text{ ms})}{(40 \ \Omega)\ln 2} = 361 \ \mu\text{F} \approx 360 \ \mu\text{F}$$

Given that the stored energy in the capacitor is 250 J, the initial voltage must be to the nearest 100 V:

$$V_0 = \sqrt{\frac{2U}{C}} = \sqrt{\frac{2(250 \text{ J})}{(361 \ \mu\text{F})}} = 1200 \text{ V}$$

ASSESS The initial current going through the chest is $I_0 = V_0/R = 30$ A. Such a huge amount of current can sometimes cause burns (see Table 24.3). But the person will likely die if this "jolt" to the heart is not applied in time.

57. **INTERPRET** This problem involves energy dissipation in an RC circuit. Given the energy dissipated in the given time, we are to find the capacitance.

DEVELOP A capacitor discharging through a resistor is described by exponential decay, with time constant RC (Equation 25.8):

$$V(t) = V(0)e^{-t/RC}$$

The energy in the capacitor is given by Equation 23.3:

$$U_C(t) = \frac{1}{2}CV(t)^2 = \frac{1}{2}CV(0)^2 e^{-2t/RC} = U_C(0)e^{-2t/RC}$$

EVALUATE If 2 J is dissipated in time t, the energy stored in the capacitor drops from $U_C(0) = 5.0$ J to $U_C(t) = 3.0$ J (assuming there are no losses due to radiation, etc.). From the equation above, the capacitance is

$$C = \frac{2t}{R\ln[U_C(0)/U_C(t)]} = \frac{2(8.6\text{ ms})}{(10\text{ k}\Omega)\ln(5.0\text{ J}/3.0\text{ J})} = 3.4\,\mu\text{F}$$

ASSESS In this problem the time constant is $RC = (10\text{ k}\Omega)(3.37\,\mu\text{F}) = 33.7$ ms. Therefore, at 8.6 ms (about 0.255 RC) the energy decreases by a factor $e^{-2(0.255)} \approx 0.6$, which is precisely what we found (i.e., from 5.0 V to $5.0 \times 0.6 = 3.0$ V).

59. **INTERPRET** This problem is about the long-term and short-term behavior of an RC circuit. For each extreme, we are to find the voltage and current in both resistors of the RC circuit of Example 25.6.
   DEVELOP In addition to the explanation in Example 25.7, we note that when the switch is closed, Kirchhoff's voltage law applied to the loop containing both resistors yields $\mathcal{E} = I_1R_1 + I_2R_2$, and Kirchhoff's law applied to the loop containing just R_2 and C is $V_C = I_2R_2$.
   EVALUATE (a) If the switch is closed at $t = 0$, the circuit behaves as if it were the circuit of Figure 25.23b, and Example 25.6 explains that $V_C(0) = 0$, $I_2(0) = 0$, so

$$I_1(0) = \frac{\mathcal{E}}{R_1} = \frac{100\text{ V}}{4.0\text{ k}\Omega} = 25\text{ mA}$$

   (b) As $t \to \infty$, the circuit behaves like the circuit of Figure 25.23c, and Example 25.7 shows that

$$I_1(\infty) = I_2(\infty) = \frac{\mathcal{E}}{R_1 + R_2} = \frac{100\text{ V}}{10\text{ k}\Omega} = 10\text{ mA}$$

   and $V_C(\infty) = I_2(\infty)R_2 = (10\text{ mA})(6.0\text{ k}\Omega) = 60\text{V}$.
   (c) Under the conditions stated, the fully charged capacitor ($V_C = 60$ V) simply discharges through R_2. (R_1 is in an open-circuit branch, so $I_1 = 0$ for the entire discharging process.) The initial discharging current is

$$I_2 = \frac{V_C}{R_2} = \frac{60\text{ V}}{6.0\text{ k}\Omega} = 10\text{ mA}$$

   (d) After a very long time, I_2 and V_C decay exponentially to zero.
   ASSESS We deduced the short-term and long-term behavior of the RC circuit without having to solve a complicated differential equation. A long time after the circuit has been closed, the capacitor becomes fully charged an no more current can cross it, so it behaves as an open circuit. When the circuit switch is reopened, the capacitor starts to discharge and eventually loses all its stored energy. It is now capable of storing charge again, and behaves like a short circuit for times much less than its RC time constant.

61. **INTERPRET** We are asked to find the voltage and internal resistance of a battery using the measured voltage values of two voltmeters with different internal resistances.
   DEVELOP The internal resistance R_i of the battery and the resistance R_m of the voltmeter are in series with the battery's emf (see circuit below), so the current is $I = \mathcal{E}/(R_i + R_m)$. The potential drop across the meter (its reading) is

$$V_m = IR_m = \frac{\mathcal{E}R_m}{R_i + R_m}$$

   From the given data, we can write

$$4.36\text{ V} = \frac{\mathcal{E}(1.00\text{ k}\Omega)}{R_i + 1.0\text{ k}\Omega} \quad \text{and} \quad 4.41\text{ V} = \frac{\mathcal{E}(1.50\text{ k}\Omega)}{R_i + 1.50\text{ k}\Omega}$$

   or $R_i + 1.00\text{ k}\Omega = \mathcal{E}(1.00\text{ k}\Omega)/(4.36\text{ V})$ and $R_i + 1.50\text{ k}\Omega = \mathcal{E}(1.5\text{ k}\Omega)/4.41\text{ V}$.

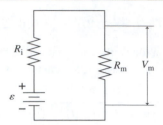

EVALUATE   Solving the simultaneous equations for ε and R_i gives

$$\varepsilon = \left(1.50 \text{ k}\Omega - 1.00 \text{ k}\Omega\right)\left(\frac{1.50 \text{ k}\Omega}{4.41 \text{ V}} - \frac{1.00 \text{ k}\Omega}{4.36 \text{ V}}\right)^{-1} = 4.51 \text{ V}$$

and $R_i = \left(4.51 \text{V}\right)1.00 \text{ k}\Omega/\left(4.36 \text{ V}\right) - 1.00 \text{ k}\Omega = 35.2 \ \Omega$.

ASSESS   An ideal voltmeter has infinite resistance. Thus, when we let $R_m \rightarrow \infty$, its reading approaches the battery voltage ε.

63.   **INTERPRET**   The electric field at the node increases due to charge accumulation and eventually reaches the breakdown field strength. We are to find how long this process will take given the rate at which charge accumulates on the sphere.

DEVELOP   The charge on the node (whether positive or negative) accumulates at a rate of $I = dq/dt = 1 \text{ A} = 1 \text{ μC/s}$, so $|q(t)| = (1 \text{ μA})t$ (where we assume that $q(0) = 0$). If the node is treated approximately as an isolated sphere, and if we assume that the charge distribution on the sphere becomes uniform at a rate much higher than the input current (so that we can treat it as a static distribution), then we can apply Gauss's law and the results of Example 21.1. Under these conditions, the electric field strength at the surface of the sphere is given as

$$E = \frac{k|q|}{r^2} = \frac{kIt}{r^2}$$

Electric breakdown occurs when $E = E_b = 3 \text{ MV/m}$.

EVALUATE   The time when the breakdown happens is

$$t = \frac{E_b r^2}{kI} = \frac{(3 \text{ MV/m})(0.5 \text{ mm})^2}{(9 \times 10^9 \text{ m/F})(1 \text{ μA})} = 80 \text{ μs}$$

to a single significant figure.

ASSESS   This problem shows that Kirchhoff's node law must hold, or else there would be a charge buildup at the node which quickly leads to an electric breakdown.

65.   **INTERPRET**   You need to specify what loudspeaker resistance is needed to get the maximum power output from an amplifier.

DEVELOP   The loudspeaker resistance will be in series with the amplifier's internal resistance. This is similar to the previous problem, where it was shown that the maximum power in the load (the loudspeaker in this case) occurs when its resistance matches the internal resistance of the power supply.

EVALUATE   From the above arguments, the optimum resistance for the loudspeaker is 8 Ω. Since this is the same as the internal resistance of the amplifier, R_{int}, the power output will be:

$$P_{max} = \frac{\varepsilon^2 R_{int}}{\left(R_{int} + R_{int}\right)^2} = \frac{\varepsilon^2}{4R_{int}}$$

If a loudspeaker with 4 Ω of resistance is connected instead, the power is reduced by

$$P = \frac{\varepsilon^2 \left(\frac{1}{2} R_{int}\right)}{\left(R_{int} + \frac{1}{2} R_{int}\right)^2} = \frac{2\varepsilon^2}{9R_{int}} = \frac{8}{9} P_{max}$$

The maximum power is specified as 100 W, so a 4-Ω loudspeaker will output 89 W.

ASSESS   A loudspeaker with half the optimum resistance still produces almost 90% of the maximum power. This shows that it's not necessary to exactly match the load to the amplifier.

67. **INTERPRET**   We're asked to determine the equivalent resistance for several complex systems of resistors.
DEVELOP   The circuit in (a) can be seen as two resistors in parallel followed in series by another pair of resistors in parallel. See the figure below. The circuit in (b) can be seen as two parallel branches, each with two resistors in series. The circuit in (c) is symmetric across a plane through the middle, so the same amount of current should flow through each side.

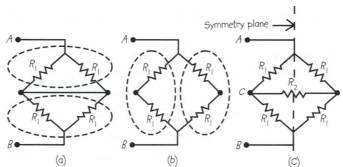

EVALUATE   (a) Each pair of parallel resistors has an equivalent resistance of $R_{\parallel} = \frac{1}{2}R_1$. Added together in series, the total resistance is (Equation 25.1):

$$R_{eq} = R_{\parallel} + R_{\parallel} = R_1$$

(b) Each branch of resistors in series has an equivalent resistance of $R_S = 2R_1$. Added together in parallel, the total resistance is (Equation 25.3b):

$$R_{eq} = \frac{R_S \cdot R_S}{R_S + R_S} = R_1$$

(c) Due to the symmetry, the potential will be the same on both sides of R_2, therefore no current will flow through this resistor. If there's no current through this branch, then the circuit is identical to the one in part (b), which means $R_{eq} = R_1$.

ASSESS   Note that the reasoning in parts (a) and (b) is easily generalized to resistances of different values; the generalization in part (c) requires the equality of ratios of resistances which are mirror images in the plane of symmetry.

69. **INTERPRET**   This problem explores the rate of increase in voltage across the capacitor of an RC circuit. We are to show that if the capacitor were to charge at its initial rate of charging (i.e., the rate at $t = 0$), then it would charge completely in a single time constant $\tau = RC$.
DEVELOP   Kirchhoff's loop law for a battery charging a capacitor through a resistor is

$$\varepsilon - IR - V_C = 0$$

Differentiate this and use Equation 25.4 to obtain

$$\frac{dV_C(t)}{dt} = \frac{d\left[\varepsilon - I(t)R\right]}{dt} = -R\left[\frac{dI(t)}{dt}\right]$$

Using $I(t) = (\varepsilon/R)e^{-t/(RC)}$ for a charging capacitor (Equation 25.5), we find

$$\frac{dV_C(t)}{dt} = -R\frac{-I(t)}{RC} = \frac{I(t)}{C}$$

For an initially uncharged capacitor, $I(t = 0) = \varepsilon/R \equiv I_0$, because an uncharged capacitor acts like a short circuit. Thus, the initial rate of increase in voltage across the capacitor is

$$\frac{dV_C(t = 0)}{dt} = \frac{\varepsilon}{RC} = \frac{\varepsilon}{\tau}$$

so we find how long it takes at this rate for the capacitor to be fully charged [i.e., to reach $V(t = \infty)$].

EVALUATE   From Equation 25.6, we see that $V(t = \infty) = \varepsilon$, so charging at the above rate, the time t it would take to reach this voltage is

$$t\left[\frac{dV_C(t=0)}{dt}\right]=t\left(\frac{\varepsilon}{\tau}\right)=V(t=\infty)=\varepsilon$$

$$t\left(\frac{\varepsilon}{\tau}\right)=\varepsilon \quad \Rightarrow \quad t=\tau$$

ASSESS The real time it takes to reach full charge is longer than one time constant because the rate of change in the voltage is not constant.

71. **INTERPRET** Using the plot provided of the capacitor voltage as a function of time, we are to find the battery voltage, time constant, and capacitance of an the *RC* circuit.

DEVELOP From Equation 25.6, $V_C = \varepsilon\left(1-e^{-t/(RC)}\right)$, we see that the voltage V_C across the capacitor asymptotically approaches the battery voltage ε as $t\to\infty$. Thus, we can read the battery voltage off the graph by finding the asymptotic limit of the capacitor voltage (see figure below). The time constant is the time it takes the capacitor voltage to reach $1-e^{-1}=63\%$ of its asymptotic value, as marked on the graph. From this estimate of the time constant τ, we can find the capacitance from using $\tau = RC$.

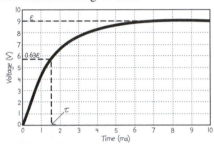

EVALUATE (a) From the asymptotic value of the capacitor voltage, we find that the battery voltage is $\varepsilon \sim 9\text{ V}$.
(b) In one time constant t, the capacitor reaches $\varepsilon\left(1-e^{-1}\right)\approx(9\text{ V})(0.63)=5.7\text{ V}$. From the graph, this occurs at approximately $\tau\sim1.5\text{ ms}$.
(c) The time constant is *RC*, so $C=\tau/R\approx(1.5\text{ ms})/4700\ \Omega\approx0.3\ \mu\text{F}$.

ASSESS From the graph, we can also see that the rate of increase of the capacitor voltage within one time constant is approximately linear, with a rate of

$$\frac{dV_C}{dt}\approx\frac{\varepsilon\left(1-e^{-1}\right)}{\tau}\approx\frac{2\varepsilon}{3\tau}$$

73. **INTERPRET** This problem is an extension of the previous problem. The emf ε_3 changes now so that it supplies the indicated current. The rest of the circuit elements remain the same and we are to find the new value of ε_3.
DEVELOP The relation between I_b and the circuit emfs and resistances, given in the solution to Problem 72, can be solved for ε_3 in Fig. 25.40, resulting in $\varepsilon_3 = \frac{1}{3}\left(8RI_b + 2\varepsilon_2 + \varepsilon_1\right)$.

EVALUATE For $I_b = 40\text{ nA}$ and with the rest of the circuit elements remaining the same,

$$\varepsilon_3 = \tfrac{1}{3}\left(8\times1.5\text{ M}\Omega\times40\text{ nA}+90\text{ mV}+75\text{ mV}\right)=220\text{ mV}$$

ASSESS Thus, ε_3 changes by an order of magnitude from 20 mV (in Problem 25.72) to 220 mV here.

75. **INTERPRET** We will use Kirchhoff's laws to write a system of equations for the circuit shown in Figure 25.23a, and from the resulting equations we are to determine the time constant of the circuit.
DEVELOP We first sketch our loops and nodes, as shown in the figure below. We have 3 unknowns, so we will need 3 equations. Nodes *A* and *B* give us duplicate information, so we will use only one of the two: our equations must then come from loops 1 and 2, and node *A*. Node *A* gives us

$$I_1 - I_2 - I_3 = 0$$

Loop 1 gives us

$$\varepsilon - I_1 R_1 - I_2 R_2 = 0$$

and loop 2 gives us

$$I_2 R_2 - V_C = 0$$

The voltage across the capacitor is given by $V_C = Q/C$, and $I_3 = dQ/dt$. We will eliminate I_1 and I_2 in our system of equations, then rearrange the results into the form of Equation 25.4, from which we can easily identify the time constant.

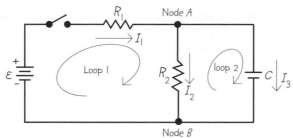

EVALUATE   From node A, $I_1 = I_2 + I_3$. Substitute this into the equation for loop 1:

$$\varepsilon - (I_2 + I_3)R_1 - I_2 R_2 = 0$$

$$I_2 = \frac{\varepsilon - I_3 R_1}{R_1 + R_2}$$

Now we substitute into the equation for loop 2:

$$\left(\frac{\varepsilon - I_3 R_1}{R_1 + R_2}\right) R_2 - V_C = 0$$

$$\varepsilon R_2 - I_3 R_1 R_2 = \frac{Q}{C}(R_1 + R_2)$$

We take the time derivative of this last equation:

$$-\frac{dI_3}{dt} R_1 R_2 = \frac{dQ}{dt}\frac{(R_1 + R_2)}{C}$$

$$-\frac{dI_3}{dt} R_1 R_2 = I_3 \frac{(R_1 + R_2)}{C}$$

Rearrange this slightly to obtain

$$\frac{dI_3}{dt} = -\frac{I_3}{R_1 R_2 C/(R_1 + R_2)}$$

Now here's a trick: rather than solve this equation, we note that it's the *same* equation as 25.4, with a different cluster of constants in the denominator. In the solution to 25.4, we found that $\tau = RC$, so here the time constant must be

$$\tau = \frac{R_1 R_2 C}{R_1 + R_2}$$

ASSESS   This trick of putting the equation in a previously solved form can save us a lot of effort. Note that we can *only* do it because all the terms in the square brackets are constants: if there was a term involving I_3 in those brackets, then it would be a different equation and we couldn't use the same solution.

77.   **INTERPRET**   We must convert a battery energy rating (in watt-hours) at a given voltage to a charge rating of ampere-hours.
   DEVELOP   Apply Equation 24.7, $P = IV$. The battery is specified at 50 watt-hours, which means that it can supply $P = 50$ W for 1 hour. We will use $P = IV$ to find I, knowing that the voltage is $V = 6$ V.
   EVALUATE $P = IV \implies I = P/V = (50\text{ W})/(6\text{ V}) = 8$ A to a single significant figure.
   ASSESS   This is an 8-A·h battery, which is sufficient for our requirements.

79.   **INTERPRET**   We're asked to analyze a situation where stray voltage passes through a dairy cow.
   DEVELOP   The voltage across the cow can be found with Ohm's law.

     EVALUATE Given the current from the previous problem, the voltage between the cow's tongue and hoof is

$$V_{\text{cow}} = IR_{\text{cow}} = (4 \text{ mA})(500 \text{ }\Omega) = 2 \text{ V}$$

The answer is (a).

ASSESS This is not a lot of voltage; it's just a little more than a D battery.

81.   **INTERPRET** We're asked to analyze a situation where stray voltage passes through a dairy cow.

     DEVELOP An ideal ammeter is one with zero resistance.

     EVALUATE If an ideal ammeter is attached from the water bowl to the ground, it will close the circuit and read the current as:

$$I = \frac{V}{R_{\text{int}}} = \frac{6.0 \text{ V}}{1 \text{ k}\Omega} = 6 \text{ mA}$$

The answer is (b).

ASSESS This gives an idea of the what the maximum current might be from the stray voltage. It also exemplifies the best way to eliminate the problem: by connecting the water bowl directly to ground. This would provide a zero resistance pathway for current to flow, so that the cow no longer gets a shock every time it goes for a drink.

26

MAGNETISM: FORCE AND FIELD

EXERCISES

Section 26.2 Magnetic Force and Field

15. **INTERPRET** This problem is about the magnetic force exerted on a moving electron.

DEVELOP The magnetic force on a charge q moving with velocity $\vec{v}$ is given by Equation 26.1: $\vec{F}_B = q\vec{v} \times \vec{B}$. The magnitude of $\vec{F}_B$ is

$$F_B = |\vec{F}_B| = |q\vec{v} \times \vec{B}| = |q|vB\sin\theta$$

EVALUATE (a) The magnetic field is a minimum when $\sin\theta = 1$ (the magnetic field perpendicular to the velocity). Thus,

$$B_{min} = \frac{F_B}{|q|v} = \frac{5.4 \times 10^{-15}\ \text{N}}{(1.6 \times 10^{-19}\ \text{C})(2.1 \times 10^7\ \text{m/s})} = 1.61 \times 10^{-3}\ \text{T} = 16\ \text{G}$$

(b) For $\theta = 45°$, the magnetic field is

$$B = \frac{F_B}{|q|v\sin\theta} = \frac{5.4 \times 10^{-15}\ \text{N}}{(1.6 \times 10^{-19}\ \text{C})(2.1 \times 10^7\ \text{m/s})\sin 45°} = \sqrt{2}\ B_{min} = 23\ \text{G}$$

ASSESS The magnetic force on the electron is very tiny. The magnetic field required to produce this force can be compared to the Earth's magnetic field, which is about 1 G.

17. **INTERPRET** In this problem we are asked to find the magnetic force on a proton moving at various angles with respect to a magnetic field.

DEVELOP The magnetic force on a charge q moving with velocity $\vec{v}$ is given by Equation 26.1: $\vec{F}_B = q\vec{v} \times \vec{B}$. The magnitude of $\vec{F}_B$ is

$$F_B = |\vec{F}_B| = |q\vec{v} \times \vec{B}| = |q|vB\sin\theta$$

The charge of the proton is $q = 1.6 \times 10^{-19}$ C.

EVALUATE (a) When $\theta = 90°$, the magnitude of the magnetic force is

$$F_B = qvB\sin(90°) = (1.6 \times 10^{-19}\ \text{C})(2.5 \times 10^5\ \text{m/s})(0.50\ \text{T}) = 2.0 \times 10^{-14}\ \text{N}$$

(b) When $\theta = 30°$, the force is

$$F_B = qvB\sin(30°) = (1.6 \times 10^{-19}\ \text{C})(2.5 \times 10^5\ \text{m/s})(0.50\ \text{T})\sin(30°) = 1.0 \times 10^{-14}\ \text{N}$$

(c) When $\theta = 0°$, the force is $F_B = qvB\sin(0°) = 0$.

ASSESS The magnetic force is a maximum $F_{B,max} = |q|vB$ when $\theta = 90°$ and a minimum $F_{B,min} = 0$ when $\theta = 0°$.

19. **INTERPRET** This problem is about the speed of a given charge if it is to pass through the velocity selector undeflected. A velocity selector contains an electric and a magnetic field that are perpendicular to each other (see Figure 26.5).

DEVELOP In the presence of both electric and magnetic fields, the force on a moving charge is (see Equation 26.2):

$$\vec{F} = \vec{F}_E + \vec{F}_B = q\left(\vec{E} + \vec{v} \times \vec{B}\right)$$

Because $\vec{E}$ is perpendicular to $\vec{B}$, as shown in Figure 26.5, the forces due to each field on a charged particle are antiparallel. Thus, we can use the scalar for of Equation 26.2,

$$\vec{F} = F_E + F_B = q\left(E + vB\sin\theta\right)$$

so the condition for a charged particle to pass undeflected through the velocity selector is that the net force on it is zero, or $F_E = -F_B$.

EVALUATE   Substituting the values given in the problem statement, we obtain

$$0 = q\left(E + vB\sin\theta\right)$$

$$v = \frac{E}{B\sin\theta} = \frac{24 \text{ kN/C}}{(0.060 \text{ T})\sin\left(90°\right)} = 400 \text{ km/s}$$

ASSESS   Only particles with this speed would pass undeflected through the mutually perpendicular fields; at any other speed, particles would be deflected. Note also that the particle velocity must be perpendicular to the magnetic field for this result to hold.

Section 26.3 Charged Particles in Magnetic Fields

21.   **INTERPRET**   This problem is about an electron undergoing circular motion in a uniform magnetic field. We want to know its period, or the time it takes to complete one revolution.

DEVELOP   Using Equation 26.3, the radius of the circular motion is $r = mv/\left(|e|B\right)$. Therefore, the period of the motion is

$$T = \frac{2\pi r}{v} = \frac{2\pi}{v}\frac{mv}{|e|B} = \frac{2\pi m}{|e|B}$$

EVALUATE   Substituting the values given in the problem statement, we find the period to be

$$T = \frac{2\pi m}{|e|B} = \frac{2\pi\left(9.11\times10^{-31} \text{ kg}\right)}{\left(1.6\times10^{-19} \text{ C}\right)\left(1.0\times10^{-4} \text{ T}\right)} = 360 \text{ ns}$$

to two significant figures.

ASSESS   The period is independent of the electron's speed and orbital radius. However, it is inversely proportional to the magnetic field strength.

23.   **INTERPRET**   This problem involves finding the magnetic field strength required for the given frequency of electrons moving in a circular path through the field. In addition, given the maximum radius of the electron path, we are to find the maximum electron energy (i.e., kinetic energy).

DEVELOP   For part (a), apply Equation 26.4, which gives the frequency of motion as a function of magnetic field. For part (b), use Equation 26.3, $r = mv/(qB)$ to find the kinetic energy $K = mv^2/2$ that corresponds to $r = 2.5$ mm.

EVALUATE   **(a)** A cyclotron frequency of 2.4 GHz for electrons implies a magnetic field strength of

$$B = \frac{2\pi fm}{e} = \frac{2\pi\left(2.4 \text{ GHz}\right)\left(9.11\times10^{-31} \text{ kg}\right)}{\left(1.6\times10^{-19} \text{ C}\right)} = 86 \text{ mT}$$

(b) Solving Equation 26.3 for v and inserting this into the expression for kinetic energy gives

$$K = \frac{1}{2}mv^2 = \frac{1}{2}m\left(\frac{rqB}{m}\right)^2 = \frac{r^2q^2B^2}{2m} = \frac{(2.5 \text{ mm}/2)^2\left(1.6\times10^{-19} \text{ C}\right)^2\left(85.9\times10^{-3} \text{ T}\right)^2}{2\left(9.11\times10^{-31} \text{ kg}\right)} = 1.6\times10^{-16} \text{ J} = 1.0 \text{ keV}$$

ASSESS   The electron's kinetic energy could also be expressed in terms of the cyclotron frequency directly, $K = \left(2\pi frm\right)^2/\left(2m\right) = 2m\left(\pi rf\right)^2$, with the same result.

Section 26.4 The Magnetic Force on a Current

25.   **INTERPRET**   This problem involves finding the force on a wire that is perpendicular to the given magnetic field and that carries the given current.

DEVELOP   Apply Equation 26.5

$$\vec{F} = I\vec{l} \times \vec{B}$$

which, in scalar form, is $F = IlB\sin\theta$.

EVALUATE   Inserting the given quantities into the expression above gives

$$F = IlB\sin\theta = (15\ \text{A})(0.50\ \text{m})(0.050\ \text{T})\sin(90°) = 0.38\ \text{N}$$

ASSESS   The direction of this force is given by the right-hand rule, crossing the current $\vec{I}$ into the magnetic field $\vec{B}$.

27.   **INTERPRET**   You need to show that a bar carrying current will need to be securely fastened inside high magnetic field experiment.

DEVELOP   You can use Equation 26.5 to find the magnitude of the magnetic force on the conducting bar:
$F = ILB\sin\theta$.

EVALUATE   Putting in the given values,

$$F = (4.1\ \text{kA})(1.3\ \text{m})(12\ \text{T})\sin 60° = 55\ \text{kN}\left[\frac{1\ \text{lb}}{4.448\ \text{N}}\right] = 12{,}000\ \text{lb}$$

Yes, you were right to suggest clamping down the bar.

ASSESS   The force will point in the direction perpendicular to the plane defined by the field and bar. With high magnetic fields such as this, it's very important to remove or secure all metal objects.

Section 26.5 Origin of the Magnetic Field

29.   **INTERPRET**   For this problem, we are given the current carried by a wire that forms a loop. Given the magnetic field strength at the loop center, we are to find the radius of the loop.

DEVELOP   This problem is dealt with in Example 26.3, so apply that result (Equation 16.9) here with $x = 0$ (since we are in the plane of the loop).

EVALUATE   Solving Equation 26.9 (with $x = 0$) for the loop radius a gives

$$B = \frac{\mu_0 I}{2a}$$

$$a = \frac{\mu_0 I}{2B} = \frac{(4\pi\times10^{-7}\ \text{N/A}^2)(15\ \text{A})}{2(80\ \mu\text{T})} = 12\ \text{cm}$$

ASSESS   The current in the loop is very high (15 A!) yet the magnetic field it produces is quite small (~ 1 G).

31.   **INTERPRET**   This problem is similar to the preceding one, except that we consider here the effect of not one, but several current-carrying loops that are positioned very close together on the same axis. We are given the current in each loop and the radius and are to find the magnetic field strength at the center of the loops.

DEVELOP   Using the principle of superposition, the total magnetic field at the center of the loops will be the sum of the magnetic field from each loop. The number n of loops involved is $n = L/(2\pi a)$ where $L = 2.2$ m and $2a = 5.0$ cm. From Problem 29, we see that the magnetic field due to a single loop at the center of the loops is $B = \mu_0 I/(2a)$.

EVALUATE   The total magnetic field is

$$B = n\frac{\mu_0 I}{a} = \left(\frac{L}{2\pi a}\right)\left(\frac{\mu_0 I}{a}\right) = \frac{\mu_0 IL}{2\pi a^2} = \frac{(4\pi\times10^{-7}\ \text{N/A}^2)(3.5\ \text{A})(2.2\ \text{m})}{2\pi(0.050\ \text{cm})^2} = 1.2\ \text{mT}$$

ASSESS   For this approximation to be valid, the loop radius must be much, much larger than the separation between the loops.

33.   **INTERPRET**   This problem involves two long parallel wires separated by 1 cm and carrying the given current (note that the current is in the same direction for both wires). We are to find the force between these wires.

DEVELOP   Apply Equation 26.11. To find the force per unit length, simply divide through by the length l of the wires:

$$\frac{F}{l} = \frac{\mu_0 I_1 I_2}{2\pi d}$$

EVALUATE   Inserting the given quantities gives

$$\frac{F}{l} = \frac{\mu_0 I_1 I_2}{2\pi d} = \frac{\left(4\pi \times 10^{-7} \text{ N/A}^2\right)\left(15 \text{ A}\right)^2}{2\pi\left(0.01 \text{ m}\right)} = 5 \text{ mN/m}$$

to a single significant figure.

ASSESS   If the currents were in the opposite directions, the force would be zero.

Section 26.6 Magnetic Dipoles

35.   **INTERPRET**   We are to find the strength of the magnetic dipole moment of the given current loop, and the magnitude of the torque it would experience when placed at 40° in the given magnetic field.

DEVELOP   To find the magnetic dipole moment, apply Equation 26.13, $\vec{\mu} = NI\vec{A}$, which in scalar form is $\mu = NIA$. The torque on this loop in a magnetic field B = 1.4 T is given by Equation 26.15.

EVALUATE   (a) The strength of the magnetic dipole moment is

$$\mu = NIA = \left(1\right)\left(0.45 \text{ A}\right)\left(5.0 \text{ cm}\right)^2 = 1.1 \times 10^{-3} \text{ A} \cdot \text{m}^2$$

(b) The torque on the current loop is

$$\tau = \left|\vec{\mu} \times \vec{B}\right| = \mu B \sin\theta = \left(1.13 \times 10^{-3} \text{ A} \cdot \text{m}^2\right)\left(1.4 \text{ T}\right)\sin\left(40°\right) = 1.0 \times 10^{-3} \text{ N} \cdot \text{m}.$$

ASSESS   The maximum torque occurs at $\theta = 90°$, as expected.

Section 26.8 Ampère's Law

37.   **INTERPRET**   This problem involves Ampère's law for magnetism, which we will use to find the current in a wire given the magnitude of the line integral of the magnetic field.

DEVELOP   Apply Equation 26.17, where the left-hand side is 8.8 mT.

EVALUATE   Inserting the given quantity for the integral and solving for the current gives

$$8.8 \text{ μT} \cdot \text{m} = \oint \vec{B} \cdot d\vec{r} = \mu_0 I_{\text{wire}}$$

$$I_{\text{wire}} = \frac{8.8 \text{ μT} \cdot \text{m}}{4\pi \times 10^{-7} \text{ N/A}^2} = 7.0 \text{ A}$$

ASSESS   The current within the area defined by the line integral is directly proportional to the value of the line integral.

39.   **INTERPRET**   This problem is similar to Example 26.7. We can apply Ampère's law to find the strength of the magnetic field inside and at the surface of the wire with the given dimensions and carrying the given current.

DEVELOP   Because the current is uniform within the wire, the fraction of current contained within 1 mm of the wire's axis is

$$I_a = I_0 \frac{\pi r_a^2}{\pi r_b^2} = I_0 \frac{r_a^2}{r_b^2}$$

where $r_a = 0.10$ mm and $r_b = 1.0$ mm. We can insert this into Ampère's law (Equation 26.17) to find the strength of the magnetic field at r_a. For part (b), we are to find the magnetic field strength at the surface of the wire (r_b), so the current enclosed is simply $I_0 = 5.0$ A.

EVALUATE (a) The magnetic field strength at 0.10 mm from the wire axis is

$$\oint \vec{B} \cdot d\vec{r} = \mu_0 I_a$$

$$B(2\pi r_a) = \mu_0 I_0 \frac{r_a^2}{r_b^2}$$

$$B = \frac{\mu_0 I_0 r_a}{2\pi r_b^2} = \frac{(4\pi\times10^{-7}\ \text{N/A}^2)(5.0\ \text{A})(0.050\ \text{mm})}{2\pi(0.50\ \text{mm})^2} = 4.0\ \text{G}$$

(b) At the surface of the wire, Ampere's law gives

$$\oint \vec{B} \cdot d\vec{r} = \mu_0 I_a$$

$$B(2\pi r_b) = \mu_0 I_0$$

$$B = \frac{\mu_0 I_0}{2\pi r_b} = \frac{(4\pi\times10^{-7}\ \text{N/A}^2)(5.0\ \text{A})}{2\pi(0.50\ \text{mm})} = 20\ \text{G}$$

ASSESS Because this problem is the same as Example 26.7, we could have directly applied Equations 26.19 and 26.18 for parts (a) and (b), respectively. The results, of course, are the same.

41. **INTERPRET** We are asked to find the magnetic field strength inside a solenoid given the current-loop density and the current.

DEVELOP Apply Equation 26.21, which gives the field inside the solenoid (i.e., many radii away from the end of the solenoid).

EVALUATE Inserting the given quantities gives

$$B = \mu_0 n I = (4\pi\times10^{-7}\ \text{N/A}^2)(3300\ \text{m}^{-1})(4.1\ \text{kA}) = 17\ \text{T}$$

ASSESS This is a very strong magnetic field.

PROBLEMS

43. **INTERPRET** The problem asks us to estimate Jupiter's magnetic dipole moment given the magnetic field strength at its poles.

DEVELOP The magnetic field on the axis of a magnetic dipole far from its center is given by $B = \mu_0 I a^2 / 2x^3$ (see Equation 26.9). If we substitute $\mu = \pi I a^2$ for the magnetic dipole moment, $B = \mu_0 \mu / 2\pi x^3$.

EVALUATE We can assume that Jupiter's poles are both one radii away from the planet's magnetic dipole: $x = 6.91\times10^7$ m (from Appendix E). Given the field at the poles, the magnetic dipole moment is

$$\mu = \frac{2\pi x^3 B}{\mu_0} = \frac{2\pi(6.91\times10^7\ \text{m})^3(14\times10^{-4}\ \text{T})}{4\pi\times10^{-7}\ \text{T}\cdot\text{m/A}} = 2.3\times10^{27}\ \text{A}\cdot\text{m}^2$$

ASSESS This is a huge dipole moment, but we can imagine it would have to be to produce a planet-wide magnetic field. The units are correct, since the magnetic dipole moment is current multiplied by area. Jupiter's magnetic field is believed to arise from currents in metallic hydrogen found deep beneath the planet's surface. If we assume that Jupiter's dipole moment were due to a single giant current loop with a radius half that of the planet, the loop would have to carry a current of $I = \frac{\mu}{A} = 6\times10^{11}$ A.

45. **INTERPRET** We are asked to approximate the amount of current flowing in the Earth's liquid core that would produce the measured magnetic field at the poles. We assume the current is confined to a single loop whose axis passes through the poles.

DEVELOP The magnetic field from a single current loop was calculated in Example 26.3 for a point on the loop's axis: $B = \mu_0 I a^2 / 2(x^2 + a^2)^{3/2}$. For the given model of the Earth's field, the radius of the loop, $a = 3000$ km, is not much smaller than the distance to the north pole: $x = 6,370$ km.

EVALUATE   Solving for the current, we arrive at

$$I = \frac{2B\left(x^2 + a^2\right)^{3/2}}{\mu_0 a^2} = \frac{2\left(62\ \mu\text{T}\right)\left[\left(6370\ \text{km}\right)^2 + \left(3000\ \text{km}\right)^2\right]^{3/2}}{\left(4\pi \times 10^{-7}\ \text{T} \cdot \text{m/A}\right)\left(3000\ \text{km}\right)^2} = 3.8\ \text{GA}$$

ASSESS   A single current loop would generate a dipole magnetic field, but the Earth's field is more complicated than that. It is believed that convection of molten iron in the liquid core creates a dynamo that sustains the planet's magnetic field.

47. INTERPRET   This problem is about a charged particle undergoing circular motion in a magnetic field, and we want to express the radius of the orbit in terms of its charge, mass, kinetic energy, and the magnetic field strength.

DEVELOP   From Equation 26.3, the radius of the circular motion is $r = mv/(qB)$. For a non-relativistic particle, $K = \frac{1}{2}mv^2$, or $v = \sqrt{2K/m}$.

EVALUATE   Therefore, the radius of the orbit is

$$r = \frac{mv}{qB} = \frac{m}{qB}\sqrt{\frac{2K}{m}} = \frac{\sqrt{2Km}}{qB}$$

ASSESS   Our result indicates that the radius is proportional to $\sqrt{K}$, or v. Thus, the greater the kinetic energy of the particle, the larger its radius.

49. INTERPRET   In this problem an electron is moving in a magnetic field with a velocity that has both parallel and perpendicular components to the magnetic field. The path is a spiral.

DEVELOP   The radius depends only on the perpendicular velocity component, $r = \frac{mv_\perp}{eB}$. On the other hand, the distance moved parallel to the field is $d = v_\parallel T$, where T is the cyclotron period.

EVALUATE   (a) The radius of the spiral path is

$$r = \frac{mv_\perp}{eB} = \frac{(9.11 \times 10^{-31}\ \text{kg})(3.1 \times 10^6\ \text{m/s})}{(1.6 \times 10^{-19}\ \text{C})(0.25\ \text{T})} = 70.6\ \mu\text{m} = 71\ \mu\text{m}$$

(b) Since $v_\parallel = v_\perp$, the distance moved parallel to the field is

$$d = v_\parallel T = v_\perp\left(\frac{2\pi m}{eB}\right) = 2\pi\left(\frac{mv_\perp}{eB}\right) = 2\pi r = 2\pi(70.6\ \mu\text{m}) = 440\ \mu\text{m}$$

ASSESS   Since motion parallel to the field is not affected by the magnetic force, with $v_\parallel = v_\perp$, the distance traveled in $t = T$ along the direction of the field is simply $d = 2\pi r$.

51. INTERPRET   You're designing a prosthetic ankle that uses an electric motor. You need to find the current necessary to achieve the desired torque.

DEVELOP   As described in Example 26.5, an electric motor consists of a current loop in a magnetic field. The torque is given by Equation 26.4: $\tau = \left|\vec{\mu} \times \vec{B}\right| = \mu B \sin\theta$. The magnetic dipole moment is equal to $\mu = NIA$ (Equation 26.12).

EVALUATE   The torque is maximum when the magnetic dipole moment is perpendicular to the field $\left(\sin\theta = 1\right)$. Solving for the current gives

$$I = \frac{\tau_{max}}{NAB} = \frac{\left(3.1\ \text{mN} \cdot \text{m}\right)}{(150)\pi\left(\frac{1}{2} \cdot 15\ \text{mm}\right)^2 (220\ \text{mT})} = 0.53\ \text{A}$$

ASSESS   Note that the units work out, since $1\ \text{T} = 1\ \text{N/A} \cdot \text{m}$. The result seems like a reasonable current for this application. But care will be needed to be sure no current leaks out into the surrounding tissue.

53. INTERPRET   This problem deals with the Hall effect, which we can use to find the number density of free electrons (i.e., mobile electrons) in copper.

DEVELOP   The geometry in this problem is the same as that in the discussion leading to Equation 26.6, which shows that the number density n of charge carriers is

$$n = IB/qV_H t$$

EVALUATE   Inserting the given quantities into this expression gives

$$n = IB/qV_H t = \frac{(6.8 \text{ A})(2.4 \text{ T})}{(1.6 \times 10^{-19} \text{ C})(1.2 \text{ μV})(1.0 \text{ mm})} = 8.5 \times 10^{22} \text{ cm}^{-3}$$

ASSESS   This is a typical number density for free electrons in a metal.

55.   **INTERPRET**   We are to find the magnetic dipole moment of a 100-turn solenoid with the given dimensions and current, and find the maximum torque this coil will experience in a 0.12-T magnetic field.

DEVELOP   Apply Equation 26.13,

$$\vec{\mu} = NI\vec{A}$$

to find the dipole moment. The maximum torque may be found by setting $\theta = 90°$ in the cross product of Equation 26.15.

EVALUATE   (a) The magnetic moment of the coil has magnitude

$$\mu = NIA = 100(5.0 \text{ A})\tfrac{1}{4}\pi(0.030 \text{ m})^2 = 0.35 \text{ A} \cdot \text{m}^2$$

(b) The maximum torque (from Equation 26.14, with $\sin\theta = 1$) is

$$\tau_{\text{max}} = \mu B = (0.353 \text{ A} \cdot \text{m}^2)(0.12 \text{ T}) = 4.2 \times 10^{-2} \text{ N} \cdot \text{m}$$

ASSESS   This is not a very strong motor.

57.   **INTERPRET**   We are to find the force on a quarter-circle of current-carrying wire in a magnetic field. We will use the equation for magnetic force on a wire, which we will express in differential form and then integrate to determine the net force.

DEVELOP   The force on a section of wire dl carrying a current I in a magnetic field B is

$$d\vec{F} = Id\vec{l} \times \vec{B} \quad \Rightarrow \quad dF = IBdl$$

where we have used $\sin\theta = 1$, because $\theta = 90°$ for this problem. Using the right-hand rule, we can see from Fig. 26.41 that the force on the wire will be everywhere radially away from the loop center (see figure below). The horizontal component of force on each segment dl of the wire is given by $dF_x = IBdl\sin\theta$, where θ is the angle between the vertical and a line between the center of curvature and the wire segment dl, as shown in the figure below. The vertical component of the force is $dF_y = IBdl\cos\theta$. The total horizontal force on the wire is then

$$\vec{F} = \hat{i} \int_{\theta=0}^{\theta=\pi/2} IB\sin\theta dl + \hat{j} \int_{\theta=0}^{\theta=\pi/2} IB\cos\theta dl$$

In terms of θ, $dl = rd\theta$. The current and field are constant: $I = 1.5 \text{ A}$ and $B = 48 \times 10^{-3} \text{ T}$. The radius of curvature is $r = 0.21 \text{ m}$.

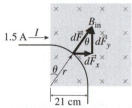

EVALUATE   Evaluating the integral gives

$$\vec{F} = \hat{i} \int_0^{\pi/2} IB\sin\theta rd\theta + \hat{j} \int_0^{\pi/2} IB\cos\theta rd\theta$$

$$= IBr\left[\hat{i} \int_0^{\pi/2} \sin\theta rd\theta + \hat{j} \int_0^{\pi/2} \cos\theta rd\theta\right] = IBr(\hat{i} + \hat{j})$$

Inserting the values gives

$$\vec{F} = (1.5 \text{ A})(48 \times 10^{-3} \text{ T})(0.21 \text{ m})(\hat{i} + \hat{j}) = (0.015 \text{ N})(\hat{i} + \hat{j})$$

The magnitude is thus $F = \sqrt{2(0.015 \text{ N})} = 0.021 \text{ N}$ and the direction is $45°$ above horizontal.

ASSESS   The symmetry of the problem makes evaluation of the integral straightforward.

59. **INTERPRET**   This problem involves finding the magnetic field of a wire that is bent into the given geometrical shape and through which flows the given current.

DEVELOP   The wire may be divided into a straight section and the loop, and the current at the loop center will be the superposition of the magnetic fields from these two components. The magnetic field due to the straight section is

$$B_{\text{straight}} = \frac{\mu_0 I}{2\pi a}$$

where a is the loop radius (see Example 26.4 and Equation 26.10). From Example 26.3, the loop contribution to the magnetic field is

$$B_{\text{loop}} = \frac{\mu_0 I}{2a}$$

where we have used $x = 0$ in Equation 26.9. Using the right-hand rule, we see that both contributions are out of the page, which we define as the $\hat{k}$ direction.

EVALUATE   Inserting the given quantities and summing the two contributions gives

$$\vec{B} = \vec{B}_{\text{straight}} + \vec{B}_{\text{loop}} = \frac{\mu_0 I}{2\pi a}\hat{k} + \frac{\mu_0 I}{2a}\hat{k} = (1+\pi)\frac{\mu_0 I}{2\pi a}\hat{k}$$

ASSESS   The superposition principle greatly simplifies this problem, both analytically and conceptually.

61. **INTERPRET**   We are to find the magnetic field at the center of a semicircular current-carrying wire by using the Biot–Savart law.

DEVELOP   Use the coordinate system shown in the figure below. The Biot-Savart law (Equation 26.7) written in a coordinate system with origin at P, gives

$$\vec{B}(P) = \frac{\mu_0 I}{4\pi}\int_{wire}\frac{d\vec{l} \times \hat{r}}{r^2}$$

where $\hat{r}$ is a unit vector from an element $d\vec{l}$ on the wire to the field point P. On the straight segments to the left and right of the semicircle, $d\vec{l}$ is parallel to $\hat{r}$ and $-\hat{r}$, respectively, so $d\vec{l} \times \hat{r} = 0$. On the semicircle, $d\vec{l}$ is perpendicular to $\hat{r}$ and the radius is constant at $r = a$.

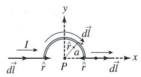

EVALUATE   Evaluating the integral gives

$$B(P) = \frac{\mu_0 I}{4\pi}\int_{semicircle}\frac{d\vec{l} \times \hat{r}}{a^2} = \int_0^\pi \frac{rd\theta}{a^2}\hat{k} = \frac{\mu_0 I}{4\pi}\left(\frac{\pi a}{a^2}\right)\hat{k} = \frac{\mu_0 I}{4a}\hat{k}$$

where $\hat{k}$ is into the page.

ASSESS   Notice that $\hat{r}$ is dimensionless, so the units work out to be $(N/A^2)(A)/m = N/(A \cdot m) = T$.

63. **INTERPRET**   This problem involves finding the force on a wire loop carrying current due to the magnetic field of a nearby straight wire also carrying current.

DEVELOP   See the figure below. At any given distance from the long, straight wire, the force on a current element in the top segment cancels that on a corresponding element in the bottom. The force on the near side (parallel currents) is attractive, and that on the far side (antiparallel currents) is repulsive. The force is given by Equation 26.5, $\vec{F} = I\vec{l} \times \vec{B}$, where the magnetic field may be found using Equation 26.10, $B = \mu_0 I/(2\pi y)$, where y is the distance from the straight wire.

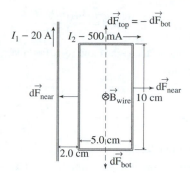

EVALUATE   Performing the sum and inserting the given quantities gives

$$F = \frac{\mu_0 I_1 I_2 l}{2\pi} \left(\frac{1}{y_{near}} - \frac{1}{y_{far}} \right)$$

$$= \frac{\mu_0 I_1 I_2 l}{2\pi} \left(\frac{1}{2.0 \text{ cm}} - \frac{1}{7.0 \text{ cm}} \right) = \frac{\left(4\pi \times 10^{-7} \text{ N/A}^2 \right) (20 \text{ A})(0.50 \text{ A})(0.10 \text{ m})}{2\pi} \left(\frac{5}{0.14 \text{ m}} \right) = 7.1 \times 10^{-6} \text{ N}$$

ASSESS   Notice that this expression reduces to Equation 26.10 for $y_{far} \rightarrow \infty$, as expected.

65.   **INTERPRET**   This problem involves Ampère's law, which we can use to find the magnetic field inside and outside a conducting pipe that carries current.

DEVELOP   Apply Ampère's law, Equation 26.17 to both situations. Inside the pipe, there is no current enclosed by the Ampèrian loop. Outside the pipe, the current enclosed is I.

EVALUATE   (a) Because there is no current enclosed inside the pipe, the magnetic field is zero ($B = 0$) inside the pipe.

(b) Outside the pipe, the field is cylindrically symmetric (provided we are far, compared to the pipe radius, from the pipe ends), so Ampères law gives

$$\oint \vec{B} \cdot d\vec{r} = B 2\pi r = \mu_0 I_{enclosed}$$

$$B = \frac{\mu_0 I}{2\pi r}$$

ASSESS   The field outside a hollow pipe is just like that outside a wire (see Example 26.7).

67.   **INTERPRET**   This problem deals with the magnetic field of a current-carrying solenoid. We are interested in the number of turns the solenoid has and the power it dissipates.

DEVELOP   A length-diameter ratio of 10 to 1 is large enough for Equation 26.20, $B = \mu_0 n I$, to be a good approximation to the field near the solenoid's center. This is the equation we shall use to calculate the number of turns. On the other hand, the power the solenoid dissipates is given by Equation 2.4.8a, $P = I^2 R$.

EVALUATE   (a) Using Equation 26.21, we find the number of turns per unit length to be

$$n = \frac{B}{\mu_0 I} = \frac{10^{-1} \text{ T}}{\left(4\pi \times 10^{-7} \text{ N/A}^2 \right) (35 \text{ A})} = 2.3 \times 10^3 \text{ m}^{-1}$$

This implies that the total number of turns is $N = nL = 2.3 \times 10^3$.

(b) A direct current is used in the solenoid, so the power dissipated (Joule heat) is

$$P = I^2 R = (35 \text{ A})^2 (2.7 \text{ }\Omega) = 3.3 \text{ kW}$$

ASSESS   That's a lot of turns in one meter. The solenoid is very tightly wound to produce such a strong field at its center.

69.   **INTERPRET**   In this problem we are asked to derive to expression for the magnetic field of a solenoid by treating it as being made up of a large number of current loops.

DEVELOP   Consider a small length of solenoid, dx, to be like a coil of radius R and current $nI \, dx$. Using Equation 26.9, the axial field is

$$dB = \frac{\mu_0(nIdx)R^2}{2(x^2+R^2)^{3/2}}$$

with direction along the axis according to the right-hand rule. For a very long solenoid, we can integrate this from $x = -\infty$ to $x = +\infty$ to find the total field.

EVALUATE Integrating over dx from $x = -\infty$ to $x = +\infty$, we find the magnetic field to be

$$B_{sol} = \frac{\mu_0 nI R^2}{2}\int_{-\infty}^{\infty}\frac{dx}{(x^2+R^2)^{3/2}} = \frac{\mu_0 nI R^2}{2}\frac{x}{R^2\sqrt{x^2+R^2}}\Big|_{-\infty}^{\infty} = \mu_0 nI$$

This is the expression given in Equation 26.20.

ASSESS For a finite solenoid, a similar integral gives the field at any point on the axis only, for example, at the center of a solenoid of length L,

$$B(0) = \frac{\mu_0 nIL}{\sqrt{L^2+4R^2}}$$

71. **INTERPRET** This problem is about the magnetic field of a coaxial cable. Ampère's law can be applied since the current distribution has line symmetry.

DEVELOP For a long, straight cable, the magnetic field can be found from Ampère's law. The field lines are cylindrically symmetric and form closed loops, hence they must be concentric circles. If we choose Ampèrian loops that correspond to these concentric circles, then $\vec{B}$ will be constant and parallel to $d\vec{r}$, and Equation 26.16 reduces to

$$\oint \vec{B}\cdot d\vec{r} = 2\pi rB = \mu_0 I_{encircled}$$

We will look at the field at different radii. As for the geometry of the coaxial cable, let $a = 0.50$ mm, $b = 5.0$ mm, and $c = 0.2$ mm, in correspondence with Figure 26.46.

EVALUATE (a) For $r = 0.10$ mm, this is within the radius, a, of the inner conductor. We will assume the current is uniformly distributed over the cross-section of the conductor, so the encircled current will be

$$I_{encircled} = I\left(\frac{\pi r^2}{\pi a^2}\right) = I\left(\frac{r}{a}\right)^2$$

Plugging this into Ampère's law gives a field magnitude of

$$B = \frac{\mu_0 I_{encircled}}{2\pi r} = \frac{\mu_0 Ir}{2\pi a^2} = \frac{\left(4\pi\times10^{-7}\,\frac{T\cdot m}{A}\right)(100\text{ mA})(0.10\text{ mm})}{2\pi(0.50\text{ mm})^2} = 8.0\ \mu T$$

(b) For $r = 5.0$ mm, this is right at the inner radius, b, of the outer conductor, so the encircled current is just the current, I, flowing in the inner conductor. The magnetic field here is

$$B = \frac{\mu_0 I_{encircled}}{2\pi r} = \frac{\left(4\pi\times10^{-7}\,\frac{T\cdot m}{A}\right)(100\text{ mA})}{2\pi(5.0\text{ mm})} = 4.0\ \mu T$$

(c) For $r = 2.0$ cm, this is beyond the outer radius, $b+c$, of the outer conductor, so the encircled current includes the two opposite flowing currents $\left(I_{encircled} = I - I = 0\right)$. Thus, the magnetic field here is zero.

ASSESS This shows that the magnetic field increases linearly with radius $(B\propto r)$ inside the inner conductor until it reaches its maximum at $r = a$. In between the conductors, the field decreases with radius as $B\propto 1/r$. Inside the outer conductor, the field will decrease to zero according to:

$$B = \frac{\mu_0 I}{2\pi r}\left[\frac{(b+c)^2 - r^2}{(b+c)^2 - b^2}\right]$$

At $r = b+c$, this gives $B = 0$ as we would expect.

73. **INTERPRET** This problem is about the magnetic field of a current-carrying conducting bar. Symmetry holds approximately in certain limits.

DEVELOP   Very near the conductor, but far from any edge, the field is like that due to a large current sheet. On the other hand, very far from the conductor, the field is like that due to a long, straight wire.

EVALUATE   (a) Approximating the bar by a large current sheet with $J_s = I/w$, Equation 26.19 gives

$$B \approx \frac{\mu_0 I}{2w}$$

(b) Approximating the bar by a long, straight wire. Equation 26.17 gives

$$B \approx \frac{\mu_0 I}{2\pi r}$$

ASSESS   The conductor exhibits different approximate symmetries, depending on where the field point is chosen.

75.   **INTERPRET**   The system is a solid conducting wire having a non-uniform current density. We are interested in the magnetic field strength both inside and outside the wire. This problem involves Ampère's law.

DEVELOP   The total current in the wire can be obtained by integrating the current density over the cross sectional area (see sketch below). The magnetic field of a long conducting wire is approximately cylindrically symmetric, as discussed in Section 26.8. The magnetic field can be found by using Ampère's law:

$$\oint \vec{B} \cdot d\vec{r} = \mu_0 I_{encircled}$$

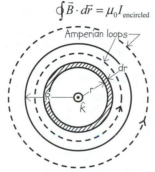

EVALUATE   (a) Using thin rings as the area elements with $dA = 2\pi r dr,$ the total current in the wire (z axis out of the page) is

$$I = \int_0^R J \, dA = \int_0^R J_0 \left(1 - \frac{r}{R}\right)(2\pi r dr) = 2\pi J_0 \left(\frac{r^2}{2} - \frac{r^3}{3R}\right)\Big|_0^R = \frac{1}{3}\pi R^2 J_0$$

(b) A concentric Ampèrian loop outside the wire encircles the total current, so Ampère's law gives

$$2\pi r B = \mu_0 I = \mu_0 \left(\frac{1}{3}\pi R^2 J_0\right)$$

$$B = \frac{\mu_0 J_0 R^2}{6r}$$

(c) Inside the wire, Ampère's law gives $2\pi r B = \mu_0 I_{encircled}.$ The calculation in part **(a)** shows that within a loop of radius $r < R,$

$$I_{encircled} = \int_0^r J dA = 2\pi J_0 \left(\frac{r^2}{2} - \frac{r^3}{3R}\right)\Big|_0^r = \pi J_0 r^2 \left(1 - \frac{2r}{3R}\right)$$

Therefore,

$$B = \frac{\mu_0 J_0 r}{2}\left(1 - \frac{2r}{3R}\right)$$

ASSESS   At $r = R$ both equations give $B = \mu_0 J_0 R/6.$ The form is the same as that shown in Equation 26.17.

77.   **INTERPRET**   You're designing a system to orient a satellite using the torque that the Earth's magnetic field will induce on a current loop. You want the maximum torque possible, but you are limited to a fixed length of wire.

DEVELOP   Let's assume that the loops are circular. The torque on such a loop is given by Equations 26.14 and 26.12:

$$\tau = \left|\vec{\mu}\times\vec{B}\right| = NIAB\sin\theta = \pi r^2 NIB\sin\theta$$

The current will be specified by the satellite's power supply. The magnetic field is that of the Earth's and the angle θ is dependent on the satellite's orientation. What you need to determine is whether one turn $(N=1)$ or many turns will give more torque, given that the total length of wire is set.

EVALUATE   The wire length is related to the size and number of loops by: $l = N(2\pi r)$. Using this to eliminate r from the torque equation gives:

$$\tau = \pi\left(\frac{l}{2\pi N}\right)^2 NIB\sin\theta = \frac{1}{N}\left(\frac{l^2 IB\sin\theta}{4\pi}\right)$$

Since $\tau \propto 1/N$, you'd get more torque from a single turn loop.

ASSESS   Although you gain by having more turns, you're losing more from reducing the area of the loop.

79.   **INTERPRET**   We are to find the force per unit length between a thin wire and a parallel ribbon, each of which carry a current I.

DEVELOP   Use the coordinate system shown in the figure below. The magnitude of the force per unit length on a thin strip of ribbon, of width dx, carrying current $I\,dx/w$, is given by Equation 26.11:

$$dF = \frac{\mu_0 I(I\,dx/w)L}{2\pi(a+x)} \quad \Rightarrow \quad \frac{dF}{L} = \frac{\mu_0 I(I\,dx/w)}{2\pi(a+x)}$$

where x is the distance from the near edge of the ribbon.

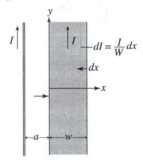

EVALUATE   Integrating the expression above from $x = 0$ to $x = w$ gives a total force of

$$\frac{F}{L} = \frac{\mu_0 I^2}{2\pi w}\int_0^w \frac{dx}{a+x} = \frac{\mu_0 I^2}{2\pi w}\ln\left(\frac{a+w}{a}\right)$$

The force is attractive since the currents are parallel.

ASSESS   Note that this expression reduces to Equation 26.11 for $w = 0$ [using $\ln(1+\varepsilon) = \varepsilon + \dots$].

81.   **INTERPRET**   We find the magnetic field at the center of a "real" solenoid of finite length, treating the solenoid as a stack of individual coils. We use the formula for the magnetic field due to a single loop, and integrate.

DEVELOP   As before in Problem 26.69, we can divide up the solenoid into infinitesimal loops with current of $nIdx$, where n is the number of turns of wire per length. Using Equation 26.9, the axial field from this infinitesimal loop is

$$dB = \frac{\mu_0(nIdx)a^2}{2(x^2+a^2)^{3/2}}$$

Now instead of integrating x from $-\infty$ to ∞, we integrate from $-l/2$ to $l/2$ to obtain the field at the center of the finite length solenoid.

EVALUATE   Performing the integration with help from the tables in Appendix A, we get

$$B = \frac{\mu_0 nIa^2}{2}\int_{-l/2}^{l/2}\frac{dx}{(x^2+a^2)^{3/2}} = \frac{\mu_0 nIa^2}{2}\left.\frac{x}{a^2\sqrt{x^2+a^2}}\right|_{-l/2}^{l/2} = \frac{\mu_0 nIl}{\sqrt{l^2+4a^2}}$$

ASSESS   We can check this formula by letting $l \to \infty$, in which case the magnetic field becomes $B = \mu_0 nI$, as was already given for an infinite solenoid in Equation 26.21.

83.   **INTERPRET**   We need to find the magnetic field necessary to create a certain force on a wire loop. We will model this as a force on a wire in a uniform magnetic field.

   DEVELOP   The force on a wire in a magnetic field is $\vec{F} = I\vec{L} \times \vec{B}$. We will optimize our speaker design by making $\vec{L}$ and $\vec{B}$ perpendicular, so $F = ILB$. From the coil diameter $d = 0.035$ m and number of turns $n = 100$, we can find the length of wire L. The current in the coil is given as $I = 2.1$ A, and the force is $F = 14.8$ N, so we will simply solve for B.

   EVALUATE

$$B = \frac{F}{IL} = \frac{F}{I(\pi n d)} = \frac{14.8 \text{ N}}{(2.1 \text{ A})\pi(100)(3.5 \text{ cm})} = 0.64 \text{ T}$$

   ASSESS   This is a fairly high field strength, but reasonable for currently available permanent magnets.

85.   **INTERPRET**   You want to consider the possible effect that magnets used in magnet therapy might have on blood flow.

   DEVELOP   You first have to estimate the typical current in a blood vessel. Each blood vessel carries a small charge, $q = 2$ pC, and is moving at a speed of $v = 12$ cm/s. There are roughly 5 billion blood cells per mL moving through a vessel of diameter 3 mm. Plugging these values into Equation 24.2, the current flowing in the vessel is

$$I = nAqv = \left(\frac{5 \times 10^9}{\text{mL}}\right)\left[\pi\left(\tfrac{1}{2} \cdot 3 \text{ mm}\right)^2\right](2 \text{ pC})(12 \text{ cm/s}) = 8.5 \text{ mA}$$

   You can compute the Hall effect that a bar magnet would cause inside a current-carrying blood vessel.

   EVALUATE   From Equation 26.6, the Hall potential is $V_H = IB/nqt$, where t is the thickness of the conducting material. In the case of blood, we can assume t is just the diameter of the blood vessel, in which case the Hall potential is

$$V_H = \frac{IB}{nqd} = \frac{(8.5 \text{ mA})(100 \text{ G})}{(5 \times 10^9 / \text{mL})(2 \text{ pC})(3 \text{ mm})} = 3 \text{ } \mu\text{V}$$

   This is roughly 10,000 times smaller than the electric potentials of bioelectric activity.

   ASSESS   A more straightforward way to calculate the Hall effect would be with $V_H \approx vBd$, which gives roughly the same answer.

87.   **INTERPRET**   We consider the magnetic field generated by a toroid.

   DEVELOP   As we explained in the previous problem, the magnetic field is symmetric around the axis of the toroid.

   EVALUATE   By the right-hand rule, it's clear that the magnetic field lines have to be in the plane of the page. That rules out choices (a) and (b). If the field lines were straight and pointing radially, that would seem to contradict Gauss's law for magnetism, Equation 26.14. One could imagine a sphere centered around where the field was radiating outward. The magnetic flux through this sphere would presumably be non-zero, as if there were a magnetic monopole at the center of the toroid. Ruling out that possibility, we're left with circular field lines, which agrees with the arguments made in the previous problem.
   The answer is (d).

   ASSESS   As described in Figure 26.8, charged particles will spiral around magnetic field lines. Therefore, inside a toroid, charged particles will orbit essentially in a circle as they spiral around the field lines. This is how the million degree fuel in a future fusion reactor will presumably be confined.

89.   **INTERPRET**   We consider the magnetic field generated by a toroid.

DEVELOP   To find the field magnitude, we can use the Ampèrian circles that were introduced in Problem 26.86.

EVALUATE   At a given radius, the magnetic field inside the coils will be

$$B = \frac{\mu_0 NI}{2\pi r}$$

The answer is (d).

ASSESS   We see here that the difference between the magnetic field in a solenoid and in a toroid is that in the former the field is uniform (Equation 26.21) but in the latter it is not $(B \propto 1/r)$.

27 ELECTROMAGNETIC INDUCTION

EXERCISES

Sections 27.2 Faraday's Law and 27.3 Induction and Energy

13. **INTERPRET**   Given a constant magnetic field, we are to find the magnetic flux that passes through the given loop.

    DEVELOP   For a stationary plane loop in a uniform magnetic field, the integral for the flux in Equation 27.1a is just $\phi_B = \vec{B} \cdot \vec{A}$.

    EVALUATE   Evaluating the dot product gives

    $$BA\cos\theta = (80 \text{ mT})\pi(2.5 \text{ cm})^2 \cos(30°) = 1.4 \times 10^{-4} \text{ Wb}.$$

    ASSESS   The SI unit of flux, $\text{T} \cdot \text{m}^2$, is also called a weber (Wb).

15. **INTERPRET**   This problem involves Faraday's law, which we can use to find the rate at which the magnetic field is changing given the current in the loop.

    DEVELOP   The flux through a stationary loop perpendicular to a magnetic field is $\phi_B = BA$, so Faraday's law (Equation 27.2) and Ohm's law (Equation 24.5) relate this to the magnitude of the induced current:

    $$I = \left| \frac{\varepsilon}{R} \right| = \left| \frac{-d\Phi_B/dt}{R} \right| = \left| \frac{-d(BA)/dt}{R} \right| = A \left| \frac{-dB/dt}{R} \right|$$

    EVALUATE   Solving this expression for the rate of change of the magnetic field gives

    $$\left| \frac{dB}{dt} \right| = \frac{IR}{A} = \frac{(0.32 \text{ A})(12 \text{ }\Omega)}{240 \times 10^{-4} \text{ m}^2} = 160 \text{ T/s}$$

    ASSESS   Whether the magnetic field is increasing or decreasing depends on the direction in which the current is circulating with respect to the magnetic field.

Section 27.4 Inductance

17. **INTERPRET**   We are to find the self inductance of the given solenoid.

    DEVELOP   This problem is treated in Example 27.6. Apply Equation 27.4.

    EVALUATE   Equation 27.4 gives

    $$L = \frac{\mu_0 N^2 A}{l} = \frac{(4\pi \times 10^{-7} \text{ H/m})(10^3)^2 \pi (2.0 \text{ cm})^2}{50 \text{ cm}} = 3.2 \text{ mH }($$

    ASSESS   Note that Equation 27.4 make use of the assumption that the solenoid length is much greater than its diameter, which holds for this problem.

19. **INTERPRET**   We are to find the induced emf in a circuit given the rate of change of the current and the circuit's inductance.

    DEVELOP   Assume that the current changes uniformly from 2.0 A to zero in 1.0 ms (or consider average values). Then $dI/dt = \Delta I/\Delta t = (-2.0 \text{ A})/(1.0 \text{ ms}) = -2.0 \times 10^3 \text{ A/s}$ and we can apply Eqution 27.5 to find the emf.

EVALUATE   The emf is

$$\varepsilon = -L\frac{dI}{dt} = -(20\ H)(2.0 \times 10^3\ \text{A/s}) = 40\ \text{kV}$$

ASSESS   The negative sign indicates that the emf opposes the change in the current.

21.   INTERPRET   We are to find the time constant of a circuit, given its resistance and its inductance.

DEVELOP   From Equations 27.6 and 27.7, we see that the time constant is $\tau = L/R$, which we can solve for the inductance given the time constant and the resistance.

EVALUATE   Inserting the given quantities gives

$$\tau_L = \frac{L}{R}$$
$$L = \tau_L R = (2.2\ \text{ms})(100\ \Omega) = 220\ \text{mH}$$

ASSESS   To verify that the units work out correctly, note that a henry is a $\text{T}\cdot\text{m}^2/\text{A}$ and an ohm is $\text{m}^2 \cdot \text{kg} \cdot \text{s}^{-3} \cdot \text{A}^{-2}$. Expressing teslas in terms of SI base units gives $(\text{kg} \cdot \text{s}^{-2} \cdot \text{A}^{-2})$

$$s \cdot \Omega = s \cdot m^2 \cdot kg \cdot s^{-3} \cdot A^{-2} = m^2 \cdot kg \cdot s^{-2} \cdot A^{-2} = T$$

Section 27.5 Magnetic Energy

23.   INTERPRET   We are to find the energy stored in the given inductor through which flows the given current.

DEVELOP   Apply Equation 27.9.

EVALUATE   Inserting the given quantities into Equation 27.9 gives

$$U = \tfrac{1}{2}LI^2 = \tfrac{1}{2}(5.0\ \text{H})(35\ \text{A})^2 = 3.1\ \text{kJ}$$

ASSESS   This is the energy it would take to lift one liter of water a height h of

$$U = mgh$$
$$h = \frac{U}{mg} = \frac{3.1\ \text{kJ}}{(1.0\ \text{kg})(9.8\ \text{m/s}^2)} = 320\ \text{m}$$

25.   INTERPRET   This problem involves the energy stored in the magnetic field of an inductor. We are to find the energy needed to raise the current of the inductor the given amount.

DEVELOP   From Equation 27.9, the energy required to raise the current from zero to I1 = 350 mA is $U_1 = LI_1^2/2$. Likewise, the energy required to raise the current from zero to I2 is $U_2 = LI_2^2/2$. The energy required to raise the current from I1 to I2 is the difference,

$$\Delta U = \frac{L}{2}\left(I_2^2 - I_1^2\right)$$

EVALUATE   Inserting the given quantities yields

$$\Delta U = \frac{L}{2}\left(I_2^2 - I_1^2\right) = \frac{220\ \text{mH}}{2}\left[(800\ \text{mA})^2 - (350\ \text{mA})^2\right] = 57\ \text{mJ}$$

ASSESS   The energy in the magnetic field is proportional to the current squared, analogous to kinetic energy, which is proportional to the velocity squared.

27.   INTERPRET   This problem is an exercise in dimensional analysis. We are to show that the given expression has units of energy density (i.e., J/m^3).

DEVELOP   The permeability constant μ_0 has units of $\text{N/A}^2 = \text{N} \cdot \text{C}^{-2} \cdot \text{s}^2$ (see discussion accompanying Equation 26.7) and the magnetic field has units of $\text{N} \cdot \text{s}/(\text{C} \cdot \text{m})$ (see discussion accompanying Equation 26.1). Combine these factors in the indicated fashion to find the units of B^2/μ_0.

EVALUATE   The units of B^2/μ_0 are

$$\left(\frac{N \cdot s}{C \cdot m}\right)^2 \left(\frac{C^2}{N \cdot s^2}\right) = \frac{N}{m^2} = \frac{N \cdot m}{m^3} = \frac{J}{m^3}$$

ASSESS   The factor 2 in the denominator does not affect the result.

29.   **INTERPRET**   We are to find the magnetic field strength in a region with the given magnetic energy density.

DEVELOP   Apply Equation 27.10.

EVALUATE   Solving Equation 27.10 for the magnetic field strength B gives

$$B = \sqrt{2\mu_0 u_B} = \sqrt{2\left(4\pi \times 10^{-7}\,\text{H/m}\right)\left(7.8\,\text{J/cm}^3\right)} = 4.4\,\text{T}$$

ASSESS   This result is for free space (i.e., empty space). If a material occupies the space, Equation 27.10 is not valid.

Section 27.6 Induced Electric Fields

31.   **INTERPRET**   This problem involves a solenoid in which the current is changing, so it has a time-varying magnetic field and thus an electric field as well. We can use Faraday's law to find the electric field as a function of r inside a solenoid.

DEVELOP   We'll use a circular Ampérian loop, of radius r, centered inside the solenoid. The flux through this loop is $\Phi = BA = \pi r^2 B$. We are told that the field in the solenoid is $B = bt$. Faraday's law, integrated around this loop, gives us

$$\oint \vec{E} \cdot d\vec{r} = -\frac{d\Phi}{dt}$$

By symmetry, the electric field is constant around any loop of a given radius, which makes the integration easy.

EVALUATE

$$\oint \vec{E} \cdot d\vec{r} = -\frac{d\Phi}{dt}$$

$$2\pi r E = -\frac{d}{dt}\left[\pi r^2 (bt)\right] = -\pi r^2 b$$

$$E = -\frac{rb}{2}$$

ASSESS   Just as in previous problems that use Gauss's and Ampère's laws, it is important to choose our symmetry to make things easy on ourselves.

PROBLEMS

33.   **INTERPRET**   This problem involves Faraday's law, which we can use to find current in the loop under the given conditions.

DEVELOP   Apply Faraday's law (Equation 27.2) and Ohm's law (Equation 27.5) to the circuit to find

$$\left.\begin{array}{c} \varepsilon = -\dfrac{d\Phi_B}{dt} \\[2mm] \varepsilon = IR \end{array}\right\} I = -\frac{1}{R}\frac{d\Phi_B}{dt} = -\frac{1}{R}\frac{d}{dt}\left(\vec{A} \cdot \vec{B}\right) - \frac{A}{R}\frac{dB_z}{dt}$$

Inserting $B_z = at^2 - b$ gives

$$I(t) = -\frac{A}{R}(2at)$$

EVALUATE   (a) Inserting $t = 3$ s gives

$$I(t = 3\,\text{s}) = -\frac{0.15\,\text{m}^2}{6.0\,\Omega}(2.0)\left(2.0\,\text{T/s}^2\right)(3.0\,\text{s}) = -0.30\,\text{A}$$

(b) $B_z = 0$ implies $at^2 = b$, or $t = \pm\sqrt{b/a}$. At this time, the current is

$$I\left(t = \sqrt{b/a}\right) = -\frac{0.15\,\text{m}^2}{6.0\,\Omega}(2.0)\left(2.0\,\text{T}/s^2\right)\left(\sqrt{(8.0\,\text{T})/(2.0\,\text{T}/s^2)}\right) = -0.20\,\text{A}$$

ASSESS   The negative sign indicates the current direction with respect to the direction of the magnetic field. If the x-y axes are as shown below and the z axis is out of the page, then $\vec{B}$ is in the same direction as $\vec{A}$ (out of the page). Using the right-hand rule, positive currents run counterclockwise and negative currents run clockwise around the loop.

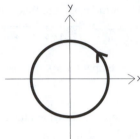

35.   **INTERPRET**   This problem is similar to Example 27.5; it involves a conducting loop rotating in a uniform, static magnetic field, so the change in the magnetic field flux through the loop results from the rotation.

DEVELOP   Use the result of Example 27.5, which shows that

$$\varepsilon = NB\pi r^2 \left[2\pi f \sin\left(2\pi t\right)\right]$$

EVALUATE   The maximum emf occurs at

$$\varepsilon = NB\pi r^2 \left(2\pi f\right)$$

$$B = \frac{\varepsilon}{N\pi r^2\left(2\pi f\right)} = \frac{360\,\mu\text{V}}{2\cdot 5\pi^2\left(0.050\,\text{m}\right)\left(10\,\text{s}^{-1}\right)} = 15\,\text{mT}$$

ASSESS   Because the maximum of the sine function is unity, we do not need to know at what time the maximum occurs.

37.   **INTERPRET**   This problem involves a magnetic field that is changing in time, so that the flux through a loop in this field changes. Thus, we can apply Faraday's law to find the induced emf.

DEVELOP   To shine at full brightness, the potential drop across the bulb must be 6 V. This is equal to the induced emf, if we neglect the resistance of the rest of the loop circuit. From Faraday's law (Equation 27.2),

$$\left|\varepsilon\right| = \left|-\frac{d\Phi_B}{dt}\right| = \left|-\frac{d\left(BA\right)}{dt}\right| = \left|\frac{A\Delta B}{\Delta t}\right|$$

EVALUATE   **(a)** Inserting the given quantities gives

$$\Delta t = \frac{A\left|\Delta B\right|}{\left|\varepsilon\right|} = \frac{\left(3.0\,\text{m}\right)^2\left(2.0\,\text{T}\right)}{6\,\text{V}} = 3\,\text{s}$$

(b) The direction of current opposes the decrease of $\vec{B}$ into the page, and thus must act to increase $\vec{B}$ into the page. From the right-hand rule, this corresponds to a clockwise current in Fig. 27.38.

ASSESS   The units of the expression work out to be units of time:

$$\frac{\left(\text{m}\right)^2\left(\text{T}\right)}{\text{V}} = \frac{\left(\text{m}\right)^2\left(\text{kg}\cdot\text{s}^2\cdot\text{A}^{-1}\right)}{\text{m}^2\cdot\text{kg}\cdot\text{s}^{-3}\cdot\text{A}^{-1}} = \text{s}$$

39.   **INTERPRET**   The solenoid current is varying in time, so the magnetic field in the solenoid varies in time and Faraday's law will be involved in finding the current induced through the wire loop in the solenoid.

DEVELOP   The magnetic field inside the solenoid is $B = \mu_0 n I$, so the flux through the loop is

$\phi_B = BA_{\text{loop}} = \mu_0 n \frac{1}{4}\pi D_{\text{loop}}^2 I$. From Faraday's and Ohm's laws (Equations 27.2 and 26.5), the magnitude of the induced current is

$$I_{\text{loop}} = \frac{|\mathcal{E}|}{R} = \frac{1}{R}\left|\frac{d\Phi_B}{dt}\right| = \mu_0\left(\frac{N}{L}\right)\frac{1}{4}\pi D_{\text{loop}}^2 \frac{1}{R}\left(\frac{dI}{dt}\right)$$

EVALUATE   (a) Inserting the given quantities gives

$$I_{\text{loop}} = \left(4\pi \times 10^{-7}\ \text{N/A}^2\right)\left(\frac{2000}{2.0\ \text{m}}\right)\left(\frac{\pi}{4}\right)(0.15\ \text{m})^2 \frac{1.0\ \text{kA/s}}{5.0\ \Omega} = 18\ \text{mA}$$

(b) If the loop encloses the solenoid, then $\Phi_B = BA_{\text{solenoid}}$, and the induced current would increase to $A_{\text{solenoid}}/A_{\text{loop}} = (1.5)^2$ times the value in part **(a)**, or 40 mA.

ASSESS   The current is greater in the outer loop because the loop encircles greater flux.

41.   **INTERPRET**   This problem involves a time-varying magnetic field that is spatially uniform. This causes a changing magnetic flux through a conducting loop, so Faraday's law will lead to an induced emf.
DEVELOP   Faraday's and Ohm's laws give the current in the loop:

$$I = \frac{\mathcal{E}}{R} = -\frac{(d\Phi_B/dt)}{R} = -\frac{A}{R}\frac{dB}{dt} = -\frac{Ab}{R}$$

EVALUATE   (a) Inserting the given values leads to

$$I = -\frac{\left(240\ \text{cm}^2\right)\left(0.35\ \text{T/s}\right)}{0.20\ \Omega} = -42\ \text{mA}$$

A normal to the loop is parallel to the z-axis and corresponds to counterclockwise positive circulation (via the right-hand rule), when viewed from above. The minus sign thus indicates a clockwise circulation when viewed from the positive z-axis.

ASSESS   The time-varying magnetic field causes a current to pass through the conducting loop.

43.   **INTERPRET**   The aim here is to find the number of turns for the rectangular coil in a generator in order to produce an alternating emf: $\mathcal{E} = \mathcal{E}_{\text{peak}} \sin(2\pi f t)$.
DEVELOP   In Example 27.5, the expression for the emf from a generator was derived. The only difference in this case is that the coil is rectangular not circular:

$$\mathcal{E} = -\frac{d\Phi_B}{dt} = NBlw(2\pi f)\sin(2\pi f t)$$

EVALUATE   Solving for N, the number of turns, gives

$$N = \frac{\mathcal{E}_{\text{peak}}}{2\pi Blwf} = \frac{(6.7\ \text{kV})}{2\pi(0.14\ \text{T})(0.75\ \text{m})(1.3\ \text{m})(60\ \text{Hz})} = 130$$

ASSESS   Notice that the alternating emf frequency is simply set by the rotation rate of the coil.

45.   **INTERPRET**   This problem involves using Lenz's law to find the sign of the voltage across the two rails in the preceding problem.
DEVELOP   As per the discussion for Problem 27.44, Lenz's law requires the current to try to circulate counterclockwise to generate an upward magnetic field to compensate for the increased downward magnetic field enclosed by the circuit.
EVALUATE   Because current flows from the positive to the negative terminal of a battery, the positive terminal will be the top bar. Thus, the positive terminal of the voltmeter must be connected to the top bar in Figure 27.39.
(b) When an ideal voltmeter replaces the resistor, no current flows (since its resistance is infinite) so no work is done moving the bar.

ASSESS   Note that work is done in accelerating the bar, because charge is separated in this process to charge the capacitor formed by the gap between the upper and lower bars. But once the bar is moving at constant velocity, n work is done.

47.   **INTERPRET**   This problem is a continuation of Problem 27.44. We are given values for the circuit elements and are asked to quantitatively characterize the circuit's response to the agent that moves the conducting bar.
DEVELOP   The situation is like that described in Example 27.4 and the solution to Problem 27.44.

EVALUATE   (a) The current is

$$I = |\varepsilon|/R = Blv/R = \frac{(0.50 \text{ T})(0.10 \text{ m})(2.0 \text{ m/s})}{4.0 \; \Omega} = 25 \text{ mA}$$

(b) The magnetic force on the conducting bar is

$$F_{mag} = IlB = (25 \text{ mA})(0.10 \text{ m})(0.50 \text{ T}) = 1.3 \times 10^{-3} \text{ N}$$

to the left.

(c) The power dissipated in the resistor is

$$P = I^2 R = (25 \text{ mA})^2 (4.0 \; \Omega) = 2.5 \text{ mW}.$$

(d) The agent pulling the bar must exert a force equal in magnitude to F_{mag} and parallel to v. Therefore, it does work at a rate

$$Fv = (1.25 \times 10^{-3} \text{ N})(2.0 \text{ m/s}) = 2.5 \text{ mW}$$

Conservation of energy requires the answers to parts (c) and (d) to be the same.

ASSESS   Note that the answer to part (d) uses the result of part (c) to three significant figures because the result of part (c) serves as an intermediate result in part (d).

49. **INTERPRET**   This problem involves a changing magnetic field that induces an electric field. Thus, Faraday's law (Equation 27.11) applies. We can use this to find the rate at which the magnetic field changes given the electric field strength. In using Faraday's law, we can make use of the line symmetry of the problem, which tells us that the electric field will be constant along circles concentric with the solenoid axis.

DEVELOP   The magnitude of the electric field is $E = F/e$. Faradays law tells us that

$$\oint \vec{E} \cdot d\vec{r} = -\frac{d\Phi_B}{dt}$$

$$2\pi r E = -\frac{d\Phi_B}{dt} = -A\frac{dB}{dt} = -\pi r^2 \frac{dB}{dt}$$

EVALUATE   Solving for the rate of change of the magnetic field gives

$$\left|\frac{dB}{dt}\right| = \frac{2E}{r} = \frac{2F}{re} = \frac{2(1.3 \times 10^{-15} \text{ N})}{(0.28 \text{ m})(1.6 \times 10^{-19} \text{ C})} = 58 \text{ T/ms}$$

ASSESS   We cannot tell if the magnetic field is increasing or decreasing because we do not have information about the direction of the force.

51. **INTERPRET**   This problem involves a coil that is moved through a magnetic field, so Faraday's law can be used to relate the changing magnetic flux through the coil to the electric field and thus to the current.

DEVELOP   Initially, the flux through the flip coil is $\phi_B = NBA$, but is reversed to $-NBA$ when the coil is rotated 180°, so $\Delta\phi_B = -2NBA$. The total charge that flows is $\Delta Q = I_{av}\Delta t$, where I_{av} is the average induced current and Δt is the time for the rotation. Use Faraday's law and Ohm's law to relate the charge to the magnetic field strength.

EVALUATE   From Faraday's and Ohm's laws,

$$I_{av} = -\frac{\Delta\Phi_B/\Delta t}{R} = \frac{2NBA}{\Delta t R}$$

so

$$\Delta Q = \frac{2NBA}{R}$$

$$B = \frac{R \; \Delta Q}{2NA}$$

ASSESS   This result agrees with that given in the problem statement.

53. **INTERPRET**   This problem involves an *RL* circuit for which we are to find the time for which the circuit has been completed (i.e., switch closed).

DEVELOP When the switch is closed, the current starts to increase, as shown in Figure 27.24. The current rise is given by Equation 27.7:

$$I(t) = \frac{\mathcal{E}_0}{R}\left(1 - e^{-Rt/L}\right)$$

which we can solve for the time t.

EVALUATE Solving for the time t gives

$$t = \frac{L}{R}\ln\left(\frac{1}{1 - RI(t)/\mathcal{E}_0}\right) = \frac{2.1\,\text{H}}{3.3\,\Omega}\ln\left(\frac{1}{1 - (3.3\,\Omega)(9.5\,\text{A})/(45\,\text{V})}\right) = 0.76\,\text{s}$$

ASSESS The time constant for this circuit is $RL = 6.9$ s, so the current in this circuit will continue to grow for about 21 s (about three time constants).

55. **INTERPRET** This problem involves the current decay in an RL circuit. We use the equation for current decay in an inductor, and energy stored in an inductor, to find the time it takes to lose 90% of the energy stored in an inductor when the circuit becomes resistive.

    DEVELOP The energy initially stored in the inductor is $U_0 = \frac{1}{2}LI_0^2$ (Equation 27.9). The decaying current through an RL circuit is given by $I = I_0 e^{-Rt/L}$. (Equation 27.8). For this problem, the initial current is $I_0 = 2.4$ kA, the inductance is $L = 0.53$ H, and the resistance is $R = 21$ mΩ. We want to calculate the time required to dissipate 90% of the initial energy.

    EVALUATE The time-dependent energy stored in the inductor is

    $$U(t) = \frac{1}{2}LI^2 = \frac{1}{2}LI_0^2 e^{-2Rt/L} = U_0 e^{-2Rt/L}$$

    so the time we're looking for is

    $$e^{-2Rt/L} = 100\% - 90\% = 0.10$$

    $$-\frac{2Rt}{L} = \ln(0.10)$$

    $$t = -\frac{L}{2R}\ln(0.10) = 20\text{ s}$$

    ASSESS The initial energy stored is 1.5 MJ, so the average power loss is nearly 69 kW! Note also that the initial current was not needed in this calculation.

57. **INTERPRET** You want to limit the voltage across elevator motors when the supplied current is suddenly switched off. Because the motors have a high inductance, they will try to keep the same current flowing through them even when the circuit is opened. A resistor placed in parallel with the motor will give a safe path for this current to flow out.

    DEVELOP In Conceptual Example 27.1, a description is given of the behavior of a circuit with a power supply connected to an inductor and resistor in parallel. You can imagine that the inductor in this example is an elevator motor and the parallel resistor is the safety element you want to install. When the elevator is in operation, a current of $I_0 = 20$ A flows through the motor. If a switch is suddenly opened, the motor's inductance will respond by driving the same current through the mini-circuit defined by the inductor and resistor (see Figure 27.26d). The voltage across the resistor, $V = I_0 R$, will be equal to the voltage across the motor (inductor).

EVALUATE (a) To limit the voltage across the motor to less than 100 V, you'll need resistors of

$$R = \frac{V}{I_0} = \frac{100 \text{ V}}{20 \text{ A}} = 5 \text{ }\Omega$$

(b) The current does not stay at the initial value. It decays exponentially according to Equation 27.8: $I = I_0 e^{-Rt/L}$. To find how much energy the resistor dissipates, you can integrate the power, $P = I^2 R$, over the time it will take for all the current to theoretically disappear (i.e., $t = \infty$).

$$\Delta U = \int_0^\infty P \, dt = \int_0^\infty I_0^2 R e^{-2Rt/L} dt = \tfrac{1}{2} L I_0^2 = \tfrac{1}{2}(2.5 \text{ H})(20 \text{ A})^2 = 500 \text{ J}$$

ASSESS Just as you might expect, the energy dissipated by the resistor is just the energy that was initially stored in the inductor (Equation 27.9).

59. **INTERPRET** We're asked to find the current in an *RL* circuit at different time points.

DEVELOP We are considering the short-term and long-term behavior of a circuit with an inductor, as was done in Conceptual Example 27.1.

EVALUATE (a) Just after the switch is closed, the inductor current is zero. We can consider this branch of the circuit as being opened $(I_3 = 0)$. Current will instead flow through R_2, which is in series with R_1 (see the figure below). The current I_2 will be

$$I_2 = I_1 = \frac{\mathcal{E}_0}{R_1 + R_2} = \frac{12 \text{ V}}{4.0 \text{ }\Omega + 8.0 \text{ }\Omega} = 1.0 \text{ A}$$

(b) After the currents have been flowing a long time, they reach steady values $(dI/dt = 0)$. This means the voltage across the inductor is zero, and we can treat it like a short-circuit. Now, R_2 and R_3 are in parallel with each other and in series with R_1 (see the figure below). This implies that the current I_1 is

$$I_1 = \frac{\mathcal{E}_0}{R_1 + R_2 R_3 / (R_2 + R_3)} = \frac{12 \text{ A}}{\left[4.0 + 8.0 \times 2.0 / (10) \right] \Omega} = 2.14 \text{ A}$$

By Kirchhoff's rules, $I_1 = I_2 + I_3$, and $I_2 R_2 = I_3 R_3$. Solving for the current I_2 gives

$$I_2 = \frac{R_3}{R_2 + R_3} I_1 = \frac{2.0}{8.0 + 2.0}(2.14 \text{ A}) = 0.43 \text{ A}$$

The current I_3 makes up for the difference: $I_3 = I_1 - I_2 = 1.71 \text{ A}$.

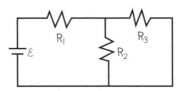

(c) When the switch is reopened, no current flows through the battery's branch, $I_1 = 0$, so we can remove it from the circuit (see the figure below). As explained in Conceptual Example 27.1, the induced emf acts to keep the current flowing through the inductor as it was before the switch was opened, i.e., $I_3 = 1.71 \text{ A}$ from part (b). The current in R_2 will be the same as in the inductor, but it will be flowing in the opposite direction as before:

$$I_2 = -I_3 = -1.7 \text{ A}$$

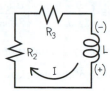

Assess Notice that the value of the inductance in L was not needed, since we are only considering the short and long term behavior of the circuit. If we wanted to calculate the currents at some intermediate time, then we would need the inductance to plug into Equation 27.7 or 27.8.

61. **Interpret** This problem involves an RL circuit with a given inductance. The energy in the inductor drops by 75% in the given time and we are to find the resistance.

Develop From Equation 27.9, $U = LI^2/2$, we find

$$U(t = 3.6 \text{ s}) = \frac{U_0}{4} = \frac{LI(t=3.6 \text{ s})^2}{2} = \frac{1}{4}\overbrace{\left(\frac{LI_0^2}{2}\right)}^{U_0} = \frac{L}{2}\left(\frac{I_0}{2}\right)^2$$

from which we deduce that $I(t = 3.6 \text{ s}) = I_0/2$. Insert this into Equation 27.8, $I(t) = I_0 e^{-Rt/L}$, to find the resistance R.

Evaluate Solving Equation 27.8 for the resistance R and inserting the given quantities gives

$$1/2 = e^{-Rt/L}$$
$$R = (L/t)\ln(2.0) = (1.0 \text{ H})/(3.6 \text{ s})\ln(2.0) = 190 \text{ m}\Omega$$

to two significant figures.

Assess The current and the resistance do not have the same time constant. Because the current is squared in the expression for energy, the time constant for the energy in the inductor is twice that for the current. Thus, the energy grows and decays at twice the rate compared to the current.

63. **Interpret** This problem involves finding the energy density in the magnetic field of a neutron star and comparing that density with other sources of energy.

Develop Apply Equation 27.10,

$$u_B = \frac{1}{2\mu_0}B^2$$

to find the energy density in the magnetic field of the neutron star.

Evaluate The energy density in a magnetic field of the neutron star is

$$u_B = \frac{1}{2\mu_0}B^2 = \frac{(1.0 \times 10^8 \text{ T})^2}{2(4\pi \times 10^{-7} \text{ H/m})} = 3.4 \times 10^{21} \text{ J/m}^3$$

This is about **(a)** 1.1×10^{11} times the energy density content of gasoline (44 MJ/kg $\times$ 800 kg /m^3 = 3.52×10^{10} J/m^3), and **(b)** 2600 times that of pure U^{235} (8×10^{13} J/kg $\times 19 \times 10^3$ kg/m^3 = 1.52×10^{18} J/m^3).

Assess The energy density in the magnetic field of the neutron star is very high compared to that of common energy sources found on Earth.

65. **Interpret** We are to find the energy per unit length within a wire that carries a given current distributed uniformly throughout the wire.

Develop Equation 26.19 $B = (\mu_0 Ir)/(2\pi R^2)$ gives the magnetic field strength inside a wire at radius r. The energy density power unit length is (Equation 27.10) $u_B = B^2/(2\mu_0)$. Combine these equations and integrate to find the energy density per unit length.

EVALUATE   Using $dV = 2\pi rL\,dr$, the energy density per unit length is

$$\frac{U}{L} = \int\frac{U_B dV}{L} = \int\frac{B^2}{2\mu_0 L}dV = \int_0^R\frac{\mu_0 I^2 r^2}{8\pi^2 R^4}2\pi r\,dr = \frac{\mu_0 I^2}{4\pi}\left.\frac{r^4}{4R^4}\right|_0^R = \frac{\mu_0 I^2}{16\pi}$$

ASSESS   The energy density, like the energy, is proportional to the current squared.

67.   **INTERPRET**   We are to compare the ratio of the electric to magnetic fields given that they have the same energy density.

DEVELOP   The energy density due to the electric field is (Equation 23.7) is $u_E = \frac{1}{2}\varepsilon_0 E^2$ and that due to the magnetic field is (Equation 27.10) $u_B = B^2/2\mu_0$.

EVALUATE   When these two energy densities are equal, their ratio is unity, which gives

$$E/B = 1/\sqrt{\mu_0 \varepsilon_0}.$$

Numerically, $\mu_0 = 4\pi\times10^{-7}$ N/A^2 and $(1/4\pi\varepsilon_0) \approx 9\times10^9$ N$\cdot$m^2/C^2, so

$$1/\sqrt{\mu_0 \varepsilon_0} \approx \sqrt{\left(9\times10^9\ \text{N}\cdot\text{m}^2/\text{C}^2\right)\left(10^{-7}\ \text{N/A}^2\right)} = 3\times10^8\ \text{m/s},$$

which is, in fact, the speed of light (see Section 29.5).

ASSESS   The speed of light is a fundamental constant of nature that can be derived from measurements of the constants of electricity and magnetism. Because these latter are the same in all inertial frames of reference, the speed of light must also be the same in all inertial frames of reference.

69.   **INTERPRET**   A conductive disk is in a changing magnetic field, and we are asked to find the current density in the disk and the rate of power dissipation in the disk. We will use Faraday's law and the resistance of the individual loops that make up the disk.

DEVELOP   We will treat the disk as a set of infinitesimal loops with radius r, thickness h, resistivity ρ, and width dr. The resistance of each such loop, using $R = \rho\frac{L}{A}$, is $R = \rho\frac{2\pi r}{hdr}$. The magnetic flux through each loop is the magnetic field dotted with the area normal, or

$$\Phi_B = B\pi r^2 = bt\pi r^2,$$

The induced emf around the loop is (Faraday's law, Equation 27.2)

$$\varepsilon = -\frac{d\Phi_B}{dt} = -b\pi r^2.$$

The current density is given by $J = \frac{I}{A} = \frac{\varepsilon/R}{hdr}$. To find the total power, we will integrate the power in each infinitesimal loop:

$$dP = \varepsilon dI \quad\Rightarrow\quad P = \int_0^a \varepsilon dI.$$

EVALUATE   **(a)** The current density is

$$J = \frac{\varepsilon}{Rhdr} = \frac{-\pi br^2}{\left(\rho\frac{2\pi r}{hdr}\right)hdr} = -\frac{br}{2\rho}$$

(b) The power dissipation is

$$P = -\int_0^a \pi br^2 dI; \quad dI = \frac{\varepsilon}{R} = \frac{-\pi br^2}{\rho\frac{2\pi r}{hdr}} - \frac{brhdr}{2\rho}$$

$$P = \int_0^a \pi br^2\frac{brh}{2\rho}dr = \frac{\pi b^2 h}{2\rho}\int_0^a r^3 dr$$

$$= \frac{\pi b^2 h a^4}{8\rho}$$

ASSESS   There are several interesting aspects of this problem. First, the current density is linear with r, and is independent of the thickness h. This makes sense: a thicker disk would have more current, but the current density would be the same. Second, the power actually depends on the fourth power of disk radius a, so increasing the size of this disk increases the power dissipation dramatically. This phenomenon is used in metal detectors, and explains why large metal objects are easier for metal detectors to find than small ones.

71. **INTERPRET** In Problem 47, we are shown a movable bar which completes a circuit in a magnetic field. That problem uses Faraday's law to find the direction of the current and the power. Here we extend the problem to find the speed of the bar as a function of time with a constant force pulling the bar.

DEVELOP There are two forces on the bar: the constant applied force, F, and the magnetic force due to the induced current in the bar: $F_m = I|\vec{l} \times \vec{B}| = IlB$. This current I follows from Faraday's law and the fact that the area of the loop is increasing:

$$I = \frac{\mathcal{E}}{R} = -\frac{1}{R}\frac{d\Phi_B}{dt} = -\frac{1}{R}\frac{d}{dt}(BA) = -\frac{1}{R}\frac{d}{dt}(Blx) = -\frac{Blv}{R}$$

According to Lenz's law, the current will flow counterclockwise in order to reduce the change in the magnetic flux. This means the magnetic force will point to the left, in the opposite direction of the applied force. Therefore, the total force on the bar is

$$F_{tot} = F - IlB = F - \frac{B^2l^2}{R}v$$

EVALUATE From Newton's second law, $F_{tot} = ma = mdv/dt$. With the above equation, we get a differential equation for $v(t)$:

$$\frac{dv}{dt} = \frac{F}{m} - \frac{B^2l^2}{Rm}v$$

We can guess that the solution will have a form of $v(t) = A - De^{-Ct}$, where A, D and C are constants. Plugging this solution into the differential equation gives

$$\frac{dv}{dt} = DCe^{-Ct} = \frac{F}{m} - \frac{B^2l^2}{Rm}\left(A - De^{-Ct}\right)$$

When $t \to \infty$, the exponential terms disappear, and we are left with

$$\frac{F}{m} - \frac{B^2l^2}{Rm}A = 0 \quad \to \quad A = \frac{FR}{B^2l^2}$$

That means the factors in front of the exponential terms must sum to zero on their own:

$$\left(DC - \frac{B^2l^2}{Rm}D\right)e^{-Ct} = 0 \quad \to \quad C = \frac{B^2l^2}{Rm}$$

Lastly, we are told that the bar starts from rest, which means $A = D$. Putting all this together:

$$v(t) = \frac{FR}{B^2l^2}\left[1 - \exp\left(-\frac{B^2l^2}{Rm}t\right)\right]$$

ASSESS Notice that there is a terminal velocity: $v(\infty) = FR/B^2l^2$, which is the highest speed reached by the bar.

73. **INTERPRET** We calculate the self-inductance per length of a coaxial cable, using the flux through the area between the two conductors.

DEVELOP The self-inductance is defined in Equation 27.3 as $L = \Phi_B/I$. To find the magnetic flux, we recall Example 26.7, where it was shown that the magnetic field lines around a single wire form concentric circles with magnitude $B = \mu_0 I/2\pi r$. With a coaxial cable, there is an outer conductor that carries the opposite current, so the encircled current is zero outside the cable, and by Ampère's law, the field is zero as well (recall Problem 26.71). Therefore, we only need to concern ourselves with the magnetic flux in between the two conductors.

EVALUATE Since the magnetic field lines wrap around the inner conductor, we imagine the flux flowing through a strip of length l and width dr, at a distance r from the center of the cable. See the figure below.

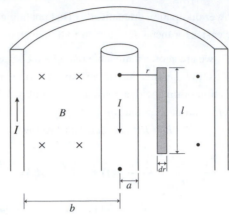

Since by construction the field is normal is normal to the strip's area, the flux through it is

$$d\Phi_B = BdA = \frac{\mu_0 Il}{2\pi r}dr$$

We now integrate this over the region where the field is nonzero, i.e. from the inner conductor's radius, a, to the outer conductor's radius, b.

$$\Phi_B = \int d\Phi_B = \int_a^b \frac{\mu_0 Il}{2\pi r}dr = \frac{\mu_0 Il}{2\pi}\ln\left(\frac{b}{a}\right)$$

The self-inductance per unit length is therefore

$$\frac{L}{l} = \frac{\mu_0}{2\pi}\ln\left(\frac{b}{a}\right)$$

ASSESS   We check this result by calculating the energy stored in the magnetic field of the coaxial cable. From Equation 27.10, the magnetic energy density is $u_B = B^2/2\mu_0$. Integrating this over the volume of a section of the cable with length l gives

$$U = \int u_B dV = \int_a^b \int_0^{2\pi} \int_0^l \frac{1}{2\mu_0}\left(\frac{\mu_0 I}{2\pi r}\right)^2 rdrd\theta dz = \frac{\mu_0 I^2 l}{4\pi}\ln\left(\frac{b}{a}\right)$$

Comparing this to Equation 27.9, $U = \frac{1}{2}LI^2$, we get the same answer for the self-inductance per unit length, L/l.

75.   **INTERPRET**   We're asked to derive the formula for the volume flow rate in a blood vessel being probed by an electromagnetic flowmeter.

   DEVELOP   The magnetic field from the flowmeter will deflect some of the moving charges in the blood, as described previously for the Hall effect. This deflection will result in an electric field across the blood vessel. Charges will continue to be deflected until the force from the electric field balances out the force from the magnetic field: $qE = qvB$, where we assume that the magnetic field is perpendicular to the blood flow. The "drift" velocity, $v = E/B$, multiplied by the cross-sectional area of the vessel, πr^2, gives the volume flow rate that we are looking for.

   EVALUATE   The flowmeter measures the voltage. If we assume the electric field is uniform, then the relationship between the voltage and the field is just $V = Ed$. Combining this with the electromagnetic force equation above, we get a volume flow rate of

$$\mathcal{F} = Av = \left[\pi\left(\tfrac{1}{2}d\right)^2\right]\left[\frac{V/d}{B}\right] = \frac{\pi d^2 V}{4Bd}$$

   ASSESS   The formula indicates that a higher voltage reading is indicative of a greater flow rate. This is because a greater flow rate results in a larger magnetic force, which requires a larger electric field to achieve equilibrium.

77.   **INTERPRET**   We consider how electric power might be "stolen" using electromagnetic induction.

   DEVELOP   The magnetic field is not uniform around the wire, so doubling the area won't necessarily double the magnetic flux. To be precise, the magnetic field is inversely proportional to the distance from the wire:

$B = \mu_0 I / 2\pi r$. Let's imagine the loop has length l and width a, and that the top of the wire is a distance y from the power line. Then, the flux through the loop is:

$$\Phi_B = \int B dA = \frac{\mu_0 I l}{2\pi} \int_y^{y+a} \frac{dr}{r} = \frac{\mu_0 I l}{2\pi} \ln\left(\frac{y+a}{y}\right)$$

EVALUATE If the loop doubles in size by extending a distance a towards the wire, the flux will increase by

$$\frac{\Phi_B'}{\Phi_B} = \frac{\ln\left(\frac{y+a}{y}\right) + \ln\left(\frac{y}{y-a}\right)}{\ln\left(\frac{y+a}{y}\right)} = 1 - \frac{\ln(1-a/y)}{\ln(1+a/y)}$$

If we assume $a \ll y$, then we can use the approximation $\ln(1+x) \approx x$, in which case $\Phi_B'/\Phi_B \approx 2$. However, if a is nearly as big as y, then $\Phi_B'/\Phi_B \to \infty$. Therefore, the flux increases by some factor greater than 2. Since the induced emf is directly related to the flux: $\mathcal{E} = -d\Phi_B/dt$, it will increase by the same factor.
The answer is (c).

ASSESS The magnetic field is greater near the wire $(B \propto 1/r)$, so the closer a farmer can place the loop to the wire, the more power he will be able to siphon off.

79. **INTERPRET** We consider how electric power might be "stolen" using electromagnetic induction.

DEVELOP You might think that power lines are always generating time-varying magnetic fields and the induced emfs that go with them, so the power company won't notice if a farmer uses some of this energy that is just being "lost" anyway. But in fact this is wrong. The magnetic field energy around a wire is not radiated away but only temporarily stored and then later given back to the power lines. During each cycle of the AC current, the magnetic fields will decrease in strength, thus inducing an emf back into the power line that helps to drive current in the next part of the cycle.

EVALUATE By the above logic, if the farmer's loop had no resistance, then current would slosh back and forth in the loop, but no energy would be expended. However, as soon as the farmer puts a load in the loop circuit (like a light bulb, for instance), some of the magnetic field energy is used to do work, and therefore less energy will cycle back from the field into the power line. As a result, more fuel must be consumed at the power plant supplying the line.
The answer is (a).

ASSESS Another way to think about this is that the loop and the wire have a mutual inductance, L. When there's no resistance, the voltage across this inductor is just: $\mathcal{E}_L = LdI/dt$. Since this voltage is 90° out of phase with the current, the total energy lost over a full cycle is zero:

$$E = \int P dt = \int I \mathcal{E} dt \propto \int \sin(2\pi ft)\cos(2\pi ft)dt = 0$$

But as soon as a resistor is added to the loop, the voltage and current will no longer be out of phase, and the energy lost over a full cycle will be non-zero.

ALTERNATING-CURRENT CIRCUITS

EXERCISES

Section 28.1 Alternating Current

15. **INTERPRET**   We are to convert from rms voltage to peak voltage and from Hz to angular frequency.

    DEVELOP   Apply Equation 28.1 to convert from rms to peak voltage and Equation 28.2 to convert from Hz to angular frequency.

    EVALUATE   (a) $V_p = \sqrt{2}\,(208\text{ V}) = 294\text{ V}$ and (b) $\omega = 2\pi\,(400\text{ Hz}) = 2.51 \times 10^3\text{ s}^{-1}$.

    ASSESS   The peak voltage, as its name implies, is greater than the rms voltage.

17. **INTERPRET**   We are to find the phase constants for a series of signals plotted as voltage versus dimensionless time.

    DEVELOP   The phase constant is a solution of Equation 28.3 for $t = 0$; that is, $V(0) = V_p \sin(\phi_V)$. Since $\sin(\phi_V) = \sin(-\phi_V \pm \pi)$, one must also consider the slope of the sinusoidal signal function at $t = 0$. In addition, the conventional range for ϕ_V usually runs from $-180°$ to $+180°$, or $-\pi \le \phi_V \le \pi$. Thus, $\phi_V = \sin^{-1}\left[V(0)/V_p\right]$ when $(dV/dt)_0 \ge 0$, but $\phi_V = -\sin^{-1}\left[V(0)/V_p \pm \pi\right]$ when $(dV/dt)_0 \le 0$.

    EVALUATE   For signal (a) in Figure 28.25, we guess that $V(0) \approx V_p/\sqrt{2}$ (since that curve next crosses zero about halfway between $\pi/2$ and π) and the slope at zero is positive, so $\phi_a = \sin^{-1}\left(1/\sqrt{2}\right) = \pi/4$ or 45°. This signal is $V_p \sin(\omega t + \phi_a) = V_p \sin(\omega t + \pi/4)$, which leads a signal with zero phase constant by 45°. For the other signals,
    (b) $V(0) = 0$ and $(dV/dt)_0 > 0$, so $\phi_b = 0$;
    (c) $V(0) = V_p$, $(dV/dt)_0 = 0$, so $\phi_c = \sin^{-1}(1) = -\sin^{-1}(1) + \pi = \pi/2$ or 90°
    (d) $V(0) = 0$ and $(dV/dt)_0 < 0$, so $\phi_d = \pm\pi$ or $\pm180°$; and
    (e) $V(0) = -V_p$ and $(dV/dt)_0 = 0$, so $\phi_e = \sin^{-1}(-1) = -\sin^{-1}(-1) - \pi = -\pi/2$ or $-90°$.

    ASSESS   We used $\left|\sin^{-1}\left[V(0)/V_p\right]\right| \le \pi/2$ or 90°, as is common on most electronic calculators, since the sine function is one-to-one only in such a restricted range.

Section 28.2 Circuit Elements in AC Circuits

19. **INTERPRET**   We are to find the rms current in each element of an RLC circuit connected across the given emf source.

    DEVELOP   Apply the equations in Table 28.1 and convert them to rms values using Equations 28.1 and 28.2.

    EVALUATE   The equations in Table 28.1 (expressed in rms values) give
    $$I_{R,\text{rms}} = V_{\text{rms}}/R = (6.3\text{ V})/(470\ \Omega) = 13\text{ mA}$$
    $$I_{C,\text{rms}} = V_{\text{rms}}\omega C = 2\pi(60\text{ Hz})(10\ \mu\text{F})(6.3\text{ V}) = 24\text{ mA}$$
    $$I_{L,\text{rms}} = V_{\text{rms}}/(\omega L) = (6.3\text{ V})/\left[2\pi(60\text{ Hz})(750\text{ mH})\right] = 22\text{ mA}$$

    ASSESS   These values are realistic for *RLC* circuits.

21.   **INTERPRET**   This problem deals with the minimum safety voltage of an capacitive circuit.

DEVELOP   Take the minimum safe voltage to be equal to the peak voltage, and use Equation 28.5 to find the peak voltage.

EVALUATE   (a) For a frequency $f = 60$ Hz, the minimum safe voltage is

$$V_p = I_p X_C = \frac{\sqrt{2}I_{rms}}{\omega C} = \frac{\sqrt{2}\,(1.4\text{ A })}{2\pi\,(60\text{ Hz})(15\ \mu\text{F})} = 250\text{ V}$$

(b) For $f = 1$ kHz, the minimum safe voltage is

$$V_p = I_p X_C = \frac{\sqrt{2}I_{rms}}{\omega C} = \frac{\sqrt{2}\,(1.0\text{ A })}{2\pi\,(1000\text{ Hz})(15\ \mu\text{F})} = 15\text{ V}$$

ASSESS   The results are given to two significant figures. The safe voltage is based on the peak voltage, which seems reasonable. Notice that the capacitor has the greatest effect (largest reactance) at low frequency.

23.   **INTERPRET**   We are to find the frequency of an inductive circuit given the rms inductance, emf, and current.

DEVELOP   Apply Equation 28.7, $I_p = V_p/(\omega L)$ and Equation 28.2, $\omega = 2\pi f$. Because

$$\frac{V_{rms}}{I_{rms}} = \frac{V_p}{I_p}$$

we can use the rms values instead of the peak values in these expressions.

EVALUATE   Combining the expressions above gives

$$f = \frac{\omega}{2\pi} = \frac{V_p}{2\pi I_p L} = \frac{V_{rms}}{2\pi I_{rms} L} = \frac{10\text{ V}}{2\pi (2.0\text{ mA})(50\text{ mH})} = 16\text{ kHz}$$

ASSESS   The inductance and frequency are inversely proportional.

Section 28.3 *LC* Circuits

25.   **INTERPRET**   Given the oscillation period of an LC circuit and its capacitance, we are to find the inductance.

DEVELOP   The inductance and capacitance are related to the frequency of an LC circuit by Equation 28.10, $\omega = (LC)^{-1/2}$. The angular frequency is related to the oscillation period T as $\omega = 2\pi f = 2\pi/T$.

EVALUATE   Solving the expression above for the inductance and inserting the given quantities gives

$$L = \frac{1}{\omega^2 C} = \left(\frac{2\pi}{2.4\text{ s}}\right)^{-2}\frac{1}{18\text{ mF}} = 8.1\text{ H}$$

ASSESS   The inductance and capacitance are inversely proportional for a given frequency.

27.   **INTERPRET**   We are to find the inductance and peak voltage of an LC circuit given its oscillation period and peak current.

DEVELOP   Using Equation 28.2, the oscillation frequency is

$$\omega = 2\pi f = \frac{2\pi}{T}$$

The inductance can be calculated from Equation 28.10: $L = 1/(\omega^2 C) = T^2/(4\pi^2 C)$

EVALUATE   (a) The inductance is

$$L = \frac{T^2}{4\pi^2 C} = \frac{(5.0\text{ ms})}{4\pi^2 (20\ \mu\text{F})} = 32\text{ mH}$$

(b) Figure 28.11 and the expressions for the electric and magnetic energies for the *LC* circuit in the text imply that $\frac{1}{2}CV_p^2 = \frac{1}{2}LI_p^2$, so

$$V_p = I_p\sqrt{L/C} = (25\text{ mA })\sqrt{(31.7\text{mH})/(20\mu\text{F})/} = 1.0\text{ V}$$

ASSESS   The results are given to two significant figures, as warranted by the data.

Section 28.4 Driven *RLC* Circuits and Resonance

29.   **INTERPRET**   We are to find the capacitance of the given *RLC* circuit, then find its impedance at the two given frequencies.

DEVELOP   The capacitance can be found from the relation between resonance frequency and the inductance and capacitance:

$$\omega_0 = \frac{1}{\sqrt{LC}}$$

Knowing the capacitance, use Equation 28.12 to find the impedance, using $X_L = \omega L$ and $X_C = 1/(\omega C)$.

EVALUATE   **(a)** From the expression for resonance in an *RLC* circuit,

$$C = 1/\left(\omega_0^2 L\right) = \left(2\pi \times 4.0 \text{ kHz}\right)^{-2}\left(20 \text{ mH}\right)^{-1} = 79 \text{ nF}$$

(b) At resonance, $X_L - X_C = 0$, so $Z = R = 75 \text{ }\Omega$.

(c) At 3 kHz, $X_L - X_C = \omega L - 1/(\omega C) = \left(2\pi\right)\left(3.0 \text{ kHz}\right)\left(20 \text{ mH}\right) - \left[\left(2\pi\right)\left(3.0 \text{ kHz}\right)\left(79.2 \text{ nF}\right)\right]^{-1} = -293 \text{ }\Omega$, so

$$Z = \sqrt{R^2 + \left(X_L - X_C\right)^2} = \sqrt{\left(75 \text{ }\Omega\right)^2 + \left(-293 \text{ }\Omega\right)^2} = 300 \text{ }\Omega$$

to two significant figures.

ASSESS   The impedance is frequency dependent, so its value is different for different frequencies.

31.   **INTERPRET**   For a series *RLC* circuit, we are to find the frequency at which the impedance is a *minimum* and the value of that impedance.

DEVELOP   From Equation 28.12, we know that the impedance Z is $Z = \sqrt{R^2 + \left(X_L - X_C\right)^2}$, where $X_C = 1/(\omega C)$ and $X_L = \omega L$. The frequency at which Z is minimum will be when $X_C = X_L$. At that resonance frequency, the impedance is $Z = R$. The component values in this circuit are $R = 18 \text{ k}\Omega$, $C = 14 \text{ }\mu\text{F}$, and $L = 0.20 \text{ H}$.

EVALUATE   **(a)** The minimum-impedance frequency is

$$X_C = X_L$$
$$\frac{1}{\omega C} = \omega L$$
$$\omega = \frac{1}{\sqrt{LC}}$$
$$f = \frac{1}{2\pi\sqrt{LC}} = \frac{1}{2\pi\sqrt{\left(0.20 \text{ H}\right)\left(14 \text{ }\mu\text{F}\right)}} = 95 \text{ Hz}$$

(b) At this frequency, the impedance is $Z = R = 18 \text{ k}\Omega$.

ASSESS   At resonance, the effects of the inductor and the capacitor cancel out, leaving only resistance.

Sections 28.5 Power in AC Circuits and 28.6 Transformers and Power Supplies

33.   **INTERPRET**   We are to find the power consumption of a device given its rms current and the current phase.

DEVELOP   The average power consumed by an AC circuit is given by Equation 28.14, $P_{ave} = V_{rms} I_{rms} \cos\phi$.

EVALUATE   Inserting the given values into the expression above gives

$$P_{ave} = V_{rms} I_{rms} \cos\phi = \left(120 \text{ V}\right)\left(4.6 \text{ A}\right)\cos\left(25°\right) = 500 \text{ W}$$

ASSESS   The maximum power for this would be (120 V)(4.6 A) = 552 W, which would require operating at a different frequency. Thus, at 25° phase, the power is about 90% of its maximum value.

35.   **INTERPRET**   We are to compare the power consumption of two circuits that have the same current and voltage; but one that is purely resistive and the other has voltage leading current. The difference in the power usage by these two circuits will be due to the difference in power factors between the two circuits.

DEVELOP   The average power consumption of a circuit is (Equation 28.14) $P_{ave} = I_{rms}V_{rms}\cos\phi$. In the first circuit, the power factor is $\cos\phi = 1$, since the circuit is purely resistive. In the second, $\phi = 20°$. In each case, $I_{rms} = 20$ A and $V_{rms} = 240$ V.

EVALUATE   For the first circuit,

$$P_{ave} = I_{rms}V_{rms}\cos\phi = I_{rms}V_{rms} = (20\text{ A})(240\text{ V}) = 4.8\text{ kW}$$

For the second circuit,

$$P_{ave} = I_{rms}V_{rms}\cos\phi = (20\text{ A})(240\text{ V})\cos(20°) = 4.5\text{ kW}.$$

ASSESS   This is a fairly direct application of a power calculation.

37. **INTERPRET**   You're trying to determine what transformer you need to run your American-bought stereo in Europe.
     DEVELOP   You need a step-down transformer that goes from Europe's 230 V to the 120 V used by your stereo. The number of turns in the primary and secondary coils are related by Equation 28.15: $N_2/N_1 = V_2/V_1$. Power, $P = IV$, is ideally conserved in the transformer, so the currents in the primary and secondary coils should be related by: $I_2/I_1 = V_1/V_2$.
     EVALUATE   (a) Given the number of turns in the primary, the number of turns in the secondary is

$$N_2 = N_1\frac{V_2}{V_1} = (460)\left(\frac{120\text{ V}}{230\text{ V}}\right) = 240$$

(b) Given the maximum primary current, the maximum secondary current will be

$$I_2 = I_1\frac{V_1}{V_2} = (1.5\text{ A})\left(\frac{230\text{ V}}{120\text{ V}}\right) = 2.9\text{ A}$$

This is below the threshold of your stereo, so the transformer will work.

ASSESS   The emf per turn in the secondary is set by the number of turns and the current in the primary. Therefore, to lower the voltage, the secondary should have less turns than the primary, as we have found. By contrast, the reduced voltage of the secondary require more current in order to conserve power. (Of course, some power will be lost in the transformer to resistive heating in the coils.)

PROBLEMS

39. **INTERPRET**   This problem is about capacitive and inductive reactances, and how they depend on the frequency.
     DEVELOP   From Equations 28.5 and 28.7, the capacitive and inductive reactances are

$$X_C = \frac{1}{\omega C} = \frac{1}{2\pi fC}, \quad X_L = \omega L = 2\pi fL$$

EVALUATE   (a) From the above equation, the frequency of the applied voltage is

$$f = \frac{1}{2\pi X_C C} = \frac{1}{2\pi(1.0\text{ k}\Omega)(2.0\text{ μF})} = 80\text{ Hz}$$

(b) Equating $X_L = X_C$ implies

$$L = \frac{X_C}{\omega} = \frac{X_C}{2\pi f} = \frac{1.0\text{ k}\Omega}{2\pi(79.6\text{ Hz})} = 2.0\text{ H}$$

(c) Doubling ω doubles X_L and halves X_C, so X_L would be four times X_C at $f = 159$ Hz.

ASSESS   Capacitive reactance X_C is inversely proportional to ω, whereas the inductive reactance X_L is proportional to ω. A larger capacitor has lower reactance and a larger inductor has higher reactance.

41. **INTERPRET**   We're asked to express the time-varying potential of an alpha wave in the human brain.
     DEVELOP   To use Equation 28.3, $V = V_p\sin(\omega t)$, we need to convert the rms voltage to the voltage amplitude, as well as the frequency to the angular frequency:

$$V_p = \sqrt{2}V_{rms} = \sqrt{2}(32\text{ μV}) = 45\text{ μV}$$
$$\omega = 2\pi f = 2\pi(10\text{ Hz}) = 63\text{ s}^{-1}$$

EVALUATE   Plugging the given values into the voltage equation gives

$$V = (45\ \mu\text{V})\sin\left[(63\ \text{s}^{-1})t\right]$$

ASSESS We may also write the angular frequency as rad/s, but radians are dimensionless, so it's not obligatory.

43. **INTERPRET** We are to find the frequency at which the given inductor and capacitor will have the same reactance given that at 10 kHz the reactance of the inductor is ten times that of the capacitor.

    DEVELOP From Equations 28.5 and 28.7, the capacitive and inductive reactances are

    $$X_C = \frac{1}{\omega C}, \quad X_L = \omega L$$

    respectively. We are given that $\omega_1 L = 10(1/\omega_1 C)$, or $LC = 10/\omega_1^2$. The reactances are equal when $LC = 1/\omega^2$, so we can solve for ω in terms of ω_1.

    EVALUATE The frequency at which the reactances are equal is

    $$\omega = \omega_1/\sqrt{10}$$
    $$f = \sqrt{10}\ \text{kH} = 3.2\ \text{kHz}$$

    ASSESS Reducing the frequency increases the capacitive reactance and decreases the inductive reactance.

45. **INTERPRET** This problem involves a capacitive circuit consisting of two capacitors connected in parallel across a emf source. We are given one capacitance and are asked to find the other, and we are also asked to find frequency at which the rms current decreases to the given value.

    DEVELOP Capacitors in parallel add, so the reactance of the combination is

    $$X_C = \left[\omega(C_1 + C_2)\right]^{-1}$$

    and, from the generalized version of Ohm's law (Equation 28.12 with $Z = X_C$) the rms current is $I_{C,\text{rms}} = \omega(C_1 + C_2)V_{\text{rms}}$, which allows us to find C_2 ($C_1 = 2.2$ nF).

    EVALUATE **(a)** At a frequency of 1.0 kHz,

    $$C_1 + C_2 = \frac{(3.4\ \text{mA})}{2\pi(1.0\ \text{kHz})(10\ \text{V})} = 54.1\ \text{nF}$$

    Thus,

    $$C_2 = (54.1 - 2.2)\ \text{nF} = 52\ \text{nF}$$

    (b) Dividing the rms currents at the two frequencies, we get $f_2/f_1 = I_{\text{rms},2}/I_{\text{rms},1}$, or

    $$f_2 = \frac{1.2\ \text{mA}}{3.4\ \text{mA}}(1.0\ \text{kHz}) = 350\ \text{Hz}$$

    ASSESS The results are reported to two significant figures, as warranted by the data.

47. **INTERPRET** This problem asks for the inductance that satisfies the resonance condition for a given range of capacitances and frequencies.

    DEVELOP Using Equations 28.2 and 28.10, the resonant frequency can be written as

    $$f = \frac{\omega}{2\pi} = \frac{1}{2\pi\sqrt{LC}}$$

    which can be solved to give

    $$L = \frac{1}{\omega^2 C} = \frac{1}{4\pi^2 f^2 C}$$

    EVALUATE Using either condition, $f_1 = 88$ MHz with $C_1 = 16.4$ pF, or $f_2 = 108$ MHz with $C_2 = 10.9$ pF, we find the inductance to be

$$L = \frac{1}{4\pi^2 f_1^2 C_1} = \frac{1}{4\pi^2 (88.0 \text{ MHz})^2 (16.4 \text{ pF})} = 0.199 \text{ } \mu\text{H}$$

$$L = \frac{1}{4\pi^2 f_2^2 C_2} = \frac{1}{4\pi^2 (108 \text{ MHz})^2 (10.9 \text{ pF})} = 0.199 \text{ } \mu\text{H}$$

ASSESS   For a given inductance L, the capacitance is inversely proportional to f^2. Thus, lower capacitance covers the higher end of the frequency band.

49. **INTERPRET**   This problem involves an LC circuit in which an oscillation occurs that transfers energy back and forth between electric and magnetic fields. One eighth of a cycle after the capacitor is charged, we are to find the fraction of their peak values of the capacitor charge, energy, and the inductor current and energy.
DEVELOP   The electric energy stored in the capacitor is given by $U_E(t) = q^2/(2C)$ where $q(t) = q_p \cos \omega t$ (see Equation 28.9). Similarly, the magnetic energy stored in the inductor is $U_B(t) = LI^2/2$, where

$$I(t) = dq/dt = -q_p \omega \sin \omega t = I_p \cos(\omega t + \pi/2)$$

The quantities are to be evaluated at $\omega t = \omega(T/8) = \frac{2\pi}{8} = \frac{\pi}{4} = 45°$ (i.e., at $\frac{1}{8}$ of a cycle). Note that phase constant zero corresponds to a fully charged capacitor at $t = 0$.

EVALUATE   **(a)** From Equation 28.9, we obtain $q(t = T/8)/q_p = \cos 45° = 1/\sqrt{2}$
(b) From the equation for electric energy, the ratio is

$$\frac{U_E(T/8)}{U_{E,p}} = \frac{q^2(T/8)/(2C)}{q_p^2/(2C)} = \frac{q^2(T/8)}{q_p^2} = \cos^2 \omega t = \cos^2 45° = \frac{1}{2}$$

(c) The ratio of the current is $I(T/8)/I_p = \cos(\omega t + \pi/2) = \cos 135° = -\frac{1}{\sqrt{2}}$. The direction of the current is away from the positive capacitor plate at $t = 0$.
(d) From the equation for magnetic energy, $U_B(T/8)/U_{B,p} = I(T/8)^2/I_p^2 = \cos^2 135° = \frac{1}{2}$
ASSESS   At one-eighth of a cycle, half of the total energy is magnetic and half is electric. This is illustrated in Figure 28.11.

51. **INTERPRET**   This problem is about an LC circuit with damping due to the resistance. We want to find the number of oscillations the circuit completes before the peak voltage is reduced by half.
DEVELOP   For a damped LC circuit, Equation 28.11 gives the charge as a function of time. Because $V(t) = q(t)/C$, the voltage as a function of time can be written as

$$V(t) = V_p e^{-Rt/2L} \cos \omega t$$

The peak voltage decays with time constant $2L/R$. Half the initial peak value is reached after a time $t = (2L/R)\ln 2$ (when $e^{-Rt/2L} = \frac{1}{2}$).

EVALUATE   Since the period of oscillation is $T = 2\pi/\omega = 2\pi\sqrt{LC}$, the number of cycles that occur within time t is

$$\frac{t}{T} = \frac{(2L/R)\ln 2}{2\pi\sqrt{LC}} = \frac{\ln 2}{\pi R}\sqrt{\frac{L}{C}} = \frac{\ln 2}{\pi(1.6 \text{ } \Omega)}\sqrt{\frac{20 \text{ mH}}{0.15 \text{ } \mu\text{F}}} = 50$$

ASSESS   This oscillation is underdamped. The larger the resistance, the more rapidly the oscillation decays.

53. **INTERPRET**   This problem is about a series RLC circuit at resonance. We want to find the smallest resistance that still keeps the capacitor voltage under its rated value when the circuit is at resonance.
DEVELOP   In a series RLC circuit at resonance, the peak capacitor voltage is

$$V_{C,p} = I_p X_C = \frac{V_p/R}{\omega_0 C} = \frac{V_p}{R}\sqrt{\frac{L}{C}}$$

where $\omega_0 = (LC)^{-1/2}$ is the resonant angular frequency.
EVALUATE   The condition that $V_{C,p} \leq 400$ V implies

$$R \geq \frac{V_p}{V_{C,p}} \sqrt{\frac{L}{C}} = \left(\frac{32 \text{ V}}{400 \text{ V}}\right) \sqrt{\frac{1.5 \text{ H}}{250 \text{ μF}}} = 6.2 \text{ Ω}$$

ASSESS Our results shows that $V_{C,p}$ is inversely proportional to R. This means that a larger resistor would be required if the capacitor has a lower voltage rating.

55. **INTERPRET** This problem involves analyzing a phasor diagram for a driven *RLC* circuit to find if the driving frequency is above or below resonance. We are also to complete the diagram and use it to find the phase difference between the applied voltage and current.

DEVELOP Our diagram has three phasors, $V_{Rp}, V_{Lp},$ and $V_{Cp},$ representing the voltages across the resistor, the inductor, and the capacitor, respectively. Because the resistor voltage is in phase with the current, I_p is in the same direction as V_{Rp}. The resonant frequency is $\omega_0 = (LC)^{-1/2}$.

EVALUATE   **(a)** From the observation that $V_{Lp} = I_p \omega L > V_{Cp} = I_p / \omega C$, we conclude that $\omega^2 > 1/(LC) = \omega_0^2$, which means the frequency is above resonance.

(b) The applied voltage phasor is the vector sum of the resistor, capacitor, and inductor voltage phasors, as shown below. The current is in phase with the voltage across the resistor, which in this case is lagging the applied voltage because

$$\phi = \tan^{-1} \left(\frac{V_{Lp} - V_{Cp}}{V_{Rp}}\right) > 0$$

by approximately 50° (as estimated from the figure).

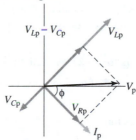

ASSESS Our circuit is inductive since $V_{Lp} > V_{Cp}$. Note that a positive ϕ means that voltage leads current, and a negative ϕ means voltage lags current. At resonance, $X_L = X_C$ and $\phi = 0$.

57. **INTERPRET** We are to find the power factor and the power dissipation in a series *RLC* circuit.

DEVELOP From the geometry of Figure 28.16, we find that the power factor of the circuit is

$$\cos\phi = \frac{V_{Rp}}{V_p} = \frac{I_p R}{I_p Z} = \frac{R}{Z}$$

The average power in the circuit is given by Equation 28.14:

$$\langle P \rangle = I_{rms} V_{rms} \cos\phi = I_{rms} (I_{rms} Z)(R/Z) = I_{rms}^2 R$$

EVALUATE   **(a)** Substituting the values given, we find the power factor is

$$\cos\phi = \frac{R}{Z} = \frac{100 \text{ Ω}}{300 \text{ Ω}} = 0.333$$

(b) The above equation gives $\langle P \rangle = I_{rms}^2 R = (200 \text{ mA})^2 (100 \text{ Ω}) = 4.00 \text{ W}$.

ASSESS   Note that the average AC power is given by the same expression as the DC power if the rms current is used. The power factor must be between zero and 1. A purely resistive circuit has a power factor of 1, while a circuit with only capacitance or inductance has a power factor of zero.

59. **INTERPRET** You want to know the percentage of power your company loses during transmission over its electric lines.

DEVELOP   For AC circuits, the average power produced is given in Equation 28.14: $\langle P \rangle = I_{rms} V_{rms} \cos\phi$, where $\cos\phi$ is the power factor. The power lost in the transmission lines is $I^2 R$ at any given time, but the average power lost will be

$$\langle P_{lost} \rangle = \langle I^2 R \rangle = \langle I^2 \rangle R = I_{rms}^2 R$$

where we have used the fact that the resistance is constant over time, as well as the definition of root-mean-squared: $I_{rms} = \sqrt{\langle I^2 \rangle}$.

EVALUATE   (a) For a power factor of 1.0, the percentage of power lost in the transmission lines is

$$\frac{\langle P_{lost} \rangle}{\langle P \rangle} = \frac{I_{rms} R}{V_{rms} \cos\phi} = \frac{(200\text{ A})(100\ \Omega)}{(365\text{ kV})(1.0)} = 5.5\%$$

(b) For a power factor of 0.60, the same percentage is

$$\frac{\langle P_{lost} \rangle}{\langle P \rangle} = \frac{I_{rms} R}{V_{rms} \cos\phi} = \frac{(200\text{ A})(100\ \Omega)}{(365\text{ kV})(0.60)} = 9.1\%$$

ASSESS   In this problem, the current is constant, so the power lost will be the same in both cases. What does change is the amount of AC power produced at the plant. For a power factor of 1.0, the current and voltage are in phase, and the power in the circuit is maximized. But for a lower power factor, the current and voltage are out of phase, so the plant is producing less power for its end-users, while still losing the same amount in the transmission lines.

61.   **INTERPRET**   This problem deals with DC power supplies. If the time constant RC is long enough, the capacitor voltage will only decrease slightly before the AC voltage from the transformer rises again to fully charge the capacitor.

DEVELOP   The scenario is depicted in Figure 28.23. From the given DC output, we find the load resistance to be $R = (22\text{ V})/(150\text{ mA}) = 147\ \Omega$. In one period of the input AC ($T = 1/f$), the capacitor voltage must decay by less than 3%, or $e^{-T/RC} \geq 0.97$.

EVALUATE   The above condition implies that

$$C \geq -\frac{T}{R\ln(0.97)} = -\frac{1}{Rf\ln(0.97)} = -\frac{1}{(60\text{ Hz})(147\ \Omega)\ln(0.97)} = 3.7\text{ mF}$$

ASSESS   If the capacitance is large enough, the load current and voltage can be made arbitrarily smooth with negligible decay.

63.   **INTERPRET**   We have an AC generator connected to a series RLC circuit, and we want to know its maximum peak voltage when the circuit is at resonance.

DEVELOP   The peak capacitor voltage is $V_{Cp} = I_p X_C$. At resonance, the impedance is $Z = R$ and $I_p = V_p / R$. The capacitive reactance is $X_C = 1/\omega_0 C = \sqrt{L/C}$.

EVALUATE   The condition that $V_{Cp} = (V_p/R)\sqrt{L/C} \leq 600$ V implies

$$V_p \leq (600\text{ V})(1.3\ \Omega)\sqrt{\frac{0.33\ \mu\text{F}}{27\text{ mH}}} = 2.7\text{ V}$$

ASSESS   The inductor voltage at resonance is

$$V_{Lp} = I_p X_L = \frac{V_p}{R}\omega_0 L = \frac{V_p}{R}\frac{L}{\sqrt{LC}} = \frac{V_p}{R}\sqrt{\frac{L}{C}}$$

which is the same as V_{Cp}. The two voltages cancel exactly at resonance. Note that V_{Cp} and V_{Lp} are both higher than V_p.

65.   **INTERPRET**   In Example 28.4, we found a frequency at which the current in an RLC circuit is half its maximum value. Here, we are to find a second frequency at which the current will be half the maximum. We shall use Equation 28.12 for Z.

DEVELOP   From Example 28.4, we have $C = 11.5$ µF, $R = 8.0$ Ω, and $L = 22$ mH. We also know that $Z = \sqrt{R^2 + (X_L - X_C)^2}$, where $X_C = 1/(\omega C)$ and $X_L = \omega L$. The current is given by Ohm's law (Equation 28.12), $I = V/Z$, and we are looking for a value of ω such that $Z = 2R$.

EVALUATE   Solving for ω gives

$$Z = \sqrt{R^2 + \left(\omega L - \frac{1}{\omega C}\right)^2} = 2R$$

$$4R^2 = R^2 + \left(\omega L - \frac{1}{\omega C}\right)^2$$

$$3R^2 = \omega^2 L^2 - 2\frac{L}{C} + \frac{1}{\omega^2 C^2}$$

$$3R^2 \omega^2 = \omega^4 L^2 - 2\frac{L\omega^2}{C} + \frac{1}{C^2}$$

$$0 = \omega^4 L^2 + \left(-2\frac{L}{C} - 3R^2\right)\omega^2 + \frac{1}{C^2}$$

$$\omega \in \{\pm 3882, \pm 10181\} \text{ rad/s}$$

The sign of ω is irrelevant. We need to convert to frequency using $f = 2\pi/\omega$, so $f \in \{618 \text{ Hz}, 1620 \text{ Hz}\}$.

ASSESS   The 618-Hz answer was given in the example, so the solution we need is $f = 1620$ Hz.

67.   **INTERPRET**   We have two capacitors connected first in series and then in parallel with an AC generator, and we want to know their capacitances given that the current drops from 30 mA to 5.5 mA upon going from parallel to series connections.

DEVELOP   Equation 28.5 gives the rms current when capacitors are connected to an AC generator,

$$I_{rms} = V_{rms}/X_C = \omega C V_{rms}$$

For the parallel connection, $C = C_1 + C_2$, (see Chapter 23), so

$$30 \text{ mA} = \left(2\pi \times 10^5 \text{ V/s}\right)(C_1 + C_2)$$

For the series connection, $C = C_1 C_2/(C_1 + C_2)$, so Equation 28.5 gives

$$5.5 \text{ mA} = \left(2\pi \times 10^5 \text{ V/s}\right)\left(\frac{C_1 C_2}{C_1 + C_2}\right)$$

The two equations can be used to solve for C_1 and C_2.

EVALUATE   Simplifying the above two equations leads to $C_1 + C_2 = 47.7$ nF and $C_1 C_2 = (20.4 \text{ nF})^2$. Eliminating C_1 from the second equation and substituting into the first equation, we obtain the following quadratic equation:

$$C_2^2 - (47.7 \text{ nF})C_2 + (20.4 \text{ nF})^2 = 0$$

Since the initial two equations are symmetric in C_1 and C_2, eliminating C_2 gives the same equation as the above, but with C_2 replaced by C_1. Thus, the solutions for C_1 and C_2 are

$$\frac{1}{2}\left[(47.7 \text{ nF}) \pm \sqrt{(47.7 \text{ nF})^2 - 4(20.4 \text{ nF})^2}\right] = 12 \text{ nF and } 36 \text{ nF}$$

ASSESS   The parallel connection yields a greater capacitance, and hence a larger current compared to the series combination. The results are reported to two significant figures, as warranted by the data.

69.   **INTERPRET**   This problem involves designing a circuit that gives the desired response to the given input signal.

DEVELOP   We want the output voltage to lead the input voltage (by 45°). By inspecting Figure 28.16, we notice that the voltage across a resistor leads the voltage across a capacitor with which it is in series. Similarly, the voltage across an inductor leads the voltage across a resistor with which it is in series. Both circuits can be adapted to the criteria of the black box in this problem.

EVALUATE Case (1): *RC* circuit with an AC input. In this circuit, $V = V_R + V_C$ and $I = I_C = I_R$. In the corresponding phasor diagram shown below, V_C lags I by 90°, V_R and I are in phase, and V is the vector sum of these (see Table 28.1). We drew I horizontally for convenience in the figure below. The impedance is thus

$$Z = \frac{V_p}{I_p} = \frac{\sqrt{V_{Rp}^2 + V_{Cp}^2}}{I_p} = \sqrt{R^2 + X_C^2} = \sqrt{R^2 + \frac{1}{\omega^2 C^2}}$$

and the phase angle is

$$\tan \phi = -\frac{V_{Cp}}{V_{Rp}} = -\frac{X_C}{R} = -\frac{1}{\omega RC}$$

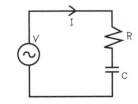

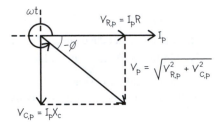

The current I always leads V, because $\phi < 0$. Recall that ϕ is defined by $I = I_p \sin \omega t$ when $V = V_p \sin(\omega t + \phi)$. The result implies that, in a series *RC* circuit, V_R leads the applied voltage, V by an angle $\tan^{-1}(1/\omega RC)$, which may be adjusted to 45° if $\omega RC = 1$. The peak voltage across the entire resistance is $V_{Rp} = V_p \cos 45° = V_p/\sqrt{2}$, so if we divide the resistance into two parts, $R_1 + R_2 = R$, with $R_2/R = 1/\sqrt{2}$, then the peak voltage across R_2 will be $V_{Rp}/\sqrt{2} = \frac{1}{2} V_p$, as desired (rms voltages have the same ratio as peak voltages).

Case (2): *RL* circuit with an AC input. When a capacitor is replaced with an inductor, the phasors for V_R and I are still parallel, but V_L leads I by 90°. The voltage V is the vector sum of V_R and V_L, so the impedance is

$$Z = \frac{V_p}{V_p} = \frac{\sqrt{V_{Rp}^2 + V_{Lp}^2}}{V_p} = \sqrt{R^2 + X_L^2} = \sqrt{R^2 + \omega^2 L^2}$$

and the phase angle is

$$\tan \phi = \frac{V_{Lp}}{V_{Rp}} = \frac{X_L}{R} = \frac{\omega L}{R}$$

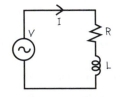

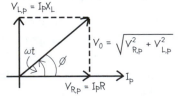

In this case, I always lags V, because $\phi > 0$ Negative ϕ is in the same sense as ωt; measured from V.

In a series RL circuit, V_L leads V by $90° - \tan^{-1}(\omega L/R)$ which equals $45°$ if $\omega L = R$. Again, $V_{Lp} = V_p/\sqrt{2}$, so if we divide L into $L_1 + L_2$, with $L_2 = L/\sqrt{2}$, the peak voltage across L_2 is $V_p/2$. Both circuits are sketched below.

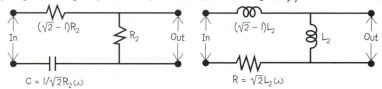

ASSESS   We have shown how the circuit can be designed in two different ways to adapt to the criteria of the black box. Our circuit conditions can be verified explicitly. Note that V_{out} is the open-circuit output voltage. If a load is connected across the output terminals, the magnitude and the phase of the voltage will be changed accordingly.

71.   **INTERPRET**   This problem involves a series RLC circuit for which we are given the current at resonance and at half the resonant frequency. We are asked to find the resistance, the inductance, and the capacitance.

DEVELOP   At resonance, the impedance is $Z = R$ and the current is $I_p = V_p/Z = V_p/R$ and $X_L - X_C = 0$. Away from resonance, $Z = V_p/I_p$ and $|X_L - X_C| = \sqrt{Z^2 - R^2}$.

EVALUATE   The resonance condition gives

$$R = \frac{V_p}{I_p} = \frac{20\text{ V}}{50\text{ mA}} = 400\ \Omega$$

On the other hand, at half the resonant frequency, $1.0\text{ kHz} = \frac{1}{2}(2.0\text{ kHz})$, the impedance is

$$Z = \frac{V_p}{I_p} = \frac{20\text{V}}{15\text{ mA}} = 1.3\text{ k}\Omega = \frac{10R}{3}$$

which gives

$$|X_L - X_C| = \sqrt{Z^2 - R^2} = R\sqrt{(10/3)^2 - 1} = \frac{\sqrt{91}}{3}R$$

With $X_L = \omega L$ and $X_C = 1/\omega C$, we obtain the following conditions:

$$\frac{1}{\omega_0 C} - \omega_0 L = 0, \qquad \frac{1}{\frac{1}{2}\omega_0 C} - \frac{1}{2}\omega_0 L = \frac{\sqrt{91}}{3}R$$

These equations can be solved for C and L, with the following result:

$$L = \frac{2\sqrt{91}\,R}{9\omega_0} = \frac{2\sqrt{91}\,(400\ \Omega)}{9(2\pi \times 2.0\text{ kHz})} = 68\text{ mH}$$

$$C = \frac{1}{\omega_0^2 L} = \frac{1}{(2\pi \times 2.0\text{ kHz})^2(67.5\text{ mH})} = 94\text{ nF}$$

ASSESS   Below resonance, capacitive reactance dominates, with $X_C > X_L$.

73.   **INTERPRET**   In this problem we are asked to derive the Q factor of an RLC circuit that satisfies the criteria given in the problem statement.

DEVELOP   To derive the expression for Q, we first need to know the power in the circuit. From Equations 28.12 and 28.14 (with rms values), and

$$\cos\phi = \frac{V_{Rp}}{V_p} = \frac{I_p R}{I_p Z} = \frac{R}{Z}$$

from Figure 28.16, the average power in a series RLC circuit can be written as

$$\langle P \rangle = I_{\text{rms}} V_{\text{rms}} \cos\phi = (V_{\text{rms}}/Z) V_{\text{rms}} (R/Z) = \frac{V_{\text{rms}}^2 R}{Z^2}$$

The above expression shows the power falls to half its resonance value (V_{rms}^2/R) when $Z = \sqrt{2}R$, or when $|X_L - X_C| = R$. In terms of the resonant frequency $\omega_0 = (LC)^{-1/2}$, this condition becomes

$$\left| \omega L - \frac{1}{\omega C} \right| = L \left| \omega - \frac{\omega_0^2}{\omega} \right| = R$$

$$\omega^2 - \omega_0^2 = \pm \frac{R}{L}\omega$$

The solutions of these quadratics, with $\omega > 0$, are

$$\omega_{\pm} = \frac{1}{2}\left[\pm \frac{R}{L} + \sqrt{\frac{R^2}{L^2} + 4\omega_0^2} \right]$$

The Q factor is then equal to $\omega_0/\Delta\omega$, where $\Delta\omega = \omega_+ - \omega_-$.

EVALUATE   If $R/L \ll \omega_0$ (or $R \ll \sqrt{L/C}$), we can neglect the first term under the square root sign compared to the second, which gives $\omega \approx \omega_0 \pm R/2L$. The difference between these two values of ω is $\Delta\omega = R/L$ from which we obtain $Q = \omega_0/\Delta\omega = \omega_0 L/R$.

ASSESS   The Q factor measures the "quality" of oscillation. The smaller the resistance, the higher the Q-factor. In the absence of resistance $(R \to 0)$, the LC circuit can oscillate indefinitely.

75.   **INTERPRET**   We are to use the equation for charge on a capacitor in an RLC circuit and the differential equation for an RLC circuit from Kirchhoff's laws to find an expression for ω.

DEVELOP   The given equations are $q(t) = q_p e^{-Rt/2L} \cos \omega t$, and $L\frac{d^2 q}{dt^2} + R\frac{dq}{dt} + \frac{q}{C} = 0$. We take the derivatives of q, substitute it into the differential equation, and solve the resulting equation for ω.

EVALUATE   We first calculate the derivatives of $q(t)$:

$$q(t) = q_p e^{-Rt/2L} \cos \omega t$$

$$q'(t) = -\frac{q_p}{2L} e^{-\frac{Rt}{2L}} \left(R\cos \omega t + 2\omega L \sin \omega t \right)$$

$$q''(t) = \frac{q_p}{4L^2} e^{-\frac{Rt}{2L}} \left[\left(R^2 - 4L^2\omega^2 \right)\cos \omega t + 4\omega RL \sin \omega t \right]$$

Substituting these into the differential equation gives us, after some algebraic steps,

$$\frac{q_p}{CL} e^{-\frac{Rt}{2L}} \left(-4L + CR^2 + 4CL^2\omega^2 \right)\cos \omega t = 0$$

For this equation to be true for all values of t, the term in parentheses must be zero.

$$-4L + CR^2 + 4CL^2\omega^2 = 0$$

$$\omega = \sqrt{\frac{1}{LC} - \frac{R^2}{4L^2}}$$

ASSESS   This reduces to $\omega = (LC)^{-1/2}$ if $R = 0$.

77.   **INTERPRET**   We are to find the maximum current in an RLC circuit at resonance.

DEVELOP   At resonance, the impedance Z is just the resistance R, and the current is the same in all series-circuit elements, so the maximum current in the inductor is just $I_{max} = V_{max}/R$. The maximum voltage is $V_{max} = 8.0$ V. The resistance is $R = 5.5\ \Omega$, and we really don't care what the inductor and capacitor values are.

EVALUATE   The maximum current is

$$I_{max} = \frac{8.0\ \text{V}}{5.5\ \Omega} = 1.5\ \text{A}$$

ASSESS   This current is within the safe limit.

79. **INTERPRET** We are analyzing a filter consisting of an RC circuit.

DEVELOP To determine which frequencies can pass through the filter, we consider the voltage across the capacitor, which will be equal to the output voltage, V_{out}. For a given frequency, the peak current through the RC circuit is given by Equation 28.12: $I_p = V_{in,p}/Z$, where in this case $Z = \sqrt{R^2 + X_C^2}$. The peak voltage across the capacitor will be $V_{Cp} = I_p X_C$. Using $X_C = 1/\omega C$, and defining $\omega_{RC} = 1/RC$, we can write the capacitor voltage as

$$V_{Cp} = \frac{V_{in,p}}{\sqrt{1 + (\omega/\omega_{RC})^2}}$$

EVALUATE If $\omega \ll \omega_{RC}$, then $V_{Cp} = V_{in,p}$, which implies that $V_{out} = V_{in}$. Therefore, low frequencies are passed from the input to the output. By contrast, for $\omega \gg \omega_{RC}$, we have $V_{Cp} = 0$. This means there is no output at high frequencies. This, then, is a low-pass filter.

The answer is (a).

ASSESS Another way to arrive at this is to recall the short-term and long-term behavior of RC circuits from Chapter 25. Over short-times ($t \ll RC$ or equivalently $\omega \gg \omega_{RC}$), the capacitor acts like a short-circuit, so current will flow through the capacitor, and there will be no voltage at the output. Over long-times ($t \gg RC$ or equivalently $\omega \ll \omega_{RC}$), the capacitor acts like an open circuit, so no current flows in the capacitor, which means the input and output have the same voltage. One might have wrongly guessed that the presence of the capacitor implies a high-pass filter, judging from the Application on loudspeakers in the text. But in that case the capacitor is in series with the output, whereas in this case it is in parallel.

81. **INTERPRET** We are analyzing a filter consisting of an RC circuit.

DEVELOP Since there is no inductance, there is technically no resonance in this circuit. The maximum output voltage is the input voltage $(V_{out} = V_{in})$, which occurs when the reactance goes to infinity. This corresponds to zero frequency, or essentially a DC signal.

EVALUATE The output voltage gradually decreases from its maximum at $\omega = 0$ to $V_{out} = 0$ at very high frequencies. Thus, there is no resonant peak at $\omega = 1/RC$, nor at $\omega = 1/\sqrt{RC}$, but the latter actually has the wrong dimensions for frequency. Since the output voltage is the same as the voltage across the capacitor, it should have the same frequency as the input, but not necessarily the same phase. Indeed, the capacitor voltage lags behind the current by 90° (see Table 28.1), and the current and input voltage have a phase difference given by Equation 28.13, which in this case is $\tan\phi = -1/\omega RC$. So the input and output voltages will differ in phase by $\phi = 90° - \tan^{-1}(\omega_{RC}/\omega)$.

The answer is (d).

ASSESS Specifically, the phase difference between the input and the output is $\phi = 90° - \tan^{-1}(\omega_{RC}/\omega)$, where $\omega_{RC} = 1/RC$. As $\omega \to \infty$, the two voltages approach 90° out of phase. Conversely, as $\omega \to 0$, the two voltages become more and more in phase.

MAXWELL'S EQUATIONS AND ELECTROMAGNETIC WAVES

EXERCISES

Section 29.2 Ambiguity in Ampère's Law

13. **INTERPRET** In this problem, we are asked to find the displacement current through a surface.

    DEVELOP As shown in Equation 29.1, Maxwell's displacement current is

    $$I_d = \epsilon_0 \frac{d\Phi_E}{dt} = \epsilon_0 \frac{d(EA)}{dt} = \epsilon_0 A \frac{dE}{dt}$$

    EVALUATE The above equation gives

    $$I_d = \epsilon_0 A \frac{dE}{dt} = \left[8.85 \times 10^{-12} \text{ C}^2/(\text{N} \cdot \text{m}^2) \right] (1.0 \text{ cm}^2) \left[1.5 \text{ V}/(\text{m} \cdot \mu\text{s}) \right] = 1.3 \text{ nA}$$

    ASSESS Displacement current arises from changing electric flux and has units of amperes (A), just like ordinary current.

Section 29.4 Electromagnetic Waves

15. **INTERPRET** We are given the electric and magnetic fields of an electromagnetic wave and asked to find the direction of propagation in terms of a unit vector.

    DEVELOP The direction of propagation of the electromagnetic wave is the same as the direction of the cross product $\vec{E} \times \vec{B}$.

    EVALUATE When $\vec{E}$ is parallel to $\hat{j}$ and $\vec{B}$ is parallel to $\hat{i}$, the direction of propagation is parallel to $\vec{E} \times \vec{B}$, or $\hat{j} \times \hat{i} = -\hat{k}$.

    ASSESS For electromagnetic waves in vacuum, the directions of the electric and magnetic fields, and of wave propagation, form a right-handed coordinate system.

Section 29.5 Properties of Electromagnetic Waves

17. **INTERPRET** This problem involves expressing the distance between the Sun and the Earth in terms of light minutes.

    DEVELOP A light minute (abbreviated as c-min) is approximately equal to

    $$1 \text{ c-min} = (3.0 \times 10^8 \text{ m/s})(60 \text{ s}) = 1.8 \times 10^{10} \text{ m}$$

    On the other hand, the mean distance of the Earth from the Sun (an astronomical unit) is about $R_{SE} = 1.5 \times 10^{11}$ m.

    EVALUATE In units of c-min, R_{SE} can be rewritten as

    $$R_{SE} = (1.5 \times 10^{11} \text{ m}) \overbrace{\frac{1 \text{ c-min}}{1.8 \times 10^{10} \text{ m}}}^{\equiv 1} = 8.3 \text{ c-min}$$

    ASSESS The result implies that it takes about 8.3 minutes for the sunlight to reach the Earth.

19. **INTERPRET** In this problem we want to deduce the airplane's altitude by measuring the travel time of a radio wave signal it sends out. The logic of this method is the same as that of the preceding problem.

    DEVELOP The speed of light is $c = 3 \times 10^8$ m/s and the total distance traveled is $\Delta r = 2h$ (neglecting the distance traveled by the plane during the transit time of the signal).

EVALUATE Since $\Delta r = 2h = c\Delta t$ (for waves traveling with speed c), the altitude h is

$$h = \frac{c\Delta t}{2} = \frac{(3.00 \times 10^8 \text{ m/s})(50 \text{ μs})}{2} = 7.5 \text{ km}$$

ASSESS The airplane is flying lower than the typical cruising altitude of 12,000 m (35,000 ft) for commercial jet airplanes.

21. INTERPRET This problem involves finding the round-trip time delay for radio signals traveling between the Earth and the Moon.

DEVELOP The time it takes to get a reply is twice the distance (out and back) divided by the speed of light, which is 3.00×10^8 m/s in vacuum.

EVALUATE From Appendix E, we find the distance between the Earth and the Moon to be $R_{ME} = 3.85 \times 10^8$ m, so the time required is

$$\Delta t = \frac{2R_{ME}}{c} = \frac{2(3.85 \times 10^8 \text{ m})}{3.00 \times 10^8 \text{ m/s}} = 2.57 \text{ s}$$

ASSESS The signal has to travel a very long distance, so a time delay of 2.57 seconds is not surprising. The time delay via geostationary satellite communication is typically between 240 ms and 280 ms (see Problem 29.18).

23. INTERPRET In this problem we are asked to find the wavelength of electromagnetic radiation that propagates through air, given its frequency.

DEVELOP The wavelength of the electromagnetic wave can be calculated using Equation 29.16c: $f\lambda = c$. Because air is not optically dense, it may be taken to be a vacuum, so the speed of light is $c = 3.00 \times 10^8$ m/s.

EVALUATE The wavelength in a vacuum (or air) is

$$\lambda = \frac{c}{f} = \frac{3.00 \times 10^8 \text{ m/s}}{60 \text{ Hz}} = 5.00 \times 10^6 \text{ m}$$

ASSESS The wavelength is almost as large as the radius of the Earth!

25. INTERPRET This problem involves finding the direction of polarization of an electromagnetic wave, which is the direction in which the electric field oscillates.

DEVELOP The direction of propagation of the electromagnetic wave is the same as the direction of the cross product $\vec{E} \times \vec{B}$. In our case, we have $\hat{E} \times \hat{B} = \hat{k}$, where $\hat{k}$ is the unit vector in the $+z$ direction.

EVALUATE Since the magnetic field points in the $+y$ direction, $\vec{B} = B\hat{j}$, we must have $\vec{E} = E\hat{i}$, so that $\hat{i} \times \hat{j} = \hat{k}$. The wave is linearly polarized.

ASSESS For electromagnetic waves in vacuum, the directions of the electric and magnetic fields, and of wave propagation, form a right-handed coordinate system. One may write $\hat{E} \times \hat{B} = \hat{n}$, where $\hat{n}$ is the unit vector in the direction of propagation.

27. INTERPRET This problem is about the intensity of a light beam that transits a polarizer. We are given the angle between the polarization of the light (i.e., its electric field) and the polarization direction of the material.

DEVELOP The intensity of the light after emerging from a polarizer is given by the Law of Malus (Equation 29.18), $S = S_0 \cos^2 \theta$, where θ is the angle between the field and the polarization direction of the material.

EVALUATE The Law of Malus gives $S/S_0 = \cos^2(70°) = 12\%$.

ASSESS The intensity depends on $\cos^2 \theta$. The limit $\theta = 0$ corresponds to the situation where the direction of polarization of the incident light is the same as the preferred direction specified by the polarizer, and $S = S_0$. On the other hand, when $\theta = 90°$, essentially no light passes through the polarizer.

Section 29.8 Energy and Momentum in Electromagnetic Waves

29. INTERPRET This problem explores the average intensity of a laser beam required for dielectric breakdown in air.

DEVELOP The average intensity of an electromagnetic wave is given by Equation 29.20:

$$\bar{S} = \frac{E_p B_p}{2\mu_0} = \frac{cB_p^2}{2\mu_0} = \frac{E_p^2}{2\mu_0 c}$$

EVALUATE With $E_p = 3 \times 10^6$ V/m, the average intensity is

$$\overline{S} = \frac{E_p^2}{2\mu_0 c} = \frac{\left(3\times10^6 \text{ V/m}\right)^2}{2\left(4\pi\times10^{-7} \text{ N/A}^2\right)\left(3.00\times10^8 \text{ m/s}\right)} = 1\times10^{10} \text{ W/m}^2$$

ASSESS   We need a very powerful laser to produce the breakdown field strength. The laser intensity can be compared to the average solar intensity which is about 1370 W/m^2.

31.   **INTERPRET**   You want to know if your new radio can pick up a signal from a remote location.

DEVELOP   Given the minimum electric field that the radio can pick up, the minimum intensity is $\overline{S} = E_p^2/2\mu_0 c$ (Equation 29.20b).

EVALUATE   The radio's intensity threshold is

$$\overline{S} = \frac{E_p^2}{2\mu_0 c} = \frac{\left(450 \text{ μV/m}\right)^2}{2\left(4\pi\times10^{-7} \frac{\text{N}}{\text{A}^2}\right)\left(3.0\times10^8 \text{ m/s}\right)} = 0.27 \text{ nW/m}^2$$

This means you will be able to hear your favorite station at your remote cabin.

ASSESS   The minimum detectable signal for a radio or other receiver is usually set by the background noise. A radio station's signal has to be significantly more powerful than stray electromagnetic waves that contribute to the "static" we hear between stations.

33.   **INTERPRET**   You want to double the range of your radio station's antenna.

DEVELOP   Listeners at a distance of $r = 15$ km from the antenna can pick up your radio station because the intensity, S, inside this perimeter is above a typical radio receiver's threshold. You want to increase the power so that the intensity at $r' = 30$ km is above threshold. The relation between intensity and power is given in Equation 29.21: $S = P/4\pi r^2$.

EVALUATE   The required power for doubling the range is

$$P' = P\left(\frac{r'}{r}\right)^2 = \left(5.0 \text{ kW}\right)\left(2\right)^2 = 20 \text{ kW}$$

ASSESS   We have assumed that the signal from the antenna radiates uniformly out in all directions, but that is not always the case. Some antennas focus their intensity in particular directions. However, no matter what the radiation pattern from the antenna, the power will have to be quadrupled to double the range in a given direction.

PROBLEMS

35.   **INTERPRET**   This problem is about the rate of change of electric field, which induces a magnetic field. Given the magnetic field strength at 50 cm from the center, we are to find the rate of change of the electric field and whether it is increasing or decreasing.

DEVELOP   The electric and magnetic fields are related by Equation 29.1. If we evaluate the integrals around the circular field line of radius r shown in Figure 29.15, and the plane area it bounds, we obtain:

$$\oint \vec{B} \cdot d\vec{r} = 2\pi r B = \mu_0 \epsilon_0 \frac{d}{dt} \int \vec{E} \cdot d\vec{A} = \frac{1}{c^2} \pi r^2 \frac{dE}{dt}$$

where we have used $c = \left(\epsilon_0 \mu_0\right)^{-1/2}$

EVALUATE   (a) Thus, the rate of change of electric field is

$$\frac{dE}{dt} = \frac{2c^2 B}{r} = \frac{2\left(3.00\times10^8 \text{ m/s}\right)^2\left(2.0 \text{ μT}\right)}{0.50 \text{ m}} = 7.2\times10^{11} \text{ V/(m·s)}$$

(b) A circulation of $\vec{B}$ clockwise around the circle gives a positive displacement current into the page, so $\vec{E}$ is increasing in this direction.

ASSESS   Any change in electric flux results in a displacement current that produces a magnetic field. The displacement current encircled by the loop of radius $r = 0.5$ m is

$$I_d = \epsilon_0 \frac{d\Phi_E}{dt} \int \vec{E} \cdot d\vec{A} = \epsilon_0 \pi r^2 \frac{dE}{dt} = 5 \text{ A}$$

This is precisely the current a long wire must carry in order to produce the same magnetic field strength at a distance $r = 0.5$ m from its center.

37.   **INTERPRET**   The problem simply asks what frequencies correspond to the UVB wavelength range.
    DEVELOP   The frequency is inversely proportional to the wavelength: $f = c/\lambda$.
    EVALUATE   The limits of the UVB band in frequency are

$$f_{min} = \frac{c}{\lambda_{max}} = \frac{3.0 \times 10^8 \text{ m/s}}{320 \text{ nm}} = 0.94 \times 10^{15} \text{ Hz} = 0.94 \text{ PHz}$$

$$f_{max} = \frac{c}{\lambda_{min}} = \frac{3.0 \times 10^8 \text{ m/s}}{290 \text{ nm}} = 1.0 \times 10^{15} \text{ Hz} = 1.0 \text{ PHz}$$

ASSESS   Since most people are more familiar with nanometers than with petahertz, the wavelength limits for UV radiation are more often given than the frequency limits.

39.   **INTERPRET**   Given the electric field strength, we are asked to find the corresponding magnetic field strength for an electromagnetic wave propagating in air, which we can treat as a vacuum.
    DEVELOP   For an EM wave in free space, Equation 29.17 gives $E = cB$.
    EVALUATE   Given that $E = 320 \text{ }\mu\text{V/m}$, the corresponding magnetic field strength is

$$B = \frac{E}{c} = \frac{320 \text{ }\mu\text{V/m}}{3.00 \times 10^8 \text{ m/s}} = 1.07 \text{ pT}$$

ASSESS   This is a very small magnetic field. Note that in an EM wave, both the field strengths E and B are not independent; once one quantity is determined, the other can be found via the relation $E = cB$.

41.   **INTERPRET**   This problem involves finding the fraction of light transmitted through a polarizer as the direction of the incident polarization changes.
    DEVELOP   The fraction of light transmitted can be found by using the Law of Malus (Equation 29.18), $S = S_0 \cos^2 \theta$, where θ is the angle between the polarization direction (i.e. the direction of the electric field) in the electromagnetic wave and the polarization direction of the polarizer. Note that, with zero voltage applied, the laser beam polarization is perpendicular to the polarizer, so zero light is transmitted. During the brown out, the electro-optic modulator manages to rotate the polarization 72°, which is 18° short of the 90° rotation needed to pass 100% of the laser light through the polarizer. In the Law of Malus, the angle θ measures the angular departure from parallel alignment (i.e., when there is 100% transmission), so we must use $\theta = 18°$ in our calculation.
    EVALUATE   From Equation 29.18, we find that $S/S_0 = \cos^2(18°) = 91\%$ is transmitted.
    ASSESS   The intensity of the laser beam depends on $\cos^2 \theta$. The limit $\theta = 0$ corresponds to the situation where the direction of polarization of the laser beam is the same as the preferred direction specified by the polarizer, and $S = S_0$.

43.   **INTERPRET**   We are to find the intensity of a light beam after passing through two polarizers that are oriented at different angles to the polarization direction of the light. Note that the second polarizer is oriented perpendicular to the initial polarization, so we may expect that no light is trasmitted.
    DEVELOP   The intensity of the light after emerging from the first polarizer is given by the Law of Malus (Equation 29.18), $S = S_0 \cos^2 \theta_1$, where θ_1 is the angle between the electric field and the first polarizer's axis. After passing through the first polarizer, the electric field is rotated to the angle $\theta 1$ and so makes an angle $\theta_2 - \theta_1$ with the axis of the second polarizer. Thus, the intensity of the light transmitted through both polarizers is

$$S/S_0 = \cos^2 \theta_1 \cos^2(\theta_2 - \theta_1)$$

EVALUATE   Two successive applications of Equation 29.18 yield

$$\frac{S}{S_0} = \cos^2(60°)\cos^2(90° - 60°) = \left(\frac{1}{4}\right)\left(\frac{3}{4}\right) = 0.1875 \approx 19\%$$

ASSESS   To see that the result makes sense, let's solve the problem in two steps. The intensity of the beam after passing the first polarizer with $\theta_1 = 60°$ is $S_1 = S_0 \cos^2 \theta_1$. Since the angle between the first and the second polarizers is $\theta_2 = 90° - 60° = 30°$, so upon emerging from the second polarizer, the intensity becomes

$$S_2 = S_1 \cos^2 \theta_2 = \left(S_0 \cos^2 \theta_1\right)\cos^2 \theta_2$$

which is the same as above.

45.   INTERPRET   We want to determine how much of a microwave oven's radiation can leak out its door and still be below regulation standards.

DEVELOP   If we assume power is leaking uniformly out the door, then the maximum power allowed is just the intensity limit multiplied by the area: $P_{max} = S_{lim} A$.

EVALUATE   The fraction of power that is allowed to leak out the oven door is

$$\frac{P_{max}}{P} = \frac{S_{lim} A}{P} = \frac{\left(5.0 \text{ mW/m}^2\right)\left(40 \text{ cm} \times 17 \text{ cm}\right)}{900 \text{ W}} = 3.8 \times 10^{-7} \cong 0.00004\%$$

ASSESS   This is less than one part in a million. It might be surprising, therefore, that a metal screen with holes in it could provide this good of protection. The holes in the metal are much smaller than the microwave wavelength $\left(\sim 12 \text{ cm}\right)$, which means very little of the radiation can pass through.

47.   INTERPRET   The problem asks for a comparison of power output between a star and a quasar. The two objects have the same brightness but are at different distances from the Earth.

DEVELOP   The average intensity of radiation received determines the apparent brightness, so $\overline{S}_{quasar} = \overline{S}_{star}$. Apply Equation 29.21 to find the relative power from each source.

EVALUATE   From Equation 29.21, $S = P/\left(4\pi r^2\right)$ we see that the above condition implies that

$$\left(\frac{P}{r^2}\right)_{quasar} = \left(\frac{P}{r^2}\right)_{star}$$

if both behave like isotropic sources (which should be a good approximation; see Problem 29.46). Thus,

$$\frac{P_{quasar}}{P_{star}} = \left(\frac{1 \times 10^{10} \text{ ly}}{5 \times 10^4 \text{ ly}}\right)^2 = 4 \times 10^{10}$$

ASSESS   The luminosity of a quasar is comparable to a galaxy of stars!

49.   INTERPRET   We are to find the transmitted power and the peak electric field in the given electromagnetic wave.

DEVELOP   Equations 29.21 and 29.20b can be combined to express the average power output of an isotropic transmitter in terms of the peak electric field at a distance r:

$$P = 4\pi r^2 \left(\frac{E_p^2}{2\mu_0 c}\right)$$

EVALUATE   (a) The transmitted power is

$$P = 4\pi r^2 \left(\frac{E_p^2}{2\mu_0 c}\right) = \frac{4\pi\left(1.5 \text{ km}\right)^2\left(350 \text{ mV/m}\right)^2}{2\left(4\pi \times 10^{-7} \text{ N/A}^2\right)\left(3.00 \times 10^8 \text{ m/s}\right)} = 4.6 \text{ kW}$$

(b) Since $r^2 E_p^2$ is a constant,

$$E_p' = \left(r/r'\right)E_p = \frac{1.5 \text{ km}}{10 \text{ km}}\left(350 \text{ mV/m}\right) = 53 \text{ mV/m}$$

at a distance of 10 km.

ASSESS   This signal is still strong enough to be captured by modern radios. Checking the units for part (a), we find

$$\frac{(km)^2 (V/m)^2}{(N/A^2)(m/s)} = \frac{(m)^2 (m \cdot kg \cdot s^{-3} \cdot A^{-1})^2 (A^2)}{(kg \cdot m/s^2)(m/s)} = m^2 \cdot kg \cdot s^{-3} = (m)\overbrace{(kg \cdot m/s^2)}^{N}\overbrace{(s^{-1})}^{J} = W$$

51. **INTERPRET**   This problem explores the influence of the symmetry of a line source of light in the distribution of intensity that it radiates. Specifically, we are to compare the variation of intensity with distance near a line source and far from the line source.

    DEVELOP   Near the lamp (i.e., at distances $\ll$ the length L of the lamp), but far from its ends, light waves travel approximately radially outwards from the tube axis. The power crossing two co-axial cylindrical patches is the same, but the area of each patch is proportional to the radius (see figure below). Very far away, the lamp appears as a point source.

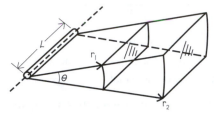

    EVALUATE   (a) Near the lamp, the intensity varies as $1/r$ because

$$S_1 A_1 = S_2 A_2 = S_1 \theta r_1 L = S_2 \theta r_2 L = \text{const}$$

$$S \propto \frac{\text{const}}{r}$$

    (b) Far from the lamp, Equation 29.21 holds, so the intensity varies like $1/r^2$

    ASSESS   Extending this reasoning, we would expect the intensity of a plane source of light to be independent of distance to the source for distances $\ll$ the dimensions of the plane source, assuming we are not near the edges of the source.

53. **INTERPRET**   We are to find the average intensity and the peak electric field of the light from a laser, given its average power and its beam diameter.

    DEVELOP   Intensity is defined as power per unit area. We know that the power of the beam is $P = 7.0$ W and the area is $A = \pi r^2$ where $r = d/2 = 5.0 \times 10^{-4}$ m, so we can find the intensity. Intensity is also given by the Poynting vector, $S = EB/\mu_0 = E^2/u_0 c$, from which we can determine the peak electric field.

    EVALUATE   (a) $S = P/A = (7.0 \text{ W})/\left[\pi (5.0 \times 10^{-4} \text{ m})^2\right] = 89 \times 10^6 \text{ W/m}^2$.

    (b) $E = \sqrt{S\mu_0 c} = \sqrt{(8.9 \times 10^6 \text{ W/m}^2)(4\pi \times 10^{-7} \text{ N/A}^2)(3.00 \times 10^8 \text{ m/s})} = 58 \times 10^3 \text{ V/m}$.

    ASSESS   This intensity is one factor in what makes a laser beam so dangerous. Seven watts is really not much power, but packed into such a small area it gives an enormous intensity.

55. **INTERPRET**   We will use Newton's third law (Chapter 4) and the radiation momentum generated by the given flashlight to find the time it takes the astronaut to accelerate from rest to 10 m/s.

    DEVELOP   By Newton's third law, the reaction force of the light emitted on the flashlight equals the rate at which momentum is carried away by the beam, or $F = dp/dt = (dU/dt)/c = P/c$. Such a force could accelerate a mass m from rest to a speed v in time $t = v/a = mv/F = mcv/P$.

    EVALUATE   For the values given for the astronaut and flashlight,

$$t = \frac{(65 \text{ kg})(3.00 \times 10^8 \text{ m/s})(10 \text{ m/s})}{1.0 \text{ W}} = 2.0 \times 10^{11} \text{ s} = 6.2 \times 10^3 \text{ y}$$

    ASSESS   This is impractically long, as one might expect.

57.   **INTERPRET**   This problem involves finding the radiation pressure at the surface of a white dwarf, given its power and radius.

DEVELOP   From Problem 29.46, the Sun radiates $P = 3.85 \times 10^{26}$ W, and the radius of the Earth is 6.37×10^6 m. Equation 29.22 gives the radiation pressure on a perfect absorber as $p_{rad} = \overline{S}/c$ and the intensity may be found using Equation 29.21, $\overline{S} = P/\left(4\pi r^2\right)$.

EVALUATE   Combining the expressions above gives a radiation pressure of

$$p_{rad} = \frac{P}{4\pi r^2 c} = \frac{3.85 \times 10^{26} \text{ W}}{4\pi \left(6.37 \times 10^6 \text{ m}\right)^2 \left(3.00 \times 10^8 \text{ m/s}\right)} = 2.52 \text{ kPa}$$

ASSESS   This is some 100 times less than the atmospheric pressure at the surface of the Earth.

59.   **INTERPRET**   This problem involves characterizing an electromagnetic wave given the relevant parameters.

DEVELOP   The average intensity of a pulse is the average power during a pulse divided by the beam area; $\overline{S} = P = \pi R^2$, and (from Equation 29.20b) the peak electric field is $E_p = \sqrt{2\mu_0 c \overline{S}}$. The wavelength may be found using Equation 29.16c, $c = f\lambda$. To find the energy in a pulse, use $U = \overline{P}_{pulse} \Delta t$, where $\Delta t = NT = N/f$, with $N = 100$ and $f = 70$ GHz. To find the average power output, calculate the power in a pulse, multiply by 1000 because there are 1000 pulses per second, and divide by 1 s to get the power (energy per unit time).

EVALUATE   (a) The peak electric field is

$$E_p = \sqrt{2\mu_0 c \overline{S}} = \frac{1}{R}\sqrt{2\mu_0 c P/\pi} = \frac{\left(8\pi \times 10^{-7} \text{ H/m}\right)\left(3.00 \times 10^8 \text{ m/s}\right)\left(45 \text{ MW}\right)^{1/2}}{0.10 \text{ m}} = 1.0 \text{ MV/m}$$

(b) The wavelength is $\lambda = c/f = \left(3.00 \times 10^8 \text{ m/s}\right)/\left(70 \text{ GHz}\right) = 4.3$ mm .
(c) The total enegy in a pulse is $U = \overline{P}_{pulse} N/f = \left(45 \text{ MW}\right)\left(100\right)\left(1/70 \text{ GHz}\right) = 64$ mJ.
(d) The momentum per pulse is given by $p = U/c = \left(64 \text{ mJ}\right)/\left(3.00 \times 10^8 \text{ m/s}\right) = 2.1 \times 10^{-10}$ kg·m/s.
(e) Every pulse carries 64.3 mJ, and there are 1000 per second, so the average power is $\overline{P} = \left(64.3 \text{ mJ}\right)\left(1000\right)/\left(1.0 \text{ s}\right) = 64$ W .

ASSESS   The average power in the beam is much less than the power per pulse because the duty cycle is

$$\text{duty cycle} = \frac{t_{on}}{t_{tot}} = \frac{\left(1000\right)\left(100\right)}{\left(70 \text{ GHz}\right)\left(1 \text{ s}\right)} = 1.4 \times 10^6$$

which explains the six-order-of-magnitude difference between $\overline{P}_{pulse}$ and $\overline{P}$.

61.   **INTERPRET**   From the transmission percentage of a stack of polarizers, we are to determine how many polarizers the stack contains. We shall use the Law of Malus.

DEVELOP   The Law of Malus (Equation 29.18) is $S = S_0 \cos^2 \theta$, where θ is the angle between the polarization of the impinging light beam and the polarization direction of the polarizer sheet. For this problem, $\theta = 14°$. The first polarizer eliminates 50% of the initially unpolarized light, and each subsequent polarizer is equivalent to multiplying the amount of light remaining by $\cos^2 \theta$, so the total percentage of the light that comes through the stack of n polarizers is $S = S_0 \frac{1}{2}\left(\cos^2 \theta\right)^{n-1}$. We are given that $S = 0.37 S_0$, so we can solve for n.

EVALUATE   The number n of polarizer sheets is

$$0.37 \, S_0 = S_0 \frac{1}{2}\left(\cos^2 \theta\right)^{n-1}$$

$$\ln\left(0.74\right) = \left(n-1\right)\ln\left(\cos^2 \theta\right)$$

$$n = \frac{\ln\left(0.74\right) + \ln\left(\cos^2 \theta\right)}{\ln\left(\cos^2 \theta\right)} = 6$$

ASSESS   The stack has six sheets. If you got 17.5 sheets, you probably forgot that the first sheet eliminates *half* the unpolarized incident light.

63. **INTERPRET**   We will use Maxwell's equations to derive the wave equation for electromagnetic radiation.

DEVELOP   We will start with the differential form of Faraday's law: $\partial E/\partial x = -\partial B/\partial t$ (Equation 29.12) and differentiate it with respect to x. Then we will take the differential form of Ampère's law: $\partial B/\partial x = -\epsilon_0\mu_0\partial E/\partial t$ (Equation 29.13) and differentiate it with respect to t. The combination of these two equations should match that of the generic wave equation: $\frac{\partial^2 y}{\partial x^2} = \frac{1}{v^2}\frac{\partial^2 y}{\partial t^2}$ (Equation 14.5).

EVALUATE   Taking the derivatives specified above, we have

$$\frac{\partial^2 E}{\partial x^2} = -\frac{\partial^2 B}{\partial x\partial t} \qquad \frac{\partial^2 B}{\partial t\partial x} = -\epsilon_0\mu_0\frac{\partial^2 E}{\partial t^2}$$

Since the order of the partial derivatives is irrelevant, the magnetic field derivatives in the two equations are equal, so we get

$$\frac{\partial^2 E}{\partial x^2} = \epsilon_0\mu_0\frac{\partial^2 E}{\partial t^2}$$

This has the same form as Equation 14.5, which implies the electric field behaves as a wave with speed $c = 1/\sqrt{\epsilon_0\mu_0}$. By reversing the differentiations, we can show the exact same thing for the magnetic field.

ASSESS   As described in Chapter 14, the solution to the wave equation is any function of the form $f(x \pm vt)$. So electromagnetic waves do not necessarily have to be sine waves, but any shape of wave can be analyzed as the sum of individual sine waves (see Figure 14.18).

65. **INTERPRET**   We are to find the quarter wavelength of the given electromagnetic radiation.

DEVELOP   We use the relationship between frequency and wavelength (Equation 29.16c) $c = f\lambda$, to find the wavelength of the signal, then divide by 4.

EVALUATE   The length of the antenna should be $L = \frac{1}{4}\lambda = c/(4f) = (3.00\times10^8\,\text{m/s})/(4\times27.3\,\text{MHz}) = 2.75$ m.

ASSESS   This seems reasonable—it's about the length of the long antennas typically seen on pickup trucks.

67. **INTERPRET**   We're asked what size of receiver dish will be needed to capture the Voyager 1 radio transmission in the future.

DEVELOP   If we assume that the spacecraft's transmitter broadcasts its radio signal uniformly in all directions, then the intensity at Earth will be $S = P_{\text{em}}/4\pi r^2$, where $P_{\text{em}} = 20\,\text{W}$ is the emitted power and r is the distance from Earth. An Earth-bound receiver of diameter, d, will be able to gather a signal with power $P_{\text{rec}} = S(\pi d^2/4)$.

EVALUATE   Given the desired receiver signal, the receiver diameter will have to be

$$d = \sqrt{\frac{4P_{\text{rec}}}{\pi S}} = 4r\sqrt{\frac{P_{\text{rec}}}{P_{\text{em}}}} = 4(25\times10^9\,\text{km})\sqrt{\frac{10^{-20}\,\text{W}}{20\,\text{W}}} = 2.2\,\text{km}$$

ASSESS   There is no single dish of this size currently. The Arecibo observatory comes closest with its 305-m diameter dish. There are plans to build a 500-m diameter single dish telescope in China, but nothing spanning 2 kilometers is in the works. However, our assumption that Voyager 1 broadcasts in all directions is not correct. The transmitter is shaped like a parabola, and therefore beams its signal towards Earth, so a smaller receiver should be sufficient.

69. **INTERPRET**   We're asked to consider the potential of solar sail technology.

DEVELOP   The solar sail is accelerated by the radiation pressure from the sun: $p_{\text{rad}} = \frac{1}{c}S$ (Equation 29.22). We don't know the area or the mass of the sail, but since the sun's intensity decreases as one over the distance squared, so will the acceleration: $a \propto 1/r^2$.

EVALUATE   The acceleration near Mars will be less than the acceleration near Earth. More specifically,

$$a_{\text{M}} = a_{\text{E}}\left(\frac{r_{\text{E}}}{r_{\text{M}}}\right)^2 = 1\,\text{m/s}^2\left(\frac{1}{1.5}\right)^2 = 0.4\,\text{m/s}^2$$

The answer is (b).

ASSESS   The acceleration of the sail is not large, but it never "shuts off" like a rocket engine does, so the speed can build up over time. Using the fact that $a\,dr = v\,dv$, we can integrate the sail's acceleration from Earth's orbital radius to that of Mars:

$$\int_{r_E}^{r_M} a_E \left(\frac{r_E}{r}\right)^2 dr = a_E r_E \left(1 - \frac{r_E}{r_M}\right) = \int v\,dv = \tfrac{1}{2}v^2$$

where we have assumed that the solar sail started at rest at Earth. Plugging in the values from above, the sail's speed once it reaches Mars would be about 300 km/s.

71. **INTERPRET**   We're asked to consider the potential of solar sail technology.

    DEVELOP   In the previous problems, we have argued that the radiation pressure and the force it applies to the sail are proportional to $1/r^2$. This is also true of the gravitational force.

    EVALUATE   Since Jupiter is roughly 5 times further from the sun than Earth, the sail force at Jupiter's position will be 25 times smaller. However, the gravitational force from the sun will also be 25 times smaller, so the sail force will still be 20 times solar gravity.

    The answer is (d).

    ASSESS   For the sail force to dominate gravity, the sail must be light-weight with a large surface area. We can write the ratio of the forces in terms of the area and the mass:

$$\frac{F_{rad}}{F_g} = \frac{P_{rad}A}{GM_s m / r^2} = \frac{P_{em}}{4\pi GM_s c}\left(\frac{A}{m}\right)$$

where we have written the sun's intensity in terms of the emitted power: $P_{em} = 3.85 \times 10^{26}\,\text{W}$. Plugging this in with the mass of the sun, $M_S = 1.99 \times 10^{30}\,\text{kg}$, we get

$$\frac{m}{A} \approx 0.8\ \text{g/m}^2 \left(\frac{F_g}{F_{rad}}\right)$$

What this says is that each gram of spacecraft requires more than a square meter of sail if the radiation force is to overcome gravity.

REFLECTION AND REFRACTION

30

EXERCISES

Section 30.1 Reflection

11. **INTERPRET**   We are to find the angle through which we must rotate a specular reflecting surface so that the reflected light rotates through 30°.

DEVELOP   Since $\theta_1 = \theta_1'$ for specular reflection, (Equation 30.1) a reflected ray is deviated by $\phi = 180° - 2\theta_1$ from the incident direction (see figure below). If rotating the mirror changes θ_1 by $\Delta\theta_1$, then the reflected ray is deviated by $\Delta\phi = \left|-2\Delta\theta_1\right|$ or twice this amount.

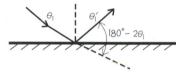

EVALUATE   Thus, if $\Delta\phi = 30°, \left|\Delta\theta_1\right| = 15°$.

ASSESS   This result can be easily verified with a small mirror.

13. **INTERPRET**   We are to find an expression for the angle of return of a light ray that is incident on a pair of perpendicularly aligned mirrors and use this expression to find the accuracy with which the mirrors must be aligned.

DEVELOP   A ray incident on the first mirror at a grazing angle α is deflected through an angle 2α (this follows from the law of reflection, see Problem 30.11). It strikes the second mirror at a grazing angle β and is deflected by an additional angle 2β. Let the total deflection be θ_f: $2\alpha + 2\beta = \theta_f$. The alignment angle of the mirrors is $\theta = \pi/2 - \alpha - \beta = \pi/2 - \Delta\theta_f/2$.

EVALUATE   Differentiating this expression gives

$$\Delta\theta = \Delta\theta_f/2$$

so for $\Delta\theta_f = 1°, \Delta\theta = 0.5°$.

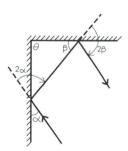

ASSESS   This is another retroreflector, but this time there is a displacement in the retro-reflected beam.

Section 30.2 Refraction

15.   **INTERPRET**   Given the speed of light in an unknown material, we are to find the index of refraction and thus identify the material.

DEVELOP   Apply Equation 30.2, $n = c/v$ and use Table 30.1 to find the corresponding material.

EVALUATE   The index of refraction of the material is $n = (3.000 \times 10^8 \, \text{m/s})/(2.292 \times 10^8 \, \text{m/s}) = 1.309$, so the material is ice.

ASSESS   This is a typical value for an index of refraction of a material that is transparent to visible light.

17.   **INTERPRET**   This problem involves Snell's law, which we can use to find the incident angle of the light beam.

DEVELOP   Apply Snell's law (Equation 30.3) to find the incident angle.

EVALUATE   Snell's law (with $n_{\text{air}} = 1$) gives $\theta_1 = \sin^{-1}(n_2 \sin \theta_2 / n_1) = \sin^{-1}[1.52 \sin(40°)/1.00] = 77.7°$.

ASSESS   The angle is between 0° and 90°, as expected.

19.   **INTERPRET**   This problem involves Snell's law, which we can use to find the angle of refraction of a light beam as it passes from glass into water.

DEVELOP   Apply Snell's law (Equation 30.3) to find θ_2, which is the angle of refraction in the water. From Table 30.1, the index of refraction of water is $n_2 = 1.333$.

EVALUATE   Equation 30.3 gives $\theta_2 = \sin^{-1}(n_1 \sin \theta_1 / n_2) = \sin^{-1}[(1.52) \sin(12.4°)/1.333] = 14.2°$.

ASSESS   This angle is greater than the incident angle (12.4°), as expected because the index of glass is larger than that of water.

21.   **INTERPRET**   We are to find the refractive index of a material given its polarizing angle in air.

DEVELOP   Apply Equation 30.4, with $n_1 = 1.0$ and $\theta_p = 62°$.

EVALUATE   Solving Equation 30.4 for n_2 gives

$$n_2 = n_1 \tan \theta_p = (1.0) \tan(62°) = 1.9.$$

ASSESS   From Table 30.1, this material could be a glass.

Section 30.3 Total Internal Reflection

23.   **INTERPRET**   We are to find the critical angle for light going from water to ice.

DEVELOP   Apply Equation 30.5, using Table 30.1 to find the indices of refraction of the media.

EVALUATE   The critical angle is

$$\theta_c = \sin^{-1}(n_2/n_1) = \sin^{-1}(1.309/1.333) = 79.1°.$$

ASSESS   This is a reasonable value for a critical angle between to similar media.

25.   **INTERPRET**   We are to find the refractive index of plastic given the critical angle for light propagation from air to plastic.

DEVELOP   Apply Equation 30.5, with $n_1 = 1.00$ $\theta_c = 37°$.

EVALUATE   At the critical angle in plastic, $\sin \theta_c = n_{\text{air}}/n_{\text{plastic}}$ (Equation 30.5), so $n_{\text{plastic}} = 1.00/\sin(37°) = 1.66$.

ASSESS   This is a reasonable value for a refractive index for a plastic.

Section 30.4 Dispersion

27.   **INTERPRET**   We are to find the angular dispersion for a beam of white light that transits an equilateral prism.

DEVELOP   The geometry for refraction through a prism is shown in the figure below. Using Snell's law (Equation 30.3) and the fact that $\theta_{\text{tot}} = \theta_1' - \alpha$ and $\theta_1 = \alpha - \theta_0'$, we find the following expression for θ_{tot}:

$$\theta_{tot} = \theta_1' - \alpha$$
$$= \text{asin}\left[n\sin\left(\alpha - \theta_0'\right)\right] - \alpha$$
$$= \text{asin}\left\{n\sin\left[\alpha - \text{asin}\left(\frac{\sin\theta_0}{n}\right)\right]\right\} - \alpha$$

where $\alpha = 60°$, n is the refractive index of the prism and we have used $n = 1$ for air. For red light, $n = n_{red} = 1.582$, whereas for violet light, $n_{violet} = 1.633$. The angular dispersion of the outgoing beam is the difference $\theta_{tot}\left(n_{violet}\right) - \theta_{tot}\left(n_{red}\right)$.

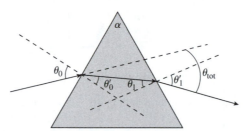

EVALUATE   Using $\theta_0 = 45°$, we find $\theta_{tot}\left(n_{violet}\right) = 7.10°$ and $\theta_{tot}\left(n_{red}\right) = 0.69°$, so the dispersion is $7.10° - 0.69° = 6.41°$.

ASSESS   We find that the violet light is deflected more than the red light, which is reasonable because the index of refraction for violet light is greater than for red light ($n_{violet} > n_{red}$).

PROBLEMS

29.   **INTERPRET**   We consider refraction in the human cornea.

DEVELOP   The angle of refraction, θ_2, can be found from Equation 30.3: $n_1 \sin\theta_1 = n_2 \sin\theta_2$. The light is coming from air $\left(n_1 = 1\right)$, so the wavelength in the cornea is $\lambda_2 = \lambda_1 / n_2$.

EVALUATE   (a) Solving for the angle of refraction gives

$$\theta_2 = \sin^{-1}\left[\frac{n_1}{n_2}\sin\theta_1\right] = \sin^{-1}\left[\frac{1}{1.40}\sin 25°\right] = 18°$$

(b) The wavelength reduces to

$$\lambda_2 = \frac{\lambda_1}{n_2} = \frac{550 \text{ nm}}{1.40} = 390 \text{ nm}$$

ASSESS   The angle of refraction is less than the angle of incidence, which is what we would expect for light entering a material of higher index of refraction.

31.   **INTERPRET**   You want to identify an unknown liquid by measuring the way that light refracts at the interface between the liquid and glass of known index of refraction.

DEVELOP   You shine the laser light so that it first passes through the unknown liquid before entering the glass $\left(n_2 = 1.52\right)$. With the measured angles, you can determine the liquid's index of refraction from Equation 30.3: $n_1 \sin\theta_1 = n_2 \sin\theta_2$.

EVALUATE   Solving for n_1 gives

$$n_1 = n_2 \frac{\sin\theta_2}{\sin\theta_1} = (1.52)\frac{\sin 27.9°}{\sin 31.5°} = 1.36$$

This agrees with the index of refraction for ethyl alcohol in Table 1.1.

ASSESS   The ethyl alcohol has smaller index of refraction than glass, so the light should bend toward the normal, as it does here. If the liquid had been benzene, which has an index of refraction very close to that of glass, the change in angle would have been nearly imperceptible.

33.   **INTERPRET**   We are to find the refractive index such that a ray impinging on the center of a cube will transect the opposing vertex (see figure below).

DEVELOP   From the figure below, we see that the angle of refraction in the glass, given by

$\sin\theta_2 = \sin(55°)/n_2$ must be less than $\operatorname{atan}\left(\frac{1}{2}\right) = 26.6°$ for the ray to emerge from the opposite face

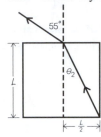

EVALUATE   Therefore, $n_2 \geq \sin(55°)/\sin(26.6°) = 1.83$.

ASSESS   The vertical face appears shorter than it really is, as is the case when one looks at legs in the 3-ft section of a swimming pool.

35.   **INTERPRET**   This problem involves the refraction of light at an air-water interface, which we shall use to find the distance at which the diver is from the lake edge.

DEVELOP   Consider the sketch below, which describes the situation. Snell's law gives the angle of refraction (θ_1) in terms of the angle of incidence ($\theta_2 = 42°$) for the light path from the flashlight to your eye. These can be related to the other given distances by means of a carefully drawn diagram. Thus,

$$\theta_1 = \operatorname{asin}\left[n_2 \sin(\theta_2)/n_1\right] = \operatorname{asin}\left[1.333\sin(42°)\right] = 63.1°,$$

where we have used indices of refraction from Table 30.1, with $n_1 = 1.00$ for air. Given this angle, we can find the desired distance from geometry.

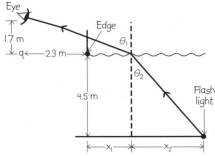

EVALUATE   The geometry of the diagram makes the horizontal distances apparent:

$$\tan\theta_1 = \frac{2.3 \text{ m} + x_1}{1.7 \text{ m}} \text{ or } x_1 = (1.7 \text{ m})\tan(63.1°) - 2.3 \text{ m} = 1.05 \text{ m},$$

and

$$\tan\theta_2 = \frac{x_2}{4.5 \text{ m}} \text{ or } x_2 = (4.5 \text{ m})\tan(42°) = 4.05 \text{ m}.$$

The total horizontal distance from the edge is $x_1 + x_2 = 5.1$ m.

ASSESS   The diver will appear to be farther from the edge of the lake, but in reality will be at the given distance.

37.   **INTERPRET**   The problem concerns laser surgery and the wavelength of UV light when it passes into the eye.

DEVELOP   The light is coming from air $(n_1 = 1)$, so the wavelength in the lens is $\lambda_2 = \lambda_1/n_2$.

EVALUATE   Using the wavelength in air and the index of refraction of the lens, the wavelength becomes

$$\lambda_2 = \frac{\lambda_1}{n_2} = \frac{193 \text{ nm}}{1.39} = 139 \text{ nm}$$

ASSESS   The laser light isn't technically supposed to enter the lens. Instead it is used to sculpt the cornea in front of the lens. The procedure helps to redirect light into the eye (through refraction) to improve vision.

39.   **INTERPRET**   We are to redo the preceding problem with different values for the refractive index of the prism, the incident angle, and the prism's apex angle.

DEVELOP   A general treatment of refraction through a prism of index of refraction $n_2 = n$, surrounded by air of index $n_1 = 1.00$, for the geometry of Figure 30.20, is given in the solution to Problem 27.

EVALUATE   For $n = 1.75$, $\alpha = 40°$, and $\theta_1 = 25°$, the other angles defined there are

$$\theta_2 = \text{asin}\big[\sin(\theta_1)/n\big] = \text{asin}\big[\sin(25°)/1.75\big] = 14.0°,$$
$$\phi_2 = \alpha - \theta_2 = 40° - 14.0° = 26.0°,$$
$$\phi_1 = \text{asin}(n\sin\phi_2) = \text{asin}(1.75)\sin(26.0°) = 50.2°,$$

and $\delta = \theta_1 + \phi_1 - \alpha = 35°$.

ASSESS   Note that ϕ_2 is less than the critical angle for this prism, which is $\text{asin}(1.00/1.75) = 34.8°$.

41.   **INTERPRET**   For this problem, we are to determine if the main beam emerges from the prism at the diagonal face or at the bottom face of the prism, and at what angle it emerges with respect to the exit face normal.

DEVELOP   The critical angle for an ice-air interface is $\theta_c = \text{asin}(n_{air}/n_{ice}) = 49.8°$, which is greater than the incident angle of 45° (see figure below). Thus, total internal reflection does not occur, and the beam emerges from the diagonal face of the prism at an angle with respect to the face normal determined by Snell's law (Equation 30.3).

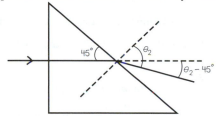

EVALUATE   From Snell's law, the exit angle (θ_2 in the sketch above) is

$$\theta_2 = \text{asin}\big[(1.309)\sin(45°)\big] = 68°$$

The deviation from the incident direction is $68° - 45° = 23°$, as shown in the sketch above.

ASSESS   Because $n_2 < n_1$, $\theta_2 > \theta_1$ to compensate and satisfy Snell's law.

43.   **INTERPRET**   We are to find the minimum refractive index for the medium surrounding the prism in Figure 30.10 for which total internal reflection does not occur.

DEVELOP   When the prism is immersed in liquid, $\theta_c = \text{asin}(n_{liquid}/1.52) \geq 45°$ for total internal reflection to occur.

EVALUATE   Therefore, $n_{liquid} \geq (1.52)\sin(45°) = 1.07$.

ASSESS   Because the liquid constitutes the second medium (i.e., the light propagates from the prism into the liquid), its refractive index must be less than that of the prism for total internal reflection to occur.

45.   **INTERPRET**   We are to show that light emanating from a point source in on material will enter a second medium in a circle with the given diameter. The critical angle at which the light is totally internally reflected will be useful for this problem.

DEVELOP   Consider the diagram below. Light from the flash will strike the water surface at the critical angle for a distance $r = h\tan\theta_c$ from a point directly over the flash. Therefore, the diameter d of the circle through which the light will emerge is $d = 2r = 2h\tan\theta_c$. But $\sin\theta_c = n_{air}/n = 1/n$ (Equation 30.5 at the water-air interface), and $\tan^2\theta = (\csc^2\theta - 1)^{-1}$ (a trigonometric identity), which we can use to show the desired relationship.

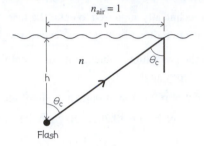

$n_{air} = 1$

Flash

EVALUATE   The diameter of the circle through which the light emerges can therefore be expressed as

$$d = 2h\tan\theta_c$$
$$= \frac{2h}{\sqrt{\csc^2\theta_c - 1}}$$
$$= \frac{2h}{\sqrt{n^2 - 1}}$$

ASSESS   The more general form of this relationship is

$$d = \frac{2h}{\sqrt{n_1^2/n_2^2 - 1}} = \frac{2hn_2}{\sqrt{n_1 - n_2}}$$

which shows that the critical angle is only relevant for $n_1 > n_2$, which is expected because $\sin\theta_c = n_2/n_1$ cannot be greater than unity.

47.   **INTERPRET**   This problem involves finding the critical angle at the interface between two given media.

   DEVELOP   Apply Equation 30.5, $\sin\theta_c = n_2/n_1$ with $n_2 = n_{glass}$ and $n_1 = n_{flint}$.

   EVALUATE   The critical angle is

$$\theta_c = \text{asin}\left(n_{glass}/n_{flint}\right) = \text{asin}(1.52/1.89) = 53.5°$$

   ASSESS   In the reverse direction, there is no critical angle because $n_{flint} > n_{glass}$.

49.   **INTERPRET**   We are to find the polarizing angle for a water-air interface.

   DEVELOP   Apply Equation 30.4, with $n_2 = n_{air}$ and $n_1 = n_{water}$.

   EVALUATE   The critical angle is

$$\theta_p = \text{atan}\left(n_{air}/n_{water}\right) = \text{atan}(1.00/1.333) = 36.9°$$

   ASSESS   This may be qualitatively verified in your neighborhood swimming pool.

51.   **INTERPRET**   This problem is about refraction of sunlight. We want to know the diameter of the tank such that sunlight can reach part of the tank bottom whenever the Sun is above the horizon.

   DEVELOP   The rays of sunlight which first hit the bottom of the tank just skim the opposite edge of the rim. The diameter and depth of the tank (d and h) are related to the angle of refraction by $\tan\theta_2 = d/h$. Combining this with Snell's law (Equation 30.3), we find (with $n_1 = 1.00$ for air and $n_2 = 1.333$ for water)

$$d = h\tan\theta_2 = h\tan\left\{\text{asin}\left[n_1\sin(\theta_1)/n_2\right]\right\}$$

   EVALUATE   If we let θ_1 approach 90° (Sun angle approaches 0°), then the tank diameter becomes

$$\lim_{\theta_1 \to 90°} d = (2.4\text{ m})\tan\left[\text{asin}\left(n_1/n_2\right)\right] = \frac{(2.4\text{ m})}{\sqrt{(n_2/n_1)^2 - 1}} = \frac{(2.4\text{ m})}{\sqrt{(1.333/1.00)^2 - 1}} = 2.7\text{ m}$$

where we have used $\tan\theta = \sin\theta/\sqrt{1 - \sin^2\theta}$.

ASSESS   If the diameter is smaller than 2.7 m then, in order for the sunlight to reach the bottom of the tank, a smaller value of θ_1 would be required. The diameter of the tank as a function of θ_1 is depicted in the figure below.

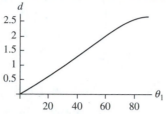

53.   **INTERPRET**   You want to center the screen so the projector's light reflects straight back towards the patient's eyes.

DEVELOP   The figure below shows the path the light should take from the projector to the patient's eyes. The lengths y and y' give the vertical distance from the center of the screen to the projector and the eyes, respectively. If you assume specular reflection, the angle of incidence, θ_1, is equal to the angle of reflection, θ_1', which means

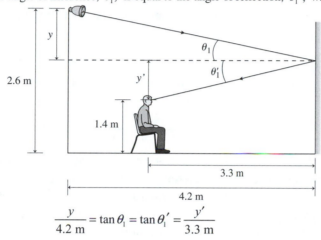

$$\frac{y}{4.2\,\mathrm{m}} = \tan\theta_1 = \tan\theta_1' = \frac{y'}{3.3\,\mathrm{m}}$$

EVALUATE   From the figure, the unknown distances must satisfy: $y + y' + 1.4\,\mathrm{m} = 2.6\,\mathrm{m}$. Combining this with the reflection criterion above, we find

$$y = \frac{2.6\,\mathrm{m} - 1.4\,\mathrm{m}}{1 + 3.3\,\mathrm{m}/4.2\,\mathrm{m}} = 0.67\,\mathrm{m}$$

This means the center of the projector should be placed $2.6\,\mathrm{m} - y = 1.9\,\mathrm{m}$ from the floor.

ASSESS   This is a little lower than the midpoint between the projector and the eyes, which makes sense since the eyes are closer to the screen than the projector.

55.   **INTERPRET**   This problem involves two refractions and a total internal reflection as the light ray passes through a spherical raindrop. Thus, we shall use Snell's law and the relationship of total internal reflection to show that the complement of the angle between the incoming and outgoing rays is as given in the problem statement.

DEVELOP   Consider the figure below. The angle ϕ can be found by summing the deflections each time the ray in Figure 30.21 is refracted or reflected. The deflection at A is $\theta - \theta'$, at B is $180° - 2\theta'$, and at C is $\theta - \theta'$. The sum is

$$\delta = 2(\theta - \theta') + 180° - 2\theta' = 180° + 2\theta - 4\theta'$$

and is related to ϕ by $180° = \phi + \delta$, so $\phi = 4\theta' - 2\theta$.

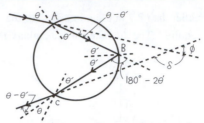

EVALUATE By eliminating θ' using Snell's law [Equation 30.3, $\sin\theta' = (\sin\theta)/n$], the desired expression

$$\phi = 4\,\text{asin}\left[\sin(\theta)/n\right] - 2\theta$$

is obtained. Note that light incident at the boundaries of the drop, at A, B, and C, is partially reflected and partially refracted; we show only the rays relevant to the formation of a rainbow.

ASSESS The angle ϕ is a complicated nonlinear function of θ, as shown on the right (with $n = 1.333$).The maximum value of ϕ is approximately equal to 42.1°. This is the average angle above the anti-solar direction that an observer sees a rainbow, because n is the average index of refraction for visible wavelengths.

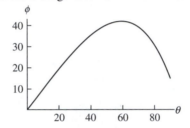

57. **INTERPRET** This problem involves two refractions and two total internal reflections as the light ray passes through a spherical raindrop. We are to use the results of the two previous problems for this problem.

DEVELOP The analysis of the secondary rainbow is similar to that of the primary rainbow (see Problems 55 and 56). The angles for an incident ray, which experiences two internal reflections in a spherical drop of water, are shown in the figure below for the emergent ray traveling downward to an observer on the ground. The total deflection for two refractions and two internal reflections,

$$\delta = 2(\theta - \theta') + 2(180° - 2\theta')$$

is related to the observation angle from the anti-solar direction ϕ by $\delta = 180° + \phi$.

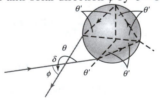

EVALUATE Combining the two equations, we obtain

$$\phi = 2\theta - 6\theta' + 180° = 2\theta - 6\,\text{asin}\left[\sin(\theta)/n\right] + 180°$$

If we differentiate Snell's law with respect to θ and substitute for θ' in terms of ϕ and θ, we get

$$\frac{d}{d\theta}(\sin\theta) = \cos\theta = \frac{d}{d\theta}(n\sin\theta') = n\cos\theta'\frac{d\theta'}{d\theta} = n\sqrt{1-\sin^2\theta'}\,\frac{d}{d\theta}\left[\frac{1}{6}(2\theta + 180° - \phi)\right]$$

$$= \sqrt{n^2 - \sin^2\theta}\left(\frac{1}{3} - \frac{1}{6}\frac{d\phi}{d\theta}\right)$$

A concentrated beam is formed for the incident angle that satisfies the condition $d\phi/d\theta = 0$. Thus,

$$\cos\theta_m = \frac{1}{3}\sqrt{n^2 - \sin^2\theta_m}$$

which implies $\cos^2\theta_m = (n^2 - 1)/8$ and $\sin^2\theta_m = (9 - n^2)/8 = n^2\sin^2\theta'_m$. Finally, the maximum value of ϕ is

$$\phi_{max} = 2\theta_m - 6\theta_m' + 180° = 2\,\mathrm{asin}\left(\sqrt{\frac{9-n^2}{8}}\right) - 6\,\mathrm{asin}\left(\sqrt{\frac{9-n^2}{8n^2}}\right) + 180°$$

For $n = 1.333$, the average angle is 50.9°. However, substituting $n_{red} = 1.330$ and $n_{violet} = 1.342$, we obtain $\phi_{max,red} = 50.10°$ and $\phi_{max,violet} = 53.22°$ for the secondary rainbow.

ASSESS   Since $\phi_{max,red} < \phi_{max,violet}$, the colors appear in the reverse order from that in the primary rainbow. Although the deflection for violet rays is always larger than that for red rays (no matter how many internal reflections are considered), the relation between ϕ and δ depends on the quadrant of δ and is different for the primary and secondary rainbows.

59.   **INTERPRET**   We will prove Snell's law starting from Fermat's principle that says light takes the path of least (or most) time when traveling between two points.

DEVELOP   We'll assume the two mediums are separated by a flat horizontal interface. Let point A be in medium 1 at a vertical distance y_A from the interface. Likewise, let point B be in medium 2 at a vertical distance y_B from the interface and a horizontal distance L from A. See the figure below.

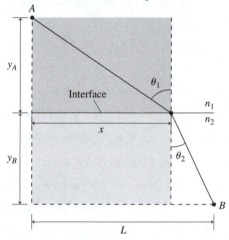

We choose an arbitrary path from A to B, characterized by the horizontal distance x between point A and the point where the path crosses the interface. This leaves a horizontal distance of $L - x$ from the crossing point to point B. As such, the time the light spends in medium 1 and medium 2 can be expressed as:

$$t_1 = \frac{\sqrt{x^2 + y_A^2}}{v_1} = n_1 \frac{\sqrt{x^2 + y_A^2}}{c}$$

$$t_2 = \frac{\sqrt{(L-x)^2 + y_B^2}}{v_2} = n_2 \frac{\sqrt{(L-x)^2 + y_B^2}}{c}$$

where we have used the speed of light in each medium: $v = c/n$. Notice that the only variable in these two equations is the distance x; the other parameters are constants. The total time for light to travel along this path is $t_1 + t_2$, which can be differentiated with respect to x to find the extremum.

EVALUATE   The derivative of the total time with respect to x is

$$\frac{dt_{tot}}{dx} = \frac{1}{c}\left[n_1 \frac{2x}{2\sqrt{x^2 + y_A^2}} - n_2 \frac{2(L-x)}{2\sqrt{(L-x)^2 + y_B^2}} \right]$$

Setting this equal to zero, we find the path that is an extremum obeys Snell's law:

$$n_1 \frac{x}{\sqrt{x^2 + y_A^2}} = n_2 \frac{(L-x)}{\sqrt{(L-x)^2 + y_B^2}} \quad \rightarrow \quad n_1 \sin\theta_1 = n_2 \sin\theta_2$$

The last step of this derivation follows from the geometry in the figure.

ASSESS   If we take the second derivative of the total time, we can show that it is always positive, which means the extremum we have found is a minimum. In other words, light chooses the fastest path from point A to point B. If light moves faster in medium 1 (i.e. $n_1 < n_2$), then the path of minimum time will extend the distance traveled in medium 1, in order to reduce the distance traveled in medium 2. An analogy would be a lifeguard rushing to reach a struggling swimmer in a lake. The fastest path may not be a straight line, but instead one in which the lifeguard runs along the shore before diving in. This is because the lifeguard moves slower through the medium of water than of air.

61.   INTERPRET   We want an expression for the time it takes light to traverse a slab of transparent material with an index of refraction that varies with depth. The light enters the material normal to the surface, so there's no angle of refraction to worry about.

DEVELOP   Inside the slab, the time it takes light to travel an infinitesimal distance, dx, is

$$dt = \frac{dx}{v} = \frac{n(x)}{c} dx$$

To find the total crossing time, we will integrate this over the slab's thickness, d.

EVALUATE   The time to traverse the slab is

$$t = \int dt = \frac{1}{c} \int_0^d \left[n_1 + (n_2 - n_1)(x/d)^2 \right] dx = \frac{d}{c} \left(\tfrac{2}{3} n_1 + \tfrac{1}{3} n_2 \right)$$

ASSESS   Since n_1 and n_2 have to be greater or equal to one, our result implies that the presence of the slab lengthens the time light takes to travel the distance d, as we would expect. If $n_1 = n_2$, then the slab has uniform index of refraction, and the time reduces to $t = \frac{d}{c} n_1$, again as we would expect.

63.   INTERPRET   We consider media with varying indices of refraction.

DEVELOP   The observer viewing the mirage is unaware of the light's curved path. He only perceives the angle at which the light enters his eye. He assumes what he sees is located in the direction implied by this angle.

EVALUATE   We redraw the figure below, extending a straight line from the observer's eye at the incident angle of the light. It's clear that the mirage will appear at the point A.

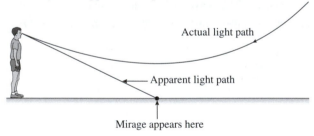

The answer is (a).

ASSESS   This answer makes sense with our experience of mirages. We think we see water on the ground in the distance, but in fact we are seeing light from the sky that is being refracted from the air near the ground.

65.   INTERPRET   We consider media with varying indices of refraction.

DEVELOP   If the refractive index of the ionosphere approaches 1 at high frequencies, that means it will have roughly the same index of refraction as the atmosphere below and above it. Thus, there will be relatively no bending of high frequency radio signals.

EVALUATE   Since higher frequencies will essentially pass through the ionosphere, an alternative method, such as satellite-based communication, will be needed to send them over long distances.

The answer is (c).

ASSESS   Most satellite-based communication is in the GHz region of the electromagnetic spectrum.

EXERCISES

Section 31.1 Images with Mirrors

17. **INTERPRET** This problem is about image formation in a plane mirror. We are to find the angle at which the mirror should be inclined so that it is possible to see one's feet when standing 50 cm from the mirror's base.

    DEVELOP Consider the sketch below. A small mirror (M) on the floor intercepts rays coming from a customer's shoes (O), which are traveling nearly parallel to the floor. The angle to the customer's eye (E) from the mirror is twice the angle of reflection α, so

    $$\tan(2\alpha) = \frac{h}{d}$$

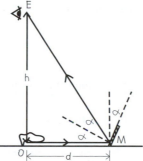

    EVALUATE Solving for the angle α, we find

    $$\alpha = \frac{1}{2}\text{atan}\left(\frac{h}{d}\right) = \frac{1}{2}\text{atan}\left(\frac{140 \text{ cm}}{50 \text{ cm}}\right) = 35°$$

    for the given distances. Therefore, the plane of the mirror should be tilted by 35° from the vertical to provide the customer with a floor-level view of her shoes.

    ASSESS The angle decreases if d is increased, and vice versa. This is consistent with common experience.

19. **INTERPRET** This problem involves image formation by a concave mirror. We want to know the orientation of the image as well as its height compared to the object.

    DEVELOP The magnification M, which is the ratio of the image height h' to object height h, is given by Equation 31.1:

    $$M = \frac{h'}{h} = -\frac{s'}{s}$$

    where s and s' are object and image distances to the mirror. The two quantities s and s' are related by the mirror equation (Equation 31.2):

    $$\frac{1}{s} + \frac{1}{s'} = \frac{1}{f}$$

    where f is the focal length of the mirror.

    EVALUATE **(a)** One can solve Equation 31.2 for s' and substitute the result into Equation 31.1 to find

    $$M = \frac{h'}{h} = -\frac{s'}{s} = -\frac{sf/(s-f)}{s} = -\frac{f}{s-f} = -\frac{f}{5f-f} = -\frac{1}{4}$$

(b) A negative magnification applies to a real, inverted image.

ASSESS   The situation corresponds to the first case shown in Table 31.1. From the ray diagram, we see that the image is real, inverted, and reduced in size. Since $s' > 0$, the image is in front of the mirror.

21.   **INTERPRET**   This problem is about image formation in a concave mirror. We want to find the object distance needed to form a full-size image.

DEVELOP   The magnification M, the ratio of the image height h' to object height h, is given by Equation 31.1:

$$M = \frac{h'}{h} = -\frac{s'}{s}$$

where s and s' are object and image distances to the mirror. A full-size image means that $|M| = 1$.

EVALUATE   **(a)** For a full-sized image, we require

$$M = \frac{h'}{h} = -\frac{s'}{s} = -\frac{sf/(s-f)}{s} = -\frac{f}{s-f} = -1$$

or $s = 2f = R$. That is, put the object at the center of curvature of the mirror (or at twice its focal length).

(b) From Equation 31.2, we have

$$s' = \frac{sf}{s-f} = \frac{(2f)f}{2f-f} = 2f > 0$$

Thus, the image is real.

ASSESS   For a spherical mirror, $|M| = 1$ applies only to a real image, $h' = -h$ unless one accepts the plane mirror, which has $f = \infty$, as a special case.

Section 31.2 Images with Lenses

23.   **INTERPRET**   This is an image-formation problem involving a converging lens. We want to find the magnification as well as the orientation of the image.

DEVELOP   The magnification of a thin lens, for paraxial rays, is given by Equation 31.4,

$$M = \frac{h'}{h} = -\frac{s'}{s} = -\frac{sf/(s-f)}{s} = -\frac{f}{s-f}$$

where we used the lens equation (Equation 31.5) to eliminate s'.

EVALUATE   If $s = 1.5f$, then $M = -f/(s-f) = -f/(1.5f - f) = -2$. The minus sign means that the image is real and inverted.

ASSESS   The lens in this problem corresponds to the second case shown in Table 31.2. With $2f > s\,(1.5f) > f$, we get a real, inverted, and enlarged image.

25.   **INTERPRET**   In this problem, we are asked to find the focal length of a magnifying glass that is a converging lens given the object and image distances.

DEVELOP   The focal length f is related to the object and image distances through the lens equation (Equation 31.5):

$$\frac{1}{f} = \frac{1}{s} + \frac{1}{s'} \quad \Rightarrow \quad f = \frac{ss'}{s+s'}$$

EVALUATE   Because the distances are given with respect to the lamp (i.e., the object), $s = 25$ cm and $s' = 1.6\text{ m} - 25\text{ cm} = 135\text{ cm}$. Substituting these into the lens equation, we find

$$f = \frac{ss'}{s+s'} = \frac{(25\text{ cm})(135\text{ cm})}{25\text{ cm} + 135\text{ cm}} = 21\text{ cm}$$

ASSESS   Since $f < s < 2f$, we expect the image to be real, inverted, and enlarged.

27.   **INTERPRET**   We are to find the focal length of a magnifying glass given the magnification and the object distance. We shall use the lens equation and the definition of magnification to solve this problem.

DEVELOP   For a single lens, magnification is defined as $M = -s'/s$. We are told that the object distance is $s = 9.0$ cm and that the magnification is $M = 1.5$ (i.e., 50% bigger) so we can find the image distance s'. The lens equation (Equation 31.5) relates the image and object distances to the focal length by $1/s + 1/s' = 1/f$, so knowing the distances we can find the focal length.

EVALUATE Using $s' = -Ms$ in the lens equation and inserting the given quantities gives

$$\frac{1}{s} - \frac{1}{Ms} = \frac{1}{f}$$

$$\frac{1}{s}\left(1 - \frac{1}{M}\right) = \frac{1}{f}$$

$$f = \frac{Ms}{M-1} = \frac{(1.5)(9.0\text{ cm})}{1.5 - 1.0} = 27\text{ cm}$$

ASSESS Be careful about the signs in problems such as these. We must remember that magnification is the *negative* of the ratio of image to object distances. If we miss this detail we will get the wrong answer of $= 5.4$ cm.

Section 31.3 Refraction in Lenses: The Details

29. **INTERPRET** This problem involves image formation by a single refracting interface between two media. Specifically, we are to find the depth of a pool of water given the image distance of an object at the bottom of the pool.

    DEVELOP The image formed by a refracting interface is described by Equation 31.6,

    $$\frac{n_1}{s} + \frac{n_2}{s'} = \frac{n_2 - n_1}{R}$$

    For the flat surface of the wading pool, the radius of curvature is $R = \infty$. The index of refraction is $n_1 = 1.333$ for water, s is the depth of the pool (your feet, the object, are on the bottom), $n_2 = 1.0$ for air, and $s' = -30$ cm (for a virtual image at the apparent depth).

    EVALUATE Thus, the above equation gives

    $$\frac{1.333}{s} + \frac{1.0}{-30\text{ cm}} = \frac{n_2 - n_1}{\infty} = 0$$

    $$s = (30\text{ cm})\frac{1.333}{1.0} = 40\text{ cm}$$

    ASSESS This problem could also be solved directly from Snell's law, without the paraxial ray approximation. Note that the image formed by a flat refracting surface is always on the same side of the surface as the object.

31. **INTERPRET** This problem involves image formation by a curved refracting interface between two media.

    DEVELOP The image formed by a refracting interface is described by Equation 31.6,

    $$\frac{n_1}{s} + \frac{n_2}{s'} = \frac{n_2 - n_1}{R}$$

    In this case, the light from the object (the insect) starts from the water in the dew drop $(n_1 = 1.333)$ and passes into the air $(n_2 = 1.0)$ through a concave surface, so the radius is negative: $R = -2.0$ mm. If you look at the surface that the insect is closest to, then $s = 2.0\,\text{mm} - 1.0\,\text{mm} = 1.0\,\text{mm}$.

    EVALUATE Solving for the image distance gives:

    $$s' = n_2\left[\frac{n_2 - n_1}{R} - \frac{n_1}{s}\right]^{-1} = \left[\frac{-0.333}{-2.0\,\text{mm}} - \frac{1.333}{1.0\,\text{mm}}\right]^{-1} = -0.86\text{ mm}$$

    The MINUS sign implies the image is virtual. The insect appears to be 0.86 mm from the surface.

    ASSESS The insect appears closer to the surface than it really is, as does the fish in Example 31.4.

Section 31.4 Optical Instruments

33. **INTERPRET** This problem is about the power of the lens required to correct vision. We are given the near point of the uncorrected eye and are asked to find the lens power needed to correct this (i.e., reduce the near point to the standard 25 cm).

    DEVELOP The uncorrected eye has a near point of 55 cm, whereas the corrected eye has a near point of 25 cm. The lens equation (Equation 31.5) relates the focal length to the object distance s and the image distance s':

$$\frac{1}{s} + \frac{1}{s'} = \frac{1}{f}$$

With the focal length f in meters, $1/f$ is the power, in diopters. The above equation gives

$$\frac{1}{s'_{retina}} = \frac{1}{f_{eye}} - \frac{1}{55\ cm} = \frac{1}{f_{eye}} = \frac{1}{f_{cor}} - \frac{1}{25\ cm}$$

EVALUATE The power of the lens is

$$P = \frac{1}{f_{cor}} = \frac{1}{0.25\ m} - \frac{1}{0.55\ m} = 2.2\ diopters$$

ASSESS Alternatively, Example 31.6 argues that the corrective lens should produce a virtual image of an object at 25 cm (the standard near point) at a distance of 55 cm (the uncorrected near point), so

$$\frac{1}{f_{cor}} = \frac{1}{0.25\ m} - \frac{1}{0.55\ m} = 2.2\ diopters$$

which is the same as above.

35. **INTERPRET** We are to find the power of a lens that would make distant objects appear to be at a distance of 80 cm, for which we shall use the lens equation.

DEVELOP The nearsighted patient described in the problem can see clearly at distances up to 80 cm. Thus, we need a lens that produces an image of distant $(s \approx \infty)$ objects at this distance $s' = -80\ cm$, where the negative sign indicates that the image will be virtual. Apply the lens equation (Equation 31.5)

$$\frac{1}{s} + \frac{1}{s'} = \frac{1}{f}$$

and report our answer in diopters, which is the inverse focal length in meters.

EVALUATE The necessary lens power is

$$\frac{1}{\cancel{s}} + \frac{1}{s'} = \frac{1}{f} = P$$

$$P = \frac{1}{s'} = \frac{1}{-0.80\ m} = -1.3\ diopters$$

ASSESS Note the sign of s' is negative, because the image is on the same side of the lens as the object.

37. **INTERPRET** We are to find the magnification of a compound microscope, given the focal lengths of the eyepiece and objective and the distance between them.

DEVELOP The magnification of a compound telescope is given by Equation 31.9:

$$M = -\frac{L}{f_o}\left(\frac{25\ cm}{f_e}\right)$$

For this problem, we are given $f_o = 6.1\ mm$, $f_e = 1.7\ cm$, and $L = 8.3\ cm$.

EVALUATE Substituting the values given, we find the overall magnification of the microscope to be

$$M = -\frac{L}{f_o}\left(\frac{25\ cm}{f_e}\right) = -\frac{83\ mm}{6.1\ mm}\left(\frac{25\ cm}{1.7\ cm}\right) = -200$$

ASSESS The image is magnified 200 times. The minus sign means that the image is inverted.

PROBLEMS

39. **INTERPRET** This is an image-formation problem involving a concave mirror. We want to know the position of the image, its height, and its orientation.

DEVELOP The magnification M, the ratio of the image height h' to object height h, is given by Equation 31.1:

$$M = \frac{h'}{h} = -\frac{s'}{s}$$

where s and s' are the object and image distances to the mirror, respectively. The two quantities s and s' are related by the mirror equation (Equation 31.2):

$$\frac{1}{s}+\frac{1}{s'}=\frac{1}{f}$$

where f is the focal length of the mirror.

EVALUATE (a) The position of the image is at

$$s'=\frac{fs}{s-f}=\frac{(17\text{ cm})(10\text{ cm})}{10\text{ cm}-17\text{ cm}}=-24\text{ cm}$$

The negative sign means that the image is behind the mirror (opposite the object).

(b) The magnification of the image is

$$M=\frac{h'}{h}=-\frac{s'}{s}=-\frac{sf/(s-f)}{s}=-\frac{f}{s-f}=-\frac{17\text{ cm}}{10\text{ cm}-17\text{ cm}}=2.43$$

Therefore, the height of the image is $h'=Mh=(2.43)12\text{ mm}=29\text{ mm}$.

(c) The image is virtual $(s'<0)$, upright $(M>0)$, and enlarged.

ASSESS The situation corresponds to the third case depicted in Table 31.1. The mirror is concave with $s<f$.

41. **INTERPRET** This problem is about image formation in a concave mirror. We want to know the position of the object, given its magnification and the focal length of the mirror.

DEVELOP Using Equations 31.1 and 31.2, the magnification of the image can be written as

$$M=\frac{h'}{h}=-\frac{s'}{s}=-\frac{sf/(s-f)}{s}=-\frac{f}{s-f}$$

The fact that the image is upright tells us that $M>0$. This equation allows us to solve for s, the position of the object.

EVALUATE With $M=+3$ and $f=27$ cm, we get

$$s=f\left(1-\frac{1}{M}\right)=(27\text{ cm})\left(1-\frac{1}{3}\right)=18\text{ cm}$$

The object is in front of the mirror.

ASSESS The image distance is $s'=-Ms=-3(18\text{ cm})=-54\text{ cm}$. The situation corresponds to the third case depicted in Table 31.1. The mirror is concave with $s<f$ and the image is virtual, upright, and enlarged.

43. **INTERPRET** This problem is about the image formed by a telescope. The main mirror of a telescope is concave, since only a concave mirror can collect light from a distant object and form a real image.

DEVELOP Inspection of Figure 31.6a (partially redrawn below) shows that the angular size of the object and image are equal

$$\frac{h}{s}=\frac{h'}{s'}\approx\theta=\frac{\text{size}}{\text{distance}}$$

Since the object distance is astronomical,

$$0\approx\frac{1}{s}=\frac{1}{f}-\frac{1}{s'}$$

$$s'\approx f$$

That is, the image distance is approximately equal to the focal length. It follows from the law of reflection that the angular size of the object and image are the same, as seen from the mirror.

EVALUATE Combining the two equations above, the image size is

$$h'=\theta s'\approx\theta f=(0.5°)\left(\frac{\pi\text{ rad}}{180°}\right)(8.5\text{ m})=7.4\text{ cm}$$

ASSESS The situation corresponds to the first case shown in Table 31.1. From the ray diagram, we see that the image is real, inverted, and reduced in size. Since $s'>0$, the image is in front of the mirror.

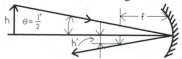

45.   **INTERPRET**   The problem asks us where to place an LCD display in front of a lens in order to project it on a distant screen.

DEVELOP   The lens is convex, so the focus is positive: $f = 7.50$ cm. Given the distance to the screen $(s' = 6.30$ m$)$, we can use the lens equation, $\frac{1}{s} + \frac{1}{s'} = \frac{1}{f}$, to find out how far from the lens the display should be put.

EVALUATE   Plugging in the values, we calculate

$$s = \left[\frac{1}{f} - \frac{1}{s'}\right]^{-1} = \left[\frac{1}{7.50\text{ cm}} - \frac{1}{6.30\text{ m}}\right]^{-1} = 7.59 \text{ cm}$$

ASSESS   The object distance satisfies: $2f > s > f$, which means the image is real, inverted and enlarged (see Table 31.2). Therefore, the display will need to be upside down in order to have the projected image right-side up.

47.   **INTERPRET**   This problem involves finding the object distance necessary for the given converging lens to form an image with the desired magnification.

DEVELOP   The magnification of a thin lens, for paraxial rays, is given by Equation 31.4,

$$M = \frac{h'}{h} = -\frac{s'}{s}$$

Combining this with the lens equation (Equation 31.5) to eliminate the image distance s' gives

$$M = -\frac{sf/(s-f)}{s} = -\frac{f}{s-f}$$

s'. Solve this equation for the object distance s.

EVALUATE   When a virtual, upright image is formed by a converging lens, the magnification is positive; M = 1.6. Therefore,

$$s = f\left(1 - \frac{1}{M}\right) = (32\text{ cm})\left(1 - \frac{1}{1.6}\right) = 12 \text{ cm}$$

ASSESS   The image formation in this problem corresponds to the third case shown in Table 31.2. With $s < f$, we get a virtual, upright, and enlarged image. Note that a diverging lens always produces a reduced image.

49.   **INTERPRET**   This problem is finding the image distance and the image type (i.e., inverted, not inverted, reduced, extended, real, virtual) formed by a converging lens with the given focal distance.

DEVELOP   Using the lens equation (Equation 31.5) and Equation 31.4 for the magnification of a thin converging (positive f) lens, we obtain

$$M = \frac{h'}{h} = -\frac{s'}{s} = -\frac{sf/(s-f)}{s} = -\frac{f}{s-f}$$

so $h' = Mh = -fh/(s-f)$ where h' is the image height.

EVALUATE   **(a)** If $f = 35$ cm and $s = f + 10$ cm $= 45$ cm, then

$$h' = Mh = -\frac{fh}{s-f} = -\frac{(35\text{ cm})(2.2\text{ cm})}{45\text{ cm} - 35\text{ cm}} = -7.7 \text{ cm}$$

The image distance is

$$-\frac{s'}{s} = \frac{h'}{h} \quad \Rightarrow \quad s' = -s\frac{h'}{h} = -(45\text{ cm})\frac{-7.7\text{ cm}}{2.2\text{ cm}} > 0$$

The negative image height signifies an inverted image and the positive image distance signifies a real image. In addition, the image is enlarged compared to the object.

(b) If $s = f - 10$ cm $= 25$ cm, then

$$h' = Mh = -\frac{fh}{s-f} = -\frac{(35\text{ cm})(2.2\text{ cm})}{25\text{ cm} - 35\text{ cm}} = +7.7 \text{ cm}$$

The image distance is

$$s' = -s\frac{h'}{h} = -(25\text{ cm})\frac{7.7\text{ cm}}{2.2\text{ cm}} < 0$$

The positive image height signifies an upright image and the negative image distance signifies a virtual image. In addition, the image is enlarged compared to the object.

ASSESS The image formed in (a) corresponds to the second case shown in Table 31.2. With $2f > s > f$, we get a real, inverted, and enlarged image. The situation in (b) corresponds to the third case shown in Table 31.2. With $s < f$, we get a virtual, upright, and enlarged image.

51. **INTERPRET** This problem involves finding the image formed by a convex lens. We are given the focal length of the lens, the distance between the object and the image, and are asked to locate the lens.

    DEVELOP Since s and s' are both positive for a real image, and the distance $a \equiv s + s' = 70$ cm is fixed, and the lens equation (Equation 31.5) can be rewritten as

    $$\frac{1}{f} = \frac{1}{s} + \frac{1}{s'} = \frac{s+s'}{ss'} = \frac{a}{ss'}$$
    $$af = (s+s')f = ss' = s(a-s) = s'(a-s')$$

    Solving the quadratic equation yields the two desired positions.

    EVALUATE The solutions for the object distance s (or the image distance s') are

    $$s = \frac{1}{2}a\left[1 \pm \sqrt{1-4f/a}\right]$$
    $$= \frac{1}{2}(70 \text{ cm})\left[1 \pm \sqrt{1-4(17 \text{ cm})/(70 \text{ cm})}\right] = 29 \text{ cm or } 41 \text{ cm}$$

    which are the desired lens locations.

    ASSESS Note that this situation has a real solution only if $0 < 4f \leq a$. Both s and s' are positive, as required for a real image.

53. **INTERPRET** We are to find the object distance needed to produce an upright image with the given magnification for the given converging lens.

    DEVELOP Using the lens equation (Equation 31.5) and Equation 31.4 for the magnification for a thin converging (positive f) lens, we obtain

    $$M = \frac{h'}{h} = -\frac{s'}{s} = -\frac{sf/(s-f)}{s} = -\frac{f}{s-f}$$

    so $s = f(1-1/M)$. Because the image is upright, the magnification is positive: $M = 1.8$.

    EVALUATE From the above equation, we get

    $$s = f\left(1-\frac{1}{M}\right) = (25 \text{ cm})\left(1-\frac{1}{1.8}\right) = 11 \text{ cm}$$

    ASSESS Since $s < f$, the situation corresponds to the third case shown in Table 31.2. The image is virtual, upright, and enlarged.

55. **INTERPRET** This problem involves image formation by a refracting plano-convex lens.

    DEVELOP The focal length of the lens is given by the lens maker's formula in Equation 31.7,

    $$\frac{1}{f} = (n-1)\left(\frac{1}{R_1} - \frac{1}{R_2}\right)$$

    With $R_1 = 26$ cm and $R_2 = \infty$ (or $R_1 = \infty$ and $R_2 = -26$ cm), the focal length is

    $$\frac{1}{f} = (n-1)\left(\frac{1}{R_1} - \frac{1}{R_2}\right) = \frac{n-1}{R_1} = \frac{1.62-1}{26 \text{ cm}} = \frac{1}{41.9 \text{ cm}} \rightarrow f = 41.9 \text{ cm}$$

    EVALUATE An object at $s = 68$ cm has its image at

    $$\frac{1}{s'} = \frac{1}{f} - \frac{1}{s} = \frac{1}{41.9 \text{ cm}} - \frac{1}{68 \text{ cm}} = \frac{1}{109 \text{ cm}} \rightarrow s' = 109 \text{ cm}$$

    This is a real, inverted image, on the opposite side of the lens from the object.

    ASSESS The image formed corresponds to the second case shown in Table 31.2. With $2f > s > f$ and $s' > 2f$, we get a real, inverted, and enlarged image.

57.   **INTERPRET**   As seen in Example 31.4, this problem involves image formation by the water in the aquarium, which forms a two-dimensional concave refracting spherical surface. The "lens" is concave toward the object, so the radius is negative.

DEVELOP   The image formed by a refracting interface is described by Equation 31.6,

$$\frac{n_1}{s} + \frac{n_2}{s'} = \frac{n_2 - n_1}{R}$$

The equation can be used to solve for the apparent distance s'.

EVALUATE   With $R = -35$ cm , $n_1 = 1.333$ for water, $s = 70$ cm $- 15$ cm $= 55$ cm (distance from the near wall), and $n_2 = 1.0$ for air, the expression above gives an image distance of

$$s' = \frac{n_2 R s}{(n_2 - n_1)s - n_1 R} = \frac{(1.0)(-35 \text{ cm})(55 \text{ cm})}{(1.0 - 1.333)(55 \text{ cm}) - (1.333)(-35 \text{ cm})} = -68 \text{ cm}$$

A negative image distance means that the image is virtual.

ASSESS   In this case, the object is closer to the refracting surface than its image (see sketch below and compare with Figure 31.23b).

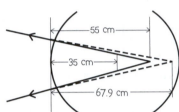

59.   **INTERPRET**   This problem involves image formation by a refracting crystal ball. We are interested in the index of refraction of the ball and are given the object and image distance of the center speck ($s_{center} = s'_{center} = R$, see Problem 31.56) and the image and object distance of the outer speck.

DEVELOP   The image distance of the outer speck is 1/3 of the distance to the center of the ball, so $s' = -|R|/3$. The object distance of the outer speck is given as $s = |R|/2$. We can then solve for Equation 31.6

$$\frac{n_1}{s} + \frac{n_2}{s'} = \frac{n_2 - n_1}{R}$$

for n_1, which is the index of the material containing the object (i.e., the ball).

EVALUATE   Equation 31.6 (with n_1 for the ball's material, $n_2 = 1.00$ for air, and $R = -|R|$ for a concave surface toward the object) gives

$$\frac{n_1}{|R|/2} + \frac{1.00}{-|R|/3} = \frac{1.00 - n_1}{-|R|}$$

This simplifies to $2n_1 - 3 = n_1 - 1$, or $n_1 = 2.00$.

ASSESS   The index of refraction of crystal indeed is about 2.0.

61.   **INTERPRET**   We have a plano-convex lens, and we want to know the relationship between its index of refraction and the radius of curvature of the curved surface. In particular, we want to find the refactive index for which the focal length is equal to the radius of curvature.

DEVELOP   The focal length of the lens is given by the lensmaker's formula (Equation 31.7),

$$\frac{1}{f} = (n-1)\left(\frac{1}{R_1} - \frac{1}{R_2}\right)$$

The plano-convex lens has $R_1 = R$ and $R_2 = \infty$ (or $R_1 = \infty$ and $R_2 = -R$). Thus, it has a focal length of

$$\frac{1}{f} = (n-1)\left(\frac{1}{R} - \frac{1}{\infty}\right) = \frac{n-1}{R}$$

$$f = \frac{R}{n-1}$$

EVALUATE   If $f = R$, then the index of refraction is $n = 2$.

ASSESS The smaller R, the more curved the lens and the more it bends light. This implies a shorter focal length. The higher refraction index n, the greater the refraction, and the shorter focal length.

63.   **INTERPRET** You are comparing two transparent materials with different indices of refraction for use in a lens. The lens must meet certain optical specifications without exceeding the given thickness.

DEVELOP You're not told the specific shape of the lens, but for simplicity you assume it is plano-convex, like the one in Example 31.5. In that case, there's only one radius of curvature to consider, and the lensmaker's formula (Equation 31.7) simplifies to $f = R/(n-1)$. It also means that you can relate the radius of curvature to the lens diameter, d, and thickness, t, by the Pythagoras formula (see figure below):

$$R^2 = \left(\frac{d}{2}\right)^2 + (R-t)^2 \quad \rightarrow \quad R = \frac{d^2}{8t} + \frac{t}{2}$$

Since $d \gg t$, the dominant term in this expression is the first term. Therefore, the upper limit on the thickness $(t_{max} = 0.8 \text{ mm})$ sets the lower limit on the radius: $R_{min} \approx d^2/8t_{max}$.

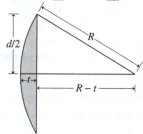

EVALUATE: The lower limit on the radius gives a lower limit on the index of refraction:

$$n > 1 + \frac{R_{min}}{f} = 1 + \frac{1}{f}\left(\frac{d^2}{8t_{max}} + \frac{t_{max}}{2}\right) = 1 + \frac{1}{17 \text{ mm}}\left(\frac{(5.5 \text{ mm})^2}{8(0.8 \text{ mm})} + \frac{(0.8 \text{ mm})}{2}\right) = 1.30$$

Both materials have indices of refractions greater than this, so you should choose plastic, because it meets the requirements and is cheaper.

ASSESS For other lens shapes, you can show that the thickness is related to the two radii of curvature by:

$$t = \left(R_1 - \sqrt{R_1^2 - \tfrac{1}{4}d^2}\right) - \left(R_2 - \sqrt{R_2^2 - \tfrac{1}{4}d^2}\right) \approx \frac{d^2}{8}\left(\frac{1}{R_1} - \frac{1}{R_2}\right)$$

The approximation is valid as long as $d \ll 2R_{1,2}$. Plugging this into the lensmaker's formula, you get $n > 1 + d^2/8ft_{max}$, which is essentially what we found above for a plano-convex lens.

65.   **INTERPRET** Diamond and glass have different indices of refraction, so this problem is about the effect on the image when the index of refraction of the lens is changed. We are to find the what type of image is formed when a glass lens is replaced by a diamond lens.

DEVELOP Using the lens equation (Equation 31.5) and Equation 31.4 for the magnification for a thin converging (positive f) lens, we obtain

$$M = \frac{h'}{h} = -\frac{s'}{s} = -\frac{sf/(s-f)}{s} = -\frac{f}{s-f}$$

Magnification is positive for a virtual image. Thus, for the crown glass lens we have

$$M_g = 2 = -\frac{f_g}{s - f_g}$$

which can be solved to give $f_g = 2s = 30$ cm. The focal length of a diamond lens with the same radius of curvature is (using Equation 31.7 and Table 30.1)

$$f_d(n_d - 1) = f_g(n_g - 1)$$

$$f_d = \left(\frac{n_g - 1}{n_d - 1}\right)f_g = \left(\frac{1.520 - 1}{2.419 - 1}\right)(30 \text{ cm}) = 11.0 \text{ cm}$$

EVALUATE   An object 15 cm from the diamond lens produces a real, inverted image (negative M) magnified by

$$M_d = -\frac{f_d}{s - f_d} = -\frac{11 \text{ cm}}{15 \text{ cm} - 11 \text{ cm}} = -2.8$$

ASSESS   In the case of a crown glass, $f_g > s$ and the image is virtual. For the diamond lens, since $s > f_d$, the image is real with $M_d < 0$.

67.   INTERPRET   We are to find the power and type of lens that must be used in order to improve the close-up capability of a camera. The original camera can focus at 60 cm, and the improved camera must be able to focus at 20 cm.

DEVELOP   For an object at 20 cm, the auxiliary lens should produce a virtual image at 60 cm from either lens (the distance between the lenses is negligible). The lens power is defined as the reciprocal of the focal length, which can be found using Equation 31.5 with $s = 20$ cm and $s' = -60$ cm.

EVALUATE   Thus, the required power is

$$P_{aux} = \frac{1}{f_{aux}} = \frac{1}{0.20 \text{ m}} + \frac{1}{-0.60 \text{ m}} = \frac{1}{0.30 \text{ m}} = 3.3 \text{ diopters}$$

Since the focal length is positive, the lens is converging.

ASSESS   The camera without the auxiliary lens can be compared to the eye, in Example 31.6, with a receding near point.

69.   INTERPRET   This problem is about the angular magnification of an astronomical telescope. We are to find the angular magnification of Jupiter given its angular diameter to the unaided eye and the focal length of a refracting telescope and that of its eyepiece.

DEVELOP   For a refracting telescope, the angular magnification is given by Equation 31.10:

$$m = \frac{\beta}{\alpha} = \frac{f_o}{f_e}$$

where α and β are the angles subtended by the actual object and the final image, respectively, while f_o and f_e are the focal lengths of the objective lens and the eye piece, respectively. Solve this equation for β.

EVALUATE   Substituting the values given, the apparent angular size is

$$\beta = \alpha\left(\frac{f_o}{f_e}\right) = 50''\left(\frac{1 \times 10^3 \text{ mm}}{40 \text{ mm}}\right) = 1250'' = 20.8' \approx 0.3°$$

ASSESS   The angular magnification in this case is $m = 25$. Note that a two-lens refracting telescope gives an inverted (real) image.

71.   INTERPRET   We are to find the diameter of a reflecting ball given that its magnification for a 6.0-cm object distance. The ball acts as a convex mirror.

DEVELOP   Using Equations 31.1 and 31.2, the magnification of the image can be written as

$$M = \frac{h'}{h} = -\frac{s'}{s} = -\frac{sf/s - f}{s} = -\frac{f}{s - f}$$

which yields $f = sM/(M - 1)$. The diameter of the ball is $D = 2R = 4|f|$, where f is the focal length (negative in this case).

EVALUATE   Using the two equations above, the diameter of the ball is

$$D = 2R = 4|f| = 4s\left|\frac{M}{M - 1}\right| = 4(6.0 \text{ cm})\left|\frac{0.75}{0.75 - 1}\right| = 72 \text{ cm}$$

ASSESS   The situation corresponds to case 4 illustrated in Table 31.1. With a convex mirror ($f < 0$) the image is virtual, upright, and reduced.

73.   INTERPRET   This problem is about the corrective power of lenses. We want to show that for closely spaced lenses, the lens powers are additive.

DEVELOP   For two closely-spaced thin lenses, distances measured from either lens are the same. We can consider that the image which would be produced by the first lens alone acts as an object for the second lens, where $s_2 = -s_1'$, because this image, if real, is on the other side of the second lens or, if virtual (i.e., $s_1' < 0$), is on the same side. The lens equations for each lens are

$$\frac{1}{f_1} = \frac{1}{s_1} + \frac{1}{s_1'} \quad \text{and} \quad \frac{1}{f_2} = \frac{1}{s_2} + \frac{1}{s_2'} = \frac{1}{-s_1'} + \frac{1}{s_2'}$$

EVALUATE   Adding the two equations yields

$$\frac{1}{f_1} + \frac{1}{f_2} = \frac{1}{s_1} + \frac{1}{s_2'} = \frac{1}{f}$$

where f is the focal length of the lens combination. Since the dioptric power of a lens is the reciprocal of its focal length in meters, the additivity of this quantity follows, under the conditions stated.

ASSESS   The additive nature of the corrective power allows an optometrist, during an eye exam, to continue to add lenses until the desired power is attained.

75.   **INTERPRET**   We are to show that objects placed equidistance on either side of the focal point of either a concave mirror or a converging lens produce equal-sized images.

DEVELOP   A concave mirror or a converging lens are both represented by positive focal lengths in the lens or mirror equations. For either, the object and image sizes are related by

$$M = \frac{h'}{h} = -\frac{s'}{s} = \frac{-f}{s-f}$$

where we have used Equations 31.4 and 31.5. The size of the image is

$$Mh = \frac{-fh}{s-f}$$

EVALUATE   The distance between the object and focal point is a constant, so $\left| s - f \right| = \pm x_0$. Thus, the two possible image heights are

$$h'_+ = \frac{-f}{x_0}h$$

$$h'_- = \frac{-f}{-x_0}h$$

Because $\left| h'_+ \right| = \left| h'_- \right|$, the image size is the same. The image types, however, are different. For $f - s = x_0$, the magnification is negative so the image is a virtual, inverted image. For $f - s = -x_0$, the image is real and upright.

ASSESS   Note that the problem statement implies that the $2f > s > 0$ (i.e., the distance from the object to the focal point is less than the focal length).

77.   **INTERPRET**   We add one more diagram to Figure 31.10, but with a ray going from the arrowhead to the center of curvature. With the resulting similar triangles, we are to show that, for a curved mirror, the focal length is half the radius of curvature.

DEVELOP   Start with a diagram, as shown in the figure below. The shaded triangles are similar, so

$$-\frac{h'}{h} = R - \frac{s'}{s} - R$$

We will also use the results from Figure 31.10a, $h'/h = -s'/s$ and the mirror equation (Equation 31.2) $1/s + 1/s' = 1/f$. We will solve for f in terms of R.

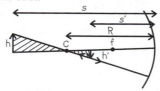

EVALUATE   From $-h'/h = R - s'/s - R$ and $h'/h = -s'/s$ we obtain

$$\frac{s'}{s} = \frac{R - s'}{s - R}$$

$$\frac{s - R}{s} = \frac{R - s'}{s'}$$

$$\frac{1}{s} = \frac{2}{R} - \frac{1}{s'}$$

Substitute this into the mirror equation:

$$\frac{1}{s} + \frac{1}{s'} = \frac{1}{f}$$

$$\left(\frac{2}{R} - \frac{1}{s'}\right) + \frac{1}{s'} = \frac{1}{f}$$

$$f = \frac{R}{2}$$

ASSESS   We have proven what was required.

79. **INTERPRET**   We are to show analytically that the maximum angular magnification of a simple magnifier is as given in the problem statement. We will use the definition of linear magnification, and the lens equation.

DEVELOP   The angular magnification is defined by the ratio of the apparent size of the image to the apparent size of the object at the near point:

$$m = \frac{h'/(25\ \text{cm})}{h/(25\ \text{cm})} = \frac{h'}{h}$$

This is just the linear magnification (Equation 31.4), $M = h'/h = -s'/s$. The image distance is $s' = 25$ cm. We will solve the lens equation for s, and substitute the result into the equation for m.

EVALUATE

$$\frac{1}{s} + \frac{1}{s'} = \frac{1}{f} \quad \Rightarrow \quad s = \frac{fs'}{s' - f}$$

$$m = \frac{h'}{h} = -\frac{s'}{s} = -s'\left(\frac{s' - f}{fs'}\right) = -\frac{s'}{f} + 1 = 1 + \frac{25\ \text{cm}}{f}$$

ASSESS   We have shown what was required.

81. **INTERPRET**   We will apply the result of Problem 31.80 to an actual lens, given the wavelength dependence of the index of refraction for the material, to find the variation in the focal length over a 10-nm range centered on the central wavelength of 550 nm.

DEVELOP   Our result from Problem 31.82 is $df/f = -dn/n - 1$. The focal length of the lens is f = 30 cm at $\lambda = 550$ nm, and the index of refraction is $n = n_0 - b\lambda$ where $n_0 = 1.546$ and $b = 4.47 \times 10^{-5}$ nm^{-1}. We want to find the total variation in focal length df over a 10-nm spread in wavelength centered on 550 nm.

EVALUATE $dn \sim b\Delta\lambda = (10\ \text{nm})(-4.47 \times 10^{-5}\ \text{nm}^{-1}) = -4.47 \times 10^{-4}$, so

$$df = -f(dn/n) - 1 = (30\ \text{cm})\frac{-4.47 \times 10^{-4}}{1.546} - 1 = 0.025\ \text{cm}$$

ASSESS   This is a small variation, because of the very small change in wavelength. For full-spectrum visible-light applications, the spread of wavelengths is about 200 nm, and the magnitude of df can be significant.

83. **INTERPRET**   We explore the meaning of a camera's f-ratio.

DEVELOP   The light admitted increases with the aperture area, but decreases with the square of the focal length. Therefore, the light admitted is inversely proportional to the square of the f-ratio.

EVALUATE: The f-ratio in this case is doubling in size, so the light admitted must be decreasing by a factor of 4. The answer is (b).

ASSESS   Many cameras have f-ratio gradations that scale by factors of the square root of 2, as for example: $f/1$, $f/1.4$, $f/2$, $f/2.8$, $f/4$, $f/5.6$, $f/8$, $f/11$, $f/16$. Each step in this sequence corresponds to a decrease in the light admitted by a factor of 2. In other words, $f/1$ is twice as fast as $f/1.4$.

85.   **INTERPRET**   We explore the meaning of a camera's f-ratio.

DEVELOP   Stopping down means reducing the lens area by squeezing closed a camera's adjustable iris. This does not change the focal length.

EVALUATE: The spherical aberration is due to the fact that the lens is not a parabola and therefore it does not focus all light rays to a single point. The problem is worst for light rays that enter at the edge of the lens, far from the center. Thus, blocking the outer edge of the lens will improve the focus, by eliminating the light rays with the most aberration.

The answer is (b).

ASSESS   Stopping down taken to its extreme leads to a pinhole camera, which has superb focus without the need of a lens. The trouble is that the image is very faint.

INTERFERENCE AND DIFFRACTION

EXERCISES

Section 32.2 Double-Slit Interference

11. **INTERPRET** This problem is about double-slit interference. We are to find the spacing between adjacent bright fringes given the wavelength of the light, the slit spacing, and the slit-screen distance.

    DEVELOP The geometrical parameters of the source, slits, and screen satisfy the conditions for which Equations 32.2a and 32.2b apply (i.e., $d \ll L$ and $\lambda \ll d$). The locations of bright fringes are given by

    $$y_{\text{bright}} = \frac{m\lambda L}{d}$$

    where m is the order number.

    EVALUATE The spacing of bright fringes is

    $$\Delta y = (m+1)\frac{\lambda L}{d} - m\frac{\lambda L}{d} = \frac{\lambda L}{d} = \frac{(550 \text{ nm})(75 \text{ cm})}{0.025 \text{ mm}} = 1.7 \text{ cm}$$

    ASSESS Since $\lambda \ll d$, the spacing between bright fringes is much smaller than L, as it should be.

13. **INTERPRET** This problem is about double-slit interference. We are interested in the wavelength of the light source.

    DEVELOP For small angles, we may approximate $\sin\theta \sim \theta$, so Equation 32.1 gives $\Delta\theta = \lambda/d$, and the interference fringes are evenly spaced.

    EVALUATE Substituting the values given, we obtain $\lambda = d\Delta\theta = (0.37 \text{ mm})(0.065°)(\pi/180°) = 420 \text{ nm}$
    ASSESS The wavelength λ is much smaller than the slit spacing d, as needed for using Equation 32.1a.

Section 32.3 Multiple-Slit Interference and Diffraction Gratings

15. **INTERPRET** The setup is a multiple-slit interference experiment. We want to know the number of minima (destructive interferences) between two adjacent maxima.

    DEVELOP In an N-slit system with slit separation d (illuminated by normally incident plane waves), the main maxima occur for angles (see Equation 32.1a) $\sin\theta = m\lambda/d$, and minima for angles (see Equation 32.4) $\sin\theta = m'\lambda/(Nd)$ (excluding $m' = 0$ or multiples of N).

    EVALUATE Between two adjacent maxima, say $m' = mN$ and $(m+1)N$, there are $N-1$ minima. The number of integers between mN and $(m+1)N$ is

    $$(m+1)N - mN - 1 = N - 1$$

    because the limits are not included. Therefore, For $N = 5$, the number of minima is 4.
    ASSESS The interference pattern resembles that shown in Figure 32.8. Note that the number of minima is independent of the order number m. Also note that our result agrees with Figure 32.8.

17. **INTERPRET** In this problem, we want to locate certain maxima and minima in a multiple-slit interference experiment. We are given the necessary parameters.

DEVELOP According to Equation 32.1a, primary maxima occur at angles $\theta = \text{asin}(m\lambda/d)$. On the other hand, minima occur at angles (see Equation 32.4)

$$\theta_{\min} = \text{asin}\left[m'\lambda/(Nd)\right], \quad m' = \pm 1, \pm 2, \pm 3, \pm 4, \pm 6, \ldots$$

where m' is an integer but not an integer multiple of N.

EVALUATE (a) Using the above equation, the first two maxima (after the central peak, $m = 0$) are for $m = 1$ and 2. The angular position for these maxima are at

$$\theta_1 = \text{asin}(1 \times 633 \text{ nm}/7.5 \text{ μm}) = 4.8°$$
$$\theta_2 = \text{asin}(2 \times 633 \text{ nm}/7.5 \text{ μm}) = 9.7°$$

(b) With $N = 5$, excluded, the third minimum is for $m' = 3$ and the sixth for $m' = 7$ (because $m' = 5$ doesn't count). Then

$$\theta_{3,\min} = \text{asin}\left[3\lambda/(5\lambda)\right] = 2.9°$$
$$\theta_{7,\min} = \text{asin}\left[7\lambda/(5\lambda)\right] = 6.8°$$

ASSESS The minima would be difficult to observe because the secondary maxima between them are faint.

19. **INTERPRET** This problem is about diffraction gratings. For a given wavelength, we are interested in the highest visible order.

DEVELOP The grating condition is $\sin\theta = m\lambda/d$ and q must be less than 90° (or $m\lambda/d < 1$) for the diffracted light to be visible. Therefore, the highest order visible is the greatest integer m less than d/λ. The grating spacing is $d = (1 \text{ cm})/10^4 = 10^3 \text{ nm}$.

EVALUATE (a) For $\lambda = 450 \text{ nm}$, the highest visible order is

$$m_{\max} < \frac{d}{\lambda} = \frac{1000 \text{ nm}}{450 \text{ nm}} = 2.22$$
$$m_{\max} = 2$$

(b) Similarly, for $\lambda = 650 \text{ nm}$, the highest visible order is

$$m_{\max} < \frac{d}{\lambda} = \frac{1000 \text{ nm}}{650 \text{ nm}} = 1.54$$
$$m_{\max} = 1$$

ASSESS Increasing wavelength lowers $m_{\max}$. This can be seen from Equation 32.1a, $d\sin\theta = m\lambda$.

Section 32.4 Interferometry

21. **INTERPRET** This problem involves interference in a thin film. We want to find the minimum film thickness that results in constructive interference for the given wavelength.

DEVELOP The condition for constructive interference from a soap film is Equation 32.7:

$$2nd = \left(m + \frac{1}{2}\right)\lambda$$

The minimum thickness corresponds to the integer $m = 0$.

EVALUATE Substituting the values given, we get

$$2t = (m + 1/2)\lambda$$

$$2nd_{\min} = \frac{1}{2}\lambda$$

$$d_{\min} = \frac{\lambda}{4n} = \frac{550 \text{ nm}}{4(1.33)} = 103 \text{ nm}$$

Note that Equation 32.7 applies to normal incidence on a thin film in air.

ASSESS The typical thickness of a thin film is on the order of 100 nm. Thin-film interference accounts for the bands of color seen in soap films or oil slicks.

23.   **INTERPRET**   The enhanced reflection is a consequence of constructive interference, so we shall look for the range of wavelengths that satisfies this condition.

DEVELOP   Equation 32.7 gives the condition for constructive interference from a given thickness of glass surrounded by air. Solving this equation for λ gives

$$\lambda = \frac{4nd}{2m+1} = \frac{4(1.65)(450 \text{ nm})}{2m+1} = \frac{2970 \text{ nm}}{2m+1}$$

EVALUATE   Integers giving wavelengths in the visible range (400 to 750 nm) are m = 2 and 3, which correspond to $\lambda = 594$ and 424 nm, respectively.

ASSESS   The wavelengths correspond to orange and blue colors, respectively.

25.   **INTERPRET**   The problem asks what portion of a soap film will appear dark because it is too thin for constructive interference in reflected light.

DEVELOP   The soap film is 20 cm high and goes from zero thickness at the top to 1-μm thick at the bottom. See figure below. White light shines on the film and the reflected light from the two soap-air surfaces will constructively interfere when the thickness of the film satisfies Equation 32.7: $2nd = \left(m + \frac{1}{2}\right)\lambda$, where the index of refraction is that of water $(n = 1.333)$ and the integer $m = 0,1,2,3....$ We are looking for the region of the film that is too thin to support constructive interference, so we will define d_{min} as the smallest thickness for a bright band and y_{min} as the distance to this first band from the top, see figure. The region defined by y_{min} will be dark.

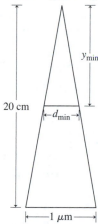

EVALUATE   White light is a combination of wavelengths from 400 to 700 nm, so there will be bright bands of different colors coming from the soap film. Near the top of film, where it is thinnest, the bands will correspond to the zeroth order $(m = 0)$. At the top of this set of bands will be the blue band for $\lambda = 400$ nm, since this corresponds to the thinnest part of the film that still supports constructive interference:

$$d_{min} = \frac{1}{2n}\left(m + \frac{1}{2}\right)\lambda = \frac{1}{2(1.333)}\left(0 + \frac{1}{2}\right)(400 \text{ nm}) = 75 \text{ nm}$$

From the figure above, we can see that this minimum thickness occurs at

$$y_{min} = d_{min}\left(\frac{20 \text{ cm}}{1 \text{ } \mu\text{m}}\right) = 1.5 \text{ cm}$$

Therefore, the top 1.5-cm of the film will be dark.

ASSESS   What happens to the light in this dark region? It is fully transmitted, so if we were to look at the backside of the soap film, the top portion would be bright, and we would see dark bands in transmission at the points corresponding to the bright bands in reflection.

Section 32.5 Huygens' Principle and Diffraction

27.   **INTERPRET**   This problem involves a single-slit diffraction of light. We are interested in the angular width of the central peak.

     DEVELOP   The condition for destructive interference in a single-slit diffraction is given by Equation 32.8:
$$a\sin\theta = m\lambda, \quad m = \pm 1, \pm 2, \ldots$$

The first minima $(m = \pm 1)$ occur at

$$\theta = a\sin\left(\pm\frac{\lambda}{a}\right)$$

$$= a\sin\left(\pm\frac{633 \text{ nm}}{2.5 \text{ μm}}\right) \pm 14.7°$$

     EVALUATE   The total angular width of the diffracted beam is $2|\theta| = 29°$.

     ASSESS   The case $m = 0$ is excluded in Equation 32.8 because it corresponds to the central maximum in which all waves are in phase.

29.   **INTERPRET**   We are to find the intensity of a diffraction maximum relative to the central peak. The second secondary maxima is the second small maxima next to the central peak.

     DEVELOP   The intensity as a function of angle in single-slit diffraction is given by Equation 32.10:

$$\overline{S} = \overline{S}_0 \left[\frac{\sin(\phi/2)}{\phi/2}\right]^2$$

The second and third minima lie at angles $\sin\theta_2 = 2\lambda/a$ and $\sin\theta_3 = 3\lambda/a$.

     EVALUATE   If we take the mid-value to be at $\sin\theta = 5\lambda/2a$, then the intensity at this angle, relative to the central intensity, is

$$\frac{\overline{S}}{\overline{S}_0} = \left(\frac{\sin(5\pi/2)}{5\pi/2}\right)^2 = \frac{4}{25\pi^2} = 1.62\times10^{-2}$$

     ASSESS   The intensity at the second secondary maximum is only about 1.62% of the central-peak intensity.

Section 32.6 The Diffraction Limit

31.   **INTERPRET**   We shall use the Rayleigh criterion to determine how large an aperture is needed on a telescope to resolve the given angle.

     DEVELOP   Apply the Rayleigh criterion for circular apertures (Equation 32.11b): $\theta_{min} = 1.22\lambda/D$. The wavelength is $\lambda = 500$ nm and the angular resolution needed is

$$\theta_{min} = 0.35 \text{ arcseconds} = 9.72\times10^{-5°} = 1.70\times10^{-6} \text{ rad}$$

Solve for D.

     EVALUATE   The diameter needed is

$$D = \frac{1.22\lambda}{\theta_{min}} = \frac{1.22(500 \text{ nm})}{1.70\times10^{-6} \text{ rad}} = 36 \text{ cm}$$

     ASSESS   Make sure that you always use radians for your angle measurements in this type of problem!

33.   **INTERPRET**   We're asked to find the diffraction limit of the eye in bright light when the pupil has contracted.

     DEVELOP   The minimum angular resolution of a circular aperture is given in Equation 32.11b: $\theta_{min} = 1.22\lambda/D$, where the result is in radians.

     EVALUATE   For the given pupil diameter and light wavelength, the resolution is

$$\theta_{min} = \frac{1.22\lambda}{D} = \frac{1.22(550\ nm)}{(2\ mm)} = 3\times10^{-4}\,rad$$

In terms of degrees, this is about 0.02°, or about 1 arcminute.

Assess This says that on a bright day our eyes would be able to distinguish objects 3 mm apart at a distance of 10 m. This is a little unrealistic. Our eyes are not only limited by the diffraction through the pupil; their resolution is also affected by the spacing of receptors (rods and cones) on the back of the retina.

PROBLEMS

35. **INTERPRET** The concept behind this problem is double-slit interference. The object of interest is the phase difference between the waves emanating from the different slits.

DEVELOP The path-length difference for waves arriving from the two different slits is

$$\Delta r = d\sin\theta \approx d\tan\theta = d\frac{y}{L}$$

since $\lambda \ll d$ and the small-angle approximation can be used (see derivation of Equations 32.2a and 32.2b). The phase difference is

$$\Delta\phi = \left(\frac{2\pi}{\lambda}\right)\Delta r = \frac{2\pi}{\lambda}\frac{yd}{L}$$

EVALUATE The phase difference is

$$\Delta\phi = \frac{2\pi(0.56\ cm)(0.035\ mm)}{(500\ nm)(1.5\ m)} = 2\pi(0.261\ cycles) = 1.6\ rad = 94°$$

Assess Constructive interference corresponds to $\Delta\phi = 2\pi m$, or $yd/L = m$, where m is an integer.

37. **INTERPRET** This problem involves a double-slit experiment. Given the slit spacing, we are asked to find the highest-order bright fringes.

DEVELOP The maximum diffraction angle for which light hits the screen is

$$\theta_{max} = atan\left(\frac{0.50\ m}{2.0\ m}\right) = 14.0°$$

From Equation 32.1a, we know that bright fringes will appear on the screen in orders of interference for which

$$\theta = asin\left(\frac{m\lambda}{d}\right) < 14.0° \Rightarrow m < \frac{d\sin(14.0°)}{633\ nm} = d(3.83\times10^5\ m^{-1})$$

EVALUATE (a) For d = 0.10 mm, $m_{max} = 38$.
(b) For $d = 10\ \mu m$, $m_{max} = 3$.
Assess The maximum order m_{max} increases as the slit spacing d decreases.

39. **INTERPRET** This problem involves a multiple-slit apparatus. We are given the number of dark fringes between two adjacent major maxima, and are asked to find the number of slits in the apparatus. We are also to find the slit separation given.

DEVELOP From Figure 32.8, we see that an N-slit system has $N-1$ minima between the major maxima. The position of the maxima is governed by Equation 32.1a, which for small angles takes the form
$$d\theta = m\lambda$$

The angular separation between adjacent maxima (i.e., between m = n and m = n + 1) is
$$d\Delta\theta = \lambda$$
The angular separation is $\Delta\theta = \pi(0.86°)/180° = 0.150\ rad$.

EVALUATE (a) For 7 minima, we have $N-1=7$, we have $N=8$ slits.

(b) The slit separation is

$$d = \frac{\lambda}{\Delta\theta} = \frac{656.3}{0.0150 \text{ rad}} = 44 \text{ μm}$$

Assess   The result is given to two significant figures because the angular separation $\Delta\theta$ is given to that precision. The angle in radians has 3 significant figures because it is an intermediate result.

41.   **Interpret**   We're asked to show that the only order where the visible spectrum doesn't overlap itself is the first order.

Develop   Within a given order, m, the spectral lines in the visible region stretch from the angle $\theta_{m,b}$ for the blue wavelength, and $\lambda_b = 400$ nm to the angle $\theta_{m,r}$ for the red wavelength $\lambda_r = 700$ nm. Each of these angles satisfies Equation 32.1a: $d\sin\theta_m = m\lambda$, so $\theta_{m,b} < \theta_{m,r}$. Overlap will occur when the blue spectral line of the $m+1$ order occurs at an angle smaller than the red spectral line of the m order: $\theta_{m+1,b} < \theta_{m,r}$.

Evaluate   If $\theta_{m+1,b} < \theta_{m,r}$, $\sin\theta_{m+1,b} < \sin\theta_{m,r}$, and

$$(m+1)\lambda_b < m\lambda_r \quad \rightarrow \quad m > \frac{\lambda_b}{\lambda_r - \lambda_b} = \frac{400 \text{ nm}}{700 \text{ nm} - 400 \text{ nm}} = 1.3$$

This says that there's no overlap for the first order, but there will be overlap for the second order and above.

Assess   We can verify that there is overlap in the second order. The $m = 2$ red line occurs at $\sin^{-1}(1400 \text{ nm}/d)$, whereas the $m = 3$ blue line occurs at $\sin^{-1}(1200 \text{ nm}/d)$. Therefore, $\theta_{3,b} < \theta_{2,r}$, which means the second and third orders overlap.

43.   **Interpret**   We are to find the diffraction order necessary to resolve (i.e., separate) two closely spaced spectral lines.

Develop   From Equation 32.5, wavelengths can be resolved if

$$\Delta\lambda > \frac{\lambda}{mN}$$

$\Delta\lambda > \lambda/mN$, or $m > \lambda/N\Delta\lambda = (648 \text{ nm})/(4500)(0.09 \text{ nm}) =$

$$m > N\frac{\lambda}{\Delta\lambda}$$

where $\Delta\lambda = 648.07 - 647.98$ nm $= 0.090$ nm and $N = 4500$.

Evaluate   The requisite order is

$$m > \frac{647.98 \text{ nm}}{(4500)(0.090 \text{ nm})} = 1.6$$

So the second or higher order is required to resolve these spectral lines.

Assess   Note that N is a dimensionless number, so the dimensions work out in the expression for the minimum order number.

45.   **Interpret**   You're assessing the feasibility of resolving Earth-sized planets with a single space telescope.

Develop   You can assume that resolving the planet means roughly that it's angular extent in the sky is at least equal to the diffraction limit of the proposed telescope. The angular extent of an Earth-sized planet at a distance of L is $2R_E/L$. Equating this to Rayleigh criterion in Equation 32.11b gives for the minimum telescope diameter:

$$D_{min} = \frac{1.22\lambda L}{2R_E}$$

Evaluate   Our equation says that the smaller the wavelength we use, the smaller the telescope has to be. So you might as well choose the lower limit of the optical wavelengths: $\lambda = 400$ nm. As such, the telescope diameter needed would be

$$D_{min} = \frac{1.22(400 \text{ nm})(5 \text{ ly})}{2(6.37\times10^6 \text{m})}\left[\frac{9.46\times10^{15}\text{m}}{1 \text{ ly}}\right] \approx 2 \text{ km}$$

A 2-km-wide telescope in space, or even on the ground, is not feasible.

ASSESS   NASA is considering ways to detect Earth-sized planets with a space telescope. However, the goal is not to resolve the planet, but merely separate its light signal from that of its host star. In this case, the angle is not set by the planet's diameter but by its orbital radius. For a planet orbiting its star at the same distance as Earth is from the sun $\left(r_E = 1.50\times10^{11}\text{m}\right)$, the minimum telescope diameter is less than 20 cm. However, this is largely irrelevant. The real challenge in getting a direct image of a distant planet is not the angular resolution, but the fact that the star is so much brighter than the planet. The starlight completely overwhelms the planet's signal, so astronomers are looking for ways to filter out the light coming from the star.

47.   **INTERPRET**   This problem is about X-ray diffraction in a crystal. We are interested in the spacing between the crystal planes, which we can find using Bragg's law.

DEVELOP   Constructive interference in X-ray diffraction is given by the Bragg condition (Equation 32.6):
$$2d\sin\theta = m\lambda, \quad m=1,2,3,...$$
Solve this for d to find the spacing between crystal planes.

EVALUATE   From the Bragg condition, one finds
$$d = \frac{m\lambda}{2\sin\theta} = \frac{(1)(97\times10^{-12}\text{ m})}{2\sin(8.5°)} = 3.3 \text{ Å}$$

ASSESS   The spacing between crystal planes is typically a few angstroms, so this result seems reasonable.

49.   **INTERPRET**   This problem involves constructive interference from a thin film. We are to find the number of times the condition for constructive interference is met for 630-nm light in a thin film that varies in thickness within the given range.

DEVELOP   In a thin film of oil between air and water ($n_{air} < n_{oil} < n_{water}$), there are 180° phase changes for reflection at both boundaries (i.e., for both rays 1 and 2 in Figure 32.7). These phase changes cancel each other, leaving only the film thickness to give the difference in path length. Therefore, for normally incident light, the term ½ in Equation 32.7 cancels due to a similar term on the left-hand side, leaving
$$2nd = m\lambda$$
The thickness d varies in the range $0.80\,\mu\text{m} \le d \le 2.1\,\mu\text{m}$, so we can find the integers m that satisfy this range for $\lambda = 630$ nm.

EVALUATE   The thickness range implies
$$0.80\,\mu\text{m} \le \frac{m\lambda}{2n} \le 2.1\,\mu\text{m}$$
$$\frac{2n(0.80\,\mu\text{m})}{\lambda} \le m \le \frac{2n(2.1\,\mu\text{m})}{\lambda}$$
$$3.17 \le m \le 8.33$$

Since m is an integer, it can range from 4 to 8, inclusive.

ASSESS   For 630 nm, this film will exhibit 5 bright fringes; for m = 4, 5, 6, 7, and 8.

51.   **INTERPRET**   This problem involves constructive interference from a thin film; in this case a film of air between two glass plates. The index of air is less than that of glass, so there is a 180° phase change at the bottom interface (air-glass interface) instead of at the top interface (glass-air).

DEVELOP   Although the phase change occurs at the second interface, as opposed to the first interface as is assumed in deriving Equation 32.6, the net effect is the same—the path difference 2d must be an odd-integer multiple of half wavelengths. The thickness of the film varies between 0 and 0.065 mm, so we can apply Equation 32.6 to find the corresponding range of m.

EVALUATE The minimum value for m is 0. The maximum value is

$$m = \frac{2nd}{\lambda} - \frac{1}{2} = \frac{2(1.00)(0.065\text{ mm})}{550\text{ nm}} - \frac{1}{2} = 235.8$$

or $m = 235$. Thus, the observer will see 236 bright bands.

ASSESS The first bright band is at zero thickness, which corresponds to $m = 0$. This band is added to the 235 remaining bands to give the total of 236 bright bands.

53. **INTERPRET** We are asked to find the distance that corresponds to the passage of 530 bright fringes in an interferometer. Thus, as one arm of the interferometer moves, the path length in that arm changes, causing alternating constructive and destructive interference (or alternating bright and dark fringes) to occur at the output of the interferometer.

DEVELOP In each arm of the interferometer, light must travel down and back, or twice the length of the arm (see, e.g., Figure 32.16). Thus, the path-length difference corresponding to 530 bright fringes is twice the distance moved by the mirror, and each successive fringe corresponds to the distance of one wavelength. This gives

$$2\Delta L = 530\lambda$$

which we can solve for ΔL.

EVALUATE Inserting the given quantities gives

$$\Delta L = \frac{530\lambda}{2} = 275(486.1\text{ nm}) = 128.8\ \mu\text{m}$$

ASSESS This distance is greater than a single wavelength, which is the minimum distance this type of apparatus can detect.

55. **INTERPRET** This problem involves an interferometer, which is used to measure the refractive index of air. Initially, one arm of the interferometer contains air. This air is gradually pumped out, which reduces the index of refraction in the arm proportionally. When no air is left, 388 bright fringes have been observed at the recombination point of the interferometer. We are to calculate the index of refraction of the air.

DEVELOP When the interferometer arm contains air, its length $2L$ in wavelengths is $N = 2L/\lambda_{air} = 2n_{air}L/\lambda$, where $\lambda_{air} = \lambda/n_{air}$ is the wavelength in air and $\lambda = 641.6$ nm (the factor 2 arises because the light must travel down and back in the interferometer arm, so measures twice the actual length L). When the interferometer arm is in vacuum, its length in wavelengths becomes $N' = 2L/\lambda$.

EVALUATE The difference between the air length and the vacuum length, in number of wavelengths, gives the number of bright fringes observed. Therefore, we can solve for the index of refraction of air as follows:

$$N - N' = 388 = \frac{2n_{air}L}{\lambda} - \frac{2L}{\lambda}$$

$$n_{air} = \left(\frac{388\lambda}{2L} + 1\right) = \left(\frac{388(641.6\text{ nm})}{2(42.5\text{ cm})}\right) = 1 + 2.93\times10^{-4}$$

ASSESS This result agrees with published results dating from 2003.

57. **INTERPRET** This problem concerns the diffraction limit of an optical system. The optical system consists of a circular concave mirror. The system has circular symmetry, so we can use the Rayleigh criterion for circular apertures.

DEVELOP To resolve a spot of 50-cm diameter at a distance of 2500 km, the necessary minimum angle to resolve is

$$\theta_{min} = \frac{(0.50\text{ m})/2}{2.5\times10^6\text{ m}} = 1.0\times10^{-7}\text{ radians}$$

The circular mirror constitutes a circular aperture for the optical system, so the minimum resolvable source separation is given by the Rayleigh criterion for circular apertures (Equation 32.11b). We can insert θ_{min} into this expression and solve for the minimum mirror diameter.

EVALUATE   The minimum mirror diameter is

$$D_{min} = \frac{1.22\lambda}{\theta_{min}} = \frac{1.22(2.8\ \mu m)}{1.0 \times 10^{-7}\ rad} = 34\ m$$

ASSESS   This is a very large mirror, especially considering the wavelength for which it is to be used. Note that we use the spot radius in determining θ_{min}. This is because this angle is defined using the distance from the maximum of the central peak to the first diffraction minimum (see Figure 32.7 and accompanying discussion).

59.   **INTERPRET**   We are to find the smallest spot that can be focused by the given lens system. Because the lens is circular, we shall apply the Rayleigh criterion for circular apertures.

     DEVELOP   The diffraction limit for a lens opening of diameter D, focusing light of wavelength λ is given by Equation 32.11b (the Rayleigh criterion for circular apertures):

$$\theta_{min} = \frac{1.22\lambda}{D}$$

The radius of a spot, at the focal length of the lens, with this angular spread, is $r = f\theta_{min}$ (the spot radius equals the distance between the central maximum and first minimum; see Figure 32.7 and accompanying discussion). The minimum spot diameter is therefore $d = 2r = 2f\theta_{min}$.

     EVALUATE   Inserting the given quantities gives

$$d = 2f\theta_{min} = \frac{2f(1.22)\lambda}{D} = 2(1.44)(1.22)(550\ nm) = 2.0\ \mu m$$

Where we have used $f/D = 1.44$, as given in the problem statement.

     ASSESS   This resolution is good enough for most commercial cameras.

61.   **INTERPRET**   We are to determine the largest distance at which humans can resolve a pair of automobile headlights. Because human pupils are circular, the Rayleigh criterion for circular apertures applies.

     DEVELOP   If we use the Rayleigh criterion (Equation 32.11b for small angles) to estimate the diffraction-limited angular resolution of the eye, at a pupil diameter of 3.1 mm and with light of wavelength 550 nm, we obtain

$$\theta_{min} = \frac{1.22\lambda}{D} = \frac{1.22(550\ nm)}{3.1\ mm} = 2.16 \times 10^{-4}\ rad$$

     EVALUATE   This angle corresponds to a linear separation of $y = 1.5$ m at a distance of

$$r = \frac{y}{\theta_{min}} = \frac{1.5\ m}{2.16 \times 10^{-4}\ rad} = 6.9\ km$$

     ASSESS   Actually, the wavelength inside the eye is different ($\lambda' = \lambda/n$) because of the average index of refraction of the eye. Even though other factors determine visual acuity, this is a reasonable ballpark estimate.

63.   **INTERPRET**   The question is whether a microscope using ultraviolet light can resolve crystallized proteins.

     DEVELOP   Suppose the minimum object size that your current optical microscope can resolve is

$$\Delta x = L\theta_{min} = \frac{1.22\lambda L}{D}$$

where L is the distance between the lens and the sample, and D is the microscope aperture. You can assume that the sales rep's UV microscope has roughly the same geometry, in which case $\Delta x_{min} \propto \lambda$.

     EVALUATE   You can assume your optical microscope uses the characteristic visible wavelength of $\lambda = 550$ nm. Therefore, the UV microscope using $\lambda = 200$ nm will have about a factor of 2 better resolution:

$$\Delta x_{UV} = \Delta x_{opt} \frac{\lambda_{UV}}{\lambda_{opt}} \approx \tfrac{1}{2} \Delta x_{opt}$$

So, yes, the sales rep is apparently correct.

ASSESS   In general, you can only resolve objects as big as the wavelength of the light that you are using. Proteins are typically only a few nanometers across, so most studies of crystallized proteins use x-ray diffraction with wavelengths between 0.1 and 10 nanometers.

65.   **INTERPRET**   We are to find an expression for the refractive index of a gas that is measured using a Michelson interferometer. We are given the difference in optical path length (i.e., the difference in the number of bright fringes) between a column of gas and an equal length of vacuum.

DEVELOP   The index of refraction in vacuum is defined to be unity. For light travelling through a gas, the wavelength of light depends on the gas through which it is travelling ($\lambda_{gas} = \lambda/n_{gas}$; λ is the vacuum wavelength). Thus, there is a difference in the number of wave cycles in the enclosed interferometer arm when the cylinder is evacuated or filled with gas. The light travels the length of the arm twice, out and back, and each cycle of difference results in one fringe shift. Thus, the number of fringes in the shift is

$$m = \frac{2L}{\lambda_{gas}} - \frac{2L}{\lambda} = \frac{2L}{\lambda}\left(n_{gas} - 1\right)$$

EVALUATE   From the above equation, the refractive index is

$$n_{gas} = 1 + \frac{m\lambda}{2L}$$

ASSESS   The interferometer allows for the determination of the refractive index of a gas.

67.   **INTERPRET**   We are asked to derive a formula for the interference pattern from Lloyd's mirror.

DEVELOP   The light reflecting off the flat mirror appears to be coming from a distance d below the surface, as shown in the figure below. Regarded in this way, the system is like a double-slit, with a slit separation of $2d$. However, the difference is that there will be a 180° phase change at the point where the light reflects off the mirror (see Figure 32.12).

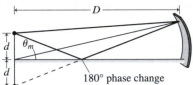

EVALUATE   Because of the extra 180° phase change, the direct and reflected beams of light will constructively interfere when their path lengths are an odd-integer multiple of half-wavelengths: $2d\sin\theta_m = \left(m + \tfrac{1}{2}\right)\lambda$. Therefore, the separation between fringes on the screen will be

$$\Delta y = D\sin\theta_{m+1} - D\sin\theta_m = D\left[\left(m + \tfrac{3}{2}\right) - \left(m + \tfrac{1}{2}\right)\right]\frac{\lambda}{2d} = \frac{D\lambda}{2d}$$

ASSESS   This is the same separation between fringes as when the light is separated by two slits. All that is different is that the bright fringes in the Lloyd's mirror setup occur at the points of the dark fringes in the double-slit experiment.

69.   **INTERPRET**   In this problem we will use the Rayleigh criterion to determine what angular spacing can be allowed between communications satellites. With this value of angle, we can find the number of satellites allowed in geosynchronous orbit before their signals begin to overlap.

DEVELOP   The Rayleigh criterion for circular apertures (Equation 32.11b) is $\theta_{min} = 1.22\lambda/D$, where the wavelength is

$$\lambda = \frac{c}{f} = \frac{3.00 \times 10^8 \text{ m/s}}{12 \text{ GHz}} = 2.5 \text{ cm}$$

and the diameter of the satellite receiver is $D = 45$ cm. The number of satellites that can fit in a circle, with this angular spacing between satellites, is $N = 2\pi/\theta_{min}$.

EVALUATE   The maximum number of satellites is

$$N = \frac{2\pi}{\theta_{min}} = \frac{2\pi D}{1.22\lambda} = \frac{2\pi(45 \text{ cm})}{1.22(2.5 \text{ cm})} = 92$$

when rounded down to the nearest integer.

ASSESS   This seems rather low, but calculating θ_{min} directly gives us an angle of 3.9°, which is consistent. We can pack more in by using shorter wavelengths or larger antennae.

71.   **INTERPRET**   We explore how interferometry can increase angular resolution in astronomy.

DEVELOP   We are told that interfering the signal of two telescopes will give the resolution of a single telescope with aperture equal to the distance between the two telescopes, i.e. $\theta_{min} = 1.22\lambda/\Delta x,$ where Δx is the telescope separation.

EVALUATE   Doubling the distance between the two telescopes should reduce by half the minimum angular separation that can be resolved.

The answer is (c).

ASSESS   Astronomers use arrays of radio telescopes with individual elements separated by 10s of meters to 1000s of kilometers. The largest arrays can obtain milliarcsecond angular resolution, which is less than a millionth of a degree.

73.   **INTERPRET**   We explore how interferometry can increase angular resolution in astronomy.

DEVELOP   For a point source directly above an interferometer, the light path to each telescope will be the same.

EVALUATE   The phase difference is proportional to the path length difference, which in this case is zero. Therefore, the electromagnetic waves will be in phase.

The answer is (a).

ASSESS   This answer is independent of the telescope separation or the wavelength being observed. The signals from a source on the bisector between two telescopes will always be in phase. It's a bit like the fact that the zeroth $(m = 0)$ order of a double-slit is always a bright fringe, since the path lengths from the two slits are equal along the bisector between them.

RELATIVITY

EXERCISES

Section 33.2 Matter, Motion, and Ether

13. **INTERPRET** In this problem we are asked to take wind speed into consideration to calculate the travel time of an airplane. Because the speeds involved are much, much less than c, we can use nonrelativistic physics.

    DEVELOP Since the velocities are small compared to c, Equation 33.5a takes the form

    $$\vec{u} = \vec{u}' + \vec{v}$$

    This is the nonrelativistic Galilean transformation of velocities as given in Equation 3.7, where $\vec{u}$ is the airplane's velocity relative to the ground, $\vec{u}'$ is the velocity of the airplane relative to the air, and $\vec{v}$ is the velocity of the air relative to the ground (i.e., the wind speed). We can use this expression to find the round-trip time t, which is

    $$t = \frac{2d}{u}$$

    With $d = 1800$ km and $u = 800$ km/h.

    EVALUATE **(a)** If $\vec{v} = 0$ (no wind), then $\vec{u} = \vec{u}'$ (ground speed equals air speed), and the round-trip travel time is

    $$t_a = \frac{2(1800 \text{ km})}{800 \text{ km/h}} = 4.50 \text{ h}$$

    (b) If $\vec{v}$ is perpendicular to $\vec{u}$, then $u'^2 = u^2 + v^2$, or

    $$u = \sqrt{u'^2 - v^2} = \sqrt{(800 \text{ km/h})^2 - (130 \text{ km/h})^2} = 789 \text{ km/h}$$

    and the round-trip travel time is

    $$t_b = \frac{2(1800 \text{ km})}{789 \text{ km/h}} = 4.56 \text{ h}$$

    (c) If $\vec{v}$ is parallel or anti-parallel to $\vec{u}$ on alternate legs of the round trip, then $u = u' \pm v$ and the travel time is (see Equation 33.2, but with c replaced by u')

    $$t_c = \frac{d}{u' + v} + \frac{d}{u' - v} = \frac{1800 \text{ km}}{800 \text{ km/h} + 130 \text{ km/h}} + \frac{1800 \text{ km}}{800 \text{ km/h} - 130 \text{ km/h}} = 4.62 \text{ h}$$

    ASSESS We find $t_a < t_b < t_c$, as mentioned in the paragraph following Equation 33.2.

Section 33.4 Space and Time in Relativity

15. **INTERPRET** This problem involves measuring a distance in two different frames of reference; the first of which is at rest with respect to the endpoints of the measurement, and the second of which is not.

    DEVELOP The distance between stars at rest in system S appears Lorentz-contracted in the spaceship's system S' according to Equation 33.4:

    $$\Delta x' = \Delta x \sqrt{1 - v^2/c^2}$$

    EVALUATE With $\Delta x = 50$ ly and $v = 0.75c$, we get

    $$\Delta x' = (50 \text{ ly}) \sqrt{1 - 0.75^2} = 33 \text{ ly}$$

    ASSESS The distance appears to be shortened or "contracted" as observed by the spaceship. Note that length contraction occurs only along the direction of motion.

17. **INTERPRET**   This problem involves measuring the length of an object in its rest frame and in a frame of reference that is moving with respect to the object. The concept of length contraction will apply.

DEVELOP   We are given the length $\Delta x' = 35$ m , which is the length measured in a frame moving at $v = 0.5c$. Equation 33.4, $\Delta x' = \Delta x\sqrt{1 - v^2/c^2}$, gives the length Δx measured in the rest system of the spaceship.

EVALUATE   Solving the equation for Δx gives

$$\Delta x = \frac{\Delta x'}{\sqrt{1 - v^2/c^2}} = \frac{35\ m}{\sqrt{1 - 1/4}} = 40\ m$$

ASSESS   The spaceship is longest in its own rest frame and is shorter to observers for whom it's moving.

19. **INTERPRET**   This is a problem about length contraction. The meter stick is measured to be shorter when it appears to be moving relative to you.

DEVELOP   The distance you measure in a frame of reference S' moving (in a direction parallel to the length of the meter stick) with speed v is $\Delta x' = 99$ cm , whereas the proper length of the meter stick is $\Delta x = 100$ cm   (in system S). These are related by Equation 33.4, $\Delta x' = \Delta x\sqrt{1 - v^2/c^2}$, which gives $0.99 = \sqrt{1 - v^2/c^2}$.

EVALUATE   Solving for v, we get $v = c\sqrt{1 - 0.99^2} = 0.14c$.

ASSESS   In order for the meter stick to measure 1% shorter, you would need to be moving at about 14% the speed of light with respect to the meter stick.

Section 33.7 Energy and Momentum in Relativity

21. **INTERPRET**   We want to know the change in momentum when we double the speed, both in the nonrelativistic and relativistic limits.

DEVELOP   The measure of momentum valid at any speed is given by Equation 33.7:

$$\vec{p} = \frac{m\vec{u}}{\sqrt{1 - u^2/c^2}} = \gamma m\vec{u}$$

Thus, doubling the speed (u becomes $2u$) increases the momentum by a factor

$$\frac{\vec{p}(2u)}{\vec{p}(u)} = \frac{2\sqrt{1 - u^2/c^2}}{\sqrt{1 - 4u^2/c^2}}$$

EVALUATE   (a) When $u = 25$ m/s, $u/c \approx 0$, so the above factor is 2.0 (to two significant figures).

(b) If $u/c = 1/3$, the factor is

$$\frac{p(2u)}{p(u)} = \frac{2\sqrt{1 - 1/9}}{\sqrt{1 - 4/9}} = 2.5$$

ASSESS   In the nonrelativistic limit, momentum $\vec{p}$ is linear in $\vec{u}$ and $\vec{p} \approx m\vec{u}$, but this no longer holds in the relativistic limit.

23. **INTERPRET**   The Newtonian expression $\vec{p} = m\vec{u}$ is valid only when $u \ll c$. and for constant mass. We want to know the speed at which the difference between this and the relativistic expression is 1%.

DEVELOP   From Equation 33.7, we find that the error in the Newtonian expression of momentum is

$$\frac{\Delta p}{p} = \frac{\gamma m u - m u}{\gamma m u} = 1 - \frac{1}{\gamma} = 1 - \sqrt{1 - u^2/c^2}$$

EVALUATE   When this factor is equal to 0.01, the speed is

$$u = c\sqrt{2\Delta p/p - \Delta p^2/p^2} = c\sqrt{2(0.01) - (0.01)^2} = 0.14c$$

ASSESS   Although $\vec{p} = m\vec{u}$ is valid at low velocity, in the relativistic limit where v/c is not negligible, Equation 33.7 should be used instead.

25. **INTERPRET**   The electron is moving at a relativistic speed, and we want to know its total energy and kinetic energy.

DEVELOP   The total energy of the electron is given by Equation 33.9:

$$E = \gamma mc^2 = \frac{mc^2}{\sqrt{1 - u^2/c^2}}$$

Its kinetic energy is given by Equation 33.8:

$$K = \gamma mc^2 - mc^2 = (\gamma - 1)mc^2$$

For this problem, the electron's speed is $v = 0.97c$, so

$$\gamma = \frac{1}{\sqrt{1 - v^2/c^2}} = \frac{1}{\sqrt{1 - (0.97)^2}} = 4.11$$

and $mc^2 = 0.511\,\text{MeV}$.

EVALUATE **(a)** From the information above, the total energy is

$$E = \gamma mc^2 = (4.11)(0.511\,\text{MeV}) = 2.1\,\text{MeV}$$

(b) The kinetic energy is

$$K = (\gamma - 1)mc^2 = (4.11 - 1.00)(0.511\,\text{MeV}) = 1.6\,\text{MeV}$$

ASSESS The total energy and kinetic energy are related by $E = K + mc^2$, where mc^2 is the rest energy of the particle. The expression demonstrates the equivalence between mass and energy. The results are reported to two significant figures to reflect the precision of the input data.

PROBLEMS

27. **INTERPRET** We are to compare the round-trip times in each branch of a Michelson–Morley interferometer. One branch is taken to be parallel to the hypothesized ether wind and the other branch is perpendicular to the ether wind. In particular, we are to show that $t_{\text{parallel}} > t_{\text{perpendicular}}$ for $0 < v < c$.

    DEVELOP The ratio of Equations 33.2 to 33.1 is

    $$\frac{t_{\text{parallel}}}{t_{\text{perpendicular}}} = \frac{2cL/(c^2 - v^2)}{2L/\sqrt{c^2 - v^2}} = \frac{1}{\sqrt{1 - v^2/c^2}}$$

    EVALUATE For $0 < v < c$, the denominator of the ratio above ranges from unity to zero, so the ratio ranges from unity to infinity. Thus, we have

    $$1 < \frac{t_{\text{parallel}}}{t_{\text{perpendicular}}} < \infty$$

    $$t_{\text{perpendicular}} < t_{\text{parallel}}\infty$$

    as claimed in the problem statement.

    ASSESS Since $t_{\text{perpendicular}} < t_{\text{parallel}}$, we conclude that the trip parallel to the ether wind always takes longer.

29. **INTERPRET** This is a problem about travel time measured in different reference frames; one reference frame is at rest with respect to the end points of the trip, whereas the other reference frame is not. Time dilation is involved.

    DEVELOP Note that the distance is given in the system S, where the Earth and the Sun are essentially at rest (the orbital speed of the Earth is very small compared to the speed of light or the speed of the spacecraft). However, the time interval is given in system S', where the spacecraft is at rest. In other words, $\Delta x = 8.3\,c \cdot \text{min}$ and $\Delta t' = 5.0\,\text{min}$. Equations 33.3 and 33.4

    $$\Delta t' = \Delta t\sqrt{1 - v^2/c^2} = \frac{\Delta t}{\gamma}$$

    $$\Delta x' = v\Delta t' = \Delta x\sqrt{1 - v^2/c^2}$$

    express time dilation and Lorentz contraction. The speed of the spacecraft can be found by using the second expression which gives

    $$v = \frac{\Delta x}{\Delta t'}\sqrt{1 - v^2/c^2}$$

    $$= \frac{c}{\sqrt{1 + (c\Delta t'/\Delta x)^2}}$$

Once we know v, we can evaluate $\gamma = \left(1 - v^2/c^2\right)^{-1/2}$ and find the time in the rest frame using the expression above for time dilation, which gives $\Delta t = \gamma \Delta t'$.

EVALUATE   (a) Inserting the given values into the expression for velocity gives

$$v = \frac{c}{\sqrt{1 + \left[\dfrac{c(5.0 \text{ min})}{8.3\, c \cdot \text{min}}\right]^2}} = 0.86c$$

(b) The time in the rest frame is

$$\Delta t = \frac{\Delta t'}{\sqrt{1 - v^2/c^2}} = \frac{5.0 \text{ min}}{\sqrt{1 - (0.857)^2}} = 9.7 \text{ min}$$

ASSESS   The result demonstrates that clocks (inside the spacecraft) moving relative to the Earth-Sun frame appear to run more slowly (5.0 min) compared to the clocks at rest (9.7 min).

31.   **INTERPRET**   This is a problem about time measured in different reference frames. The first frame is the rest frame of the end points of the journey (separated by N light years), and the second frame is the moving frame of the spaceship (in which the time spent traveling is to be N years). Time dilation is involved.

DEVELOP   The distance is given in the system S, where the Earth and the distant star are essentially at rest, but the time interval is given in the moving system S', where the spacecraft is at rest. Thus, $\Delta x = N$ ly and $\Delta t' = N$ y . One ly, or light-year, is the distance light travels in one year, and equals c multiplied by one year. Equations 33.3 and 33.4,

$$\Delta t' = \Delta t \sqrt{1 - v^2/c^2} = \frac{\Delta t}{\gamma}$$

$$\Delta x' = v\Delta t' = \Delta x \sqrt{1 - v^2/c^2}$$

for time dilation and Lorentz contraction, relate the given quantities to $\Delta x'$ and $\Delta t'$. We use the second expression (i.e., the expression for length contraction) to find

$$v = \frac{\Delta x}{\Delta t'} \sqrt{1 - v^2/c^2}$$

$$= \frac{c}{\sqrt{1 + \left(c\Delta t'/\Delta x\right)^2}}$$

EVALUATE   Inserting the given quantities gives

$$v = \frac{c}{\sqrt{1 + \left[\dfrac{c(N \text{ y})}{N\, c \cdot \text{y}}\right]^2}} = \frac{c}{\sqrt{2}} = 0.71c$$

ASSESS   To show that the result is consistent, we note that in the reference frame of the spacecraft the distance $\Delta x'$ is contracted:

$$\Delta x' = \Delta x \sqrt{1 - v^2/c^2} = \Delta x \sqrt{1 - 1/2} = N/\sqrt{2} \text{ ly}$$

So it will take

$$\Delta t' = \frac{\Delta x'}{v} = \frac{N/\sqrt{2}\, c \cdot \text{y}}{c/\sqrt{2}} = N \text{ y}$$

of the traveler's life to get there.

33.   **INTERPRET**   This is the well-known "twin paradox" problem involving time dilation. We want to know the ages of the twins after one undergoes space travel and returns. The two reference frames are that of the Earth and the distant star, which is the rest frame, and that of the spaceship, which is the moving frame.

DEVELOP   In the rest frame S, twin A must wait $\Delta t = \Delta x/v = 2(30\, c \cdot \text{y})/0.95c = 63.2$ y for twin B to return. Using Equation 33.3 for time dilation, twin B (who is in the frame S') ages for

$$\Delta t' = \Delta t\sqrt{1 - v^2/c^2} = (63.2\ \text{y})\sqrt{1 - (0.95)^2} = 19.7\ \text{y}$$

during the trip.

EVALUATE   Therefore, twin A is 83.2 y old (63.2 + 20), while twin B returns at age 39.7 y (19.7 + 20).

ASSESS   This is an intriguing consequence of time dilation. Note that only twin A's reference frame is inertial.

35.   **INTERPRET**   This is a problem involving relativistic velocity addition.

DEVELOP   Our galaxy S' is moving with speed $v = 0.75c$ relative to one of the galaxies S mentioned in the question, and the other galaxy is moving with speed $u' = 0.75c$ relative to us. All velocities are assumed to be along a common x-x' axis. The speed of one galaxy as measured by an observer in the other galaxy can be obtained by using the relativistic velocity addition formula (Equation 33.5a):

$$u = \frac{u' + v}{1 + u'v/c^2}$$

EVALUATE   Substituting the values given, we get

$$u = \frac{0.75c + 0.75c}{1 + (0.75c)^2} = 0.96c$$

ASSESS   The naïve answer, 0.75c + 0.75c = 1.5c, is inconsistent with relativity.

37.   **INTERPRET**   In this problem we are to show that if the speed of an object is less than c in one inertial reference frame, then the same conclusion holds in any other inertial frame.

DEVELOP   The problem is equivalent to showing that, if $u' < c$ and $v < c$ in the relativistic velocity addition formula (Equation 33.5a)

$$u = \frac{u' + v}{1 + u'v/c^2}$$

then $u < c$. Note that the two initial conditions above may be written as

$$u' < c \qquad v < c$$
$$u'/c < 1 \qquad v/c < 1$$
$$0 < 1 - u'/c \qquad 0 < 1 - v/c$$

EVALUATE   The conclusion follows almost immediately, because if $0 < 1 - u'/c$ and $0 < 1 - v/c$, then

$$0 < (1 - u'/c)(1 - v/c) = 1 - u'/c - v/c + u'v/c^2$$
$$u'/c + v/c < 1 + u'v/c^2$$
$$\frac{u'/c + v/c}{1 + u'v/c^2} < 1$$
$$\frac{u' + v}{1 + u'v/c^2} < c$$

but the left-hand side is just u (compare with Equation 33.5a), so we have shown that $u < c$.

ASSESS   Equation 33.5a applies to the special case where all the velocities are collinear, but the conclusion is true in general.

39.   **INTERPRET**   We are given two events (interstellar spacecraft launching) that take place at different instants as observed in the rest frame of the common galaxy. We want to know the order of occurrence as observed in the frame of reference of a spaceship moving at $0.99c$ with respect to the galaxy.

DEVELOP   Denote the frame of the galaxy by S. The spacecraft are launched at $t_A = 0$ and $t_B = 5 \times 10^5$ y from $x_A = 0$ and $x_B = 10^5$ ly (we choose the origin of S at civilization A for simplicity). Let S' be the frame of the traveler from C. The Lorentz transformation between S and S' is summarized in Table 33.1, where $v = 0.99c$ (positive x and x' axes from A to B), and

$$\gamma = \frac{1}{\sqrt{1 - v^2/c^2}} = \frac{1}{\sqrt{1 - (0.99)^2}} = 7.089$$

To see which event happened first in S', use the Lorentz transformation to transform t_A and t_B into S'.
In S', $t'_A = \gamma(t_A - x_A v/c^2) = 0$, and

$$t'_B = \gamma(t_B - x_B v/c^2) = (7.089)[5 \times 10^4 \text{ y} - (10^5 \text{ ly})(0.99c)/c^2] = -3.47 \times 10^5 \text{ y}$$

which is earlier than $t'_A = 0$.

EVALUATE   In S',

$$t'_A = \gamma\left(t_A - x_A v/c^2\right) = 0$$

and

$$t'_B = \gamma\left(t_B - x_B v/c^2\right) = (7.089)\left[5.0 \times 10^4 \text{ y} - \left(1.0 \times 10^5 \, c \cdot \text{y}\right)(0.99/c)\right] = -3.5 \times 10^5 \text{ y}$$

so the observer from C assigns priority to civilization B by $3.5 \times 10^5 \text{ y}$.

ASSESS   This problem demonstrates that two observers may see the occurrence of two events differently, depending on their frame of reference.

41.   **INTERPRET**   In special relativity the concept of simultaneity is not absolute but relative; it depends on the reference frame of the observer. We want to know whether or not there exists a reference frame in which the two events described in Problem 33.39 are seen to take place simultaneously.

DEVELOP   Refer to Problem 33.39 for a description of the frames of reference. In frame S, the light travel time between A and B is $10^5 c \cdot \text{y}/c = 10^5 \text{ y}$. In Problem 33.39, t_B is less than this, so the two launchings cannot be causally related. Therefore, there is a frame S' moving from A to B with speed v relative to S in which the events are simultaneous.

EVALUATE   Simultaneity requires that (see Table 33.1) $0 = t'_A = t'_B = \gamma\left(t_B - x_B v/c^2\right)$ or

$$\frac{v}{c} = \frac{ct_B}{x_B} = \frac{\left(1 \, c \cdot \text{y/y}\right)\left(5.0 \times 10^4 \text{ y}\right)}{1.0 \times 10^5 \text{ ly}}$$

$$v = 0.50c$$

ASSESS   If the launchings are causally related, then they cannot be simultaneous in any frame ($t'_A < t'_B$ always).

43.   **INTERPRET**   In this problem we want to use the "light box" to derive the time dilation formula given in Equation 33.3.

DEVELOP   The reference frame of the box S' is moving with speed v in the x direction relative to the frame S, which is at rest in Figure 33.6b. Let the S coordinates of event A be t_A and x_A, and those of event B be t_B and $x_B = x_A + v(t_B - t_A)$. To find $\Delta t'$, we apply the Lorentz transformation from S to S' (see Table 33.1).

EVALUATE   With $t' = \gamma\left(t - xv/c^2\right)$, we get

$$\Delta t' = t'_B - t'_A = \gamma\left[t_B - t_A - \left(x_B - x_A\right)v/c^2\right] = \gamma\left(t_B - t_A\right)\left(1 - v^2/c^2\right) = \Delta t\sqrt{1 - v^2/c^2}$$

which is Equation 33.3.

ASSESS   The equation shows that $\Delta t' < \Delta t$. That is, the time interval measured in the spaceship frame S' is shorter than that measured in S.

45.   **INTERPRET**   This problem involves time dilation. We are given the time limit of 75 years in the reference frame of the human who is traveling, and asked how fast she should go to reach a star 200 ly away as measured in the reference frame of the Earth.

DEVELOP   The distance is given in the system S, where the Earth and the star are essentially at rest, but the time interval is given in the system S' where the spacecraft is at rest. Thus, $\Delta x = 200 \text{ ly}$ and $\Delta t' = 75 \text{ y}$. Equations 33.3 and 33.4,

$$\Delta t' = \Delta t\sqrt{1 - v^2/c^2} = \frac{\Delta t}{\gamma}$$

$$\Delta x' = v\Delta t' = \Delta x\sqrt{1 - v^2/c^2}$$

for time dilation and Lorentz contraction, relate the given quantities to $\Delta x'$ and $\Delta t'$. We use the second expression (i.e., the expression for length contraction) to find

$$v = \frac{\Delta x}{\Delta t'}\sqrt{1-v^2/c^2}$$

$$= \frac{c}{\sqrt{1+\left(c\Delta t'/\Delta x\right)^2}}$$

EVALUATE Inserting the given values into the expression above for velocity gives

$$v = \frac{c}{\sqrt{1+\left(75\,c\cdot y/200\,c\cdot y\right)^2}} = 0.94c$$

ASSESS The time elapsed on the Earth may be found from Equation 33.3 for time dilation. The result is

$$\Delta t = \frac{\Delta t'}{\sqrt{1-v^2/c^2}} = \frac{75\,y}{\sqrt{1-(0.936)^2}} = 213\,y$$

Thus, none of her colleagues would be alive when she arrives at the star (not to mention that it would take another 200 years for her to signal her arrival to those on Earth!).

47. **INTERPRET** This is a problem about calculating the distance and time between two events, as measured in different reference frames. The first reference frame S is essentially stationary with respect to the Earth and the star, whereas the second reference frame S' is that of the spaceship moving at velocity v with respect to S.
DEVELOP We shall follow the Problem-Solving Strategy 33.1 for Lorentz transformation. In the Earth-star frame (system S), we choose $x_A = 0$ and $t_A = 0$. In system S', events A and B both occur at the spaceship, for which we can choose $x'_A = x'_B = 0$ and $t'_A = 0$.
EVALUATE (a) In system S, we are given $x_B - x_A = 10$ ly, so

$$\Delta t = t_B - t_A = \frac{x_B - x_A}{v} = \frac{10\,ly}{0.80c} = 13\,y$$

(b) In system S', we have $x'_B - x'_A = 0$. However, $\Delta t' = t'_B - t'_A = (t_B - t_A)/\gamma$ from time dilation (Equation 33.3), so $\Delta t' = (12.5\,y)\sqrt{1-(0.80)^2} = 7.5\,y$.
(c) For the space-time interval, one has

$$(\Delta s)^2 = c^2(\Delta t)^2 - (\Delta x)^2 = (12.5\,ly)^2 - (10\,ly)^2 = 56.25\,ly^2$$
$$(\Delta s')^2 = c^2(\Delta t')^2 - (\Delta x')^2 = (7.5\,ly)^2 - 0 = 56.25\,ly^2$$

as required by invariance.
ASSESS Our result shows that $\Delta t' < \Delta t$. In other words, the time interval measured in the spaceship frame S' is shorter than that measured in S. The space-time interval, however, remains the same in both reference frames; $\Delta s^2 = \Delta s'^2$.

49. **INTERPRET** This is a problem about the spacetime interval between two events. The events are connected by a light signal.
DEVELOP Choose the x axis along the line separating the positions of the events. Since A and B are connected by the passage of a light beam,

$$|\Delta x| = |x_B - x_A| = c|t_B - t_A| = c|\Delta t|$$

EVALUATE From Equation 33.6, one sees that the spacetime interval between them is zero:

$$(\Delta s)^2 = c^2(\Delta t)^2 - (\Delta x)^2 = 0$$

ASSESS An event with zero spacetime interval relative to A is said to lie on the light cone of A.

51. **INTERPRET** We're given the time and distance between two distant events observed in a particular reference frame S, and are to find the time and distance in another reference frame S' that is moving at the given speed with respect to S.
DEVELOP The coordinates of the events in S and S' are related by the Lorentz transformation in Table 33.1, with $v/c = 0.8$ and $\gamma = (1-v^2/c^2)^{-1/2} = 5/3$.

EVALUATE   (a) The distance between A and B measured by an observer in S' is

$$x'_B - x'_B = \gamma \left[x_B - x_A - v\left(t_B - t_A\right) \right]$$

$$= \frac{5}{3}\left[3.8 \text{ ly} - (0.80c)(1.6 \text{ y}) \right] = 4.2 \text{ ly}$$

(b) Similarly, the time between A and B measured by an observer in S' is

$$t'_A - t'_B = \gamma \left[t_B - t_A - \left(v/c^2\right)\left(x_B - x_A\right) \right]$$

$$= \frac{5}{3}\left[1.6 \text{ y} - (0.80/c)(3.8 \text{ ly}) \right] = -2.4 \text{ ly}$$

Thus, B occurs before A in S'.

ASSESS   Since the light travel time from the position of A to that of B is greater than the magnitude of the time difference (3.8 y versus 1.6 y in S, or 4.2 y versus 2.4 y in S'), the events are not causally connected.

53.   **INTERPRET**   We're given the kinetic energy of a proton and asked to find its speed and momentum.

DEVELOP   For the proton, $mc^2 = 938 \text{ MeV}$, so $K = (\gamma - 1)mc^2 = 500 \text{ MeV}$ implies

$$\gamma = 1 + \frac{K}{mc^2} = 1 + \frac{500 \text{ MeV}}{938 \text{ MeV}} = 1.53$$

EVALUATE   (a) Since $\gamma = \left(1 - v^2/c^2\right)^{-1/2}$, the speed of the proton is

$$v = c\sqrt{1 - 1/\gamma^2} = 0.758c$$

(b) Using Equation 33.7, we find the momentum to be

$$p = \gamma mv = (1.53)\left(938 \text{ MeV}/c^2\right)(0.758c) = 1.09 \text{ GeV}/c = 5.82 \times 10^{-19} \text{ kg} \cdot \text{m/s}$$

ASSESS   Since the kinetic energy is not negligible compared to the rest energy, the Newtonian expression of momentum ($p = mv$) is not applicable.

55.   **INTERPRET**   This is a problem about mass-energy conversion using $E = mc^2$.

DEVELOP   The energy-equivalent of 1 g is

$$E = mc^2 = \left(1 \times 10^{-3} \text{ kg}\right)\left(3 \times 10^8 \text{ m/s}\right)^2 = 9 \times 10^{13} \text{ J}$$

EVALUATE   This amount of energy could supply a large city, with a power consumption of 10^9 W, for a period of time

$$t = \frac{E}{P} = \frac{9 \times 10^{13} \text{ J}}{10^9 \text{ W}} = 9 \times 10^4 \text{ s} = 25 \text{ h}$$

ASSESS   This is an enormous amount of energy harnessed from just 1 g of raisin (or of any other matter).

57.   **INTERPRET**   We are asked about the kinetic energy of an electron, with its speed given. The relativistic formula is needed when v/c is not negligible.

DEVELOP   The kinetic energy of the electron is given by Equation 33.8:

$$K = \frac{m_e c^2}{\sqrt{1 - u^2/c^2}} - m_e c^2 = m_e c^2 \left[\frac{1}{\sqrt{1 - u^2/c^2}} - 1 \right]$$

where $m_e c^2 = 0.511 \text{ MeV}$ is the rest energy for an electron.

EVALUATE   (a) Since $u^2/c^2 = 1.0 \times 10^{-6} \ll 1$, we expand the square root and obtain

$$K \approx m_e c^2 \left(\frac{u^2}{2c^2} \right) = \frac{1}{2}(0.511 \text{ MeV})\left(1.0 \times 10^{-6}\right) = 0.26 \text{ eV}$$

(b) When $v/c = 0.60$, we use the exact expression above for kinetic energy. The result is

$$K = \frac{0.511 \text{ MeV}}{\sqrt{1 - (0.60)^2}} - 0.511 \text{ MeV} = 1.3 \text{ keV}$$

(c) Similarly, when $v/c = 0.99$,

$$K = \frac{0.511\,\text{MeV}}{\sqrt{1-(0.99)^2}} - 0.511\,\text{MeV} = 3.1\,\text{MeV}$$

ASSESS The Newtonian result ($K = m_e u^2/2$) is valid only when $u \ll c$.

59. **INTERPRET** In this problem we are to show that the kinetic energy in Equation 33.8 reduces to the Newtonian result $K = m_e c^2/2$ when $u \ll c$.

DEVELOP The binomial expansion valid for $|x| < 1$ is

$$(1+x)^p = 1 + px + \frac{p(p-1)}{2!}x^2 + \cdots$$

EVALUATE For $u/c \ll 1$, Equation 33.8 can be expanded to yield

$$K = mc^2 \left[\frac{1}{\sqrt{1-u^2/c^2}} - 1 \right] = mc^2 \left[1 + \frac{1}{2}\frac{u^2}{c^2} + \frac{3}{8}\left(\frac{u^2}{c^2}\right)^2 + \cdots - 1 \right] = \frac{1}{2}mu^2 \left(1 + \frac{3u^2}{4c^2} + \cdots \right) \approx \frac{1}{2}mu^2$$

ASSESS We indeed recover the Newtonian expression for kinetic energy when $u/c \ll 1$.

61. **INTERPRET** In this problem, we want to prove that the spacetime interval is relativistically invariant.

DEVELOP Consider two frames S and S' that are related by the Lorentz transformations of Table 33.1. (Since the equations are linear, they also apply to differences between coordinates.) Forming the spacetime interval in S' and transforming it using the Lorentz transformations gives

$$(\Delta s')^2 = c^2\Delta t'^2 - \Delta x'^2 - \Delta y'^2 - \Delta z'^2 = c^2\gamma^2\left(\Delta t - v\Delta x/c^2\right)^2 - \gamma^2\left(\Delta x - v\Delta t\right)^2 - \Delta y^2 - \Delta z^2$$

$$= \gamma^2\left(c^2\Delta t^2 - 2v\Delta x\Delta t + v^2\Delta x^2/c^2 - \Delta x^2 + 2v\Delta x\Delta t - v^2\Delta t^2\right) - \Delta y^2 - \Delta z^2$$

$$= \gamma^2\left(1 - v^2/c^2\right)\left(c^2\Delta t^2 - \Delta x^2\right) - \Delta y^2 - \Delta z^2 = c^2\Delta t^2 - \Delta x^2 - \Delta y^2 - \Delta z^2 = (\Delta s)^2$$

EVALUATE Therefore, $(\Delta s')^2 = (\Delta s)^2$, and the spacetime interval is invariant.

ASSESS The spacetime interval $(\Delta s)^2 = c^2(\Delta t)^2 - (\Delta x)^2$ describes the relationship between two events in a manner that is independent of the chosen reference frame.

63. **INTERPRET** In this problem we are asked to find the speed at which $K = mc^2$.

DEVELOP The kinetic energy of a particle is given by Equation 33.8:

$$K = \frac{mc^2}{\sqrt{1-u^2/c^2}} - mc^2 = mc^2\left(\frac{1}{\sqrt{1-u^2/c^2}} - 1\right) = mc^2(\gamma - 1)$$

When the kinetic energy is equal to the rest energy, $(\gamma - 1)mc^2 = mc^2$, or $\gamma = 2$.

EVALUATE From $\gamma = \left(1 - v^2/c^2\right)^{-1/2}$, the speed of the particle is

$$v = c\sqrt{1 - 1/\gamma^2} = c\sqrt{1 - 1/4} = c\frac{\sqrt{3}}{2} = 0.866c$$

ASSESS The speed of the particle is about $0.866c$ when $K = mc^2$. This is in the relativistic regime.

65. **INTERPRET** This problem explores the change in momentum when the speed of an object is changed.

DEVELOP The expression for momentum valid at any speed is given by Equation 33.7:

$$\vec{p} = \gamma m\vec{v}$$

Since $v_2 = 1.05v_1$ and $p_2 = 5p_1$, we write

$$\frac{v_2}{p_2} = \frac{1}{\gamma_2 m} = \frac{1.05v_1}{5p_1} = \frac{1.05}{5\gamma_1 m}$$

$$\frac{1.05}{\gamma_1} = \frac{5}{\gamma_2}$$

EVALUATE   To find the original speed, we rewrite the above equation as

$$(1.05)^2 \left(1 - v_1^2/c^2\right) = 5^2 \left(1 - v_2^2/c^2\right) = 5^2 \left[1 - (1.05)^2 v_1^2/c^2\right]$$

which yields

$$v = c\sqrt{\frac{25 - (1.05)^2}{24(1.05)^2}} = 0.95c$$

ASSESS   The particle must be moving at a relativistic speed. In the Newtonian limit where $v \ll c$, $p = mv$, when the speed increases by 5%, the corresponding momentum also must increase by 5%.

67.   **INTERPRET**   This problem is about the Doppler shift for light. Given a source moving toward us at a given speed, we are to derive the given expression for the Doppler-shifted frequency. For a source speed much, much less than the speed of light, we are to show that this result simplifies to the classical result (Equation 14.13).

DEVELOP   Consider the sketch below. Let S be the rest system of a source of light waves (with frequency and wavelength $\lambda f = c$) that moves with speed u towards an observer in S' (who measures $\lambda' f' = c$). Suppose that N waves are emitted in S in a time interval Δt. The first wavefront has traveled a distance $c\Delta t$ in S, so the wavelength (i.e., the distance between surfaces of constant phase) is $\lambda = c\Delta t/N$. In S', however, the wavefronts are "piled up" in a smaller distance due to the motion of S so the wavelength is

$$\lambda' = \frac{c\Delta t' - u\Delta t'}{N} = \frac{(c - u)\Delta t'}{N}$$

which gives

$$\frac{\lambda'}{\lambda} = \frac{(1 - u/c)\Delta t'}{\Delta t}$$

Now, Δt (the proper time interval in the source's rest system) is related to $\Delta t'$ (the time interval measured in a system where the source is moving) by time dilation, $\Delta t' = \gamma \Delta t$ (Equation 33.3 with altered notation), so $\lambda'/\lambda = \gamma(1 - u/c)$ or, in terms of frequency, $f/f' = \gamma(1 - u/c)$.

EVALUATE   Since $\gamma = 1/\sqrt{1 - u^2/c^2}$, this can be written as

$$\frac{f'}{f} = \frac{1 - u/c}{\sqrt{1 - u^2/c^2}} = \frac{\sqrt{(1 - u/c)^2}}{\sqrt{(1 - u/c)(1 + u/c)}} = \sqrt{\frac{1 + u/c}{1 - u/c}}$$

which is the radial Doppler shift (i.e., along the line of sight) in special relativity, with u positive for approach and negative for recession (note the difference in signs with Equation 14.13). For $u/c \ll 1$,

$$\sqrt{\frac{1 + u/c}{1 - u/c}} \approx \left(1 + \frac{1}{2}\frac{u}{c}\right)\left(1 - \frac{1}{2}\frac{-u}{c}\right) \approx 1 + \frac{u}{c}$$

which, allowing for the difference in signs, is the same as the limit of Equation 14.13 for $u/c \ll 1$.

ASSESS   Note that it is more customary to write this limit as

$$\frac{\Delta f}{f} = \frac{f' - f}{f} = \frac{u}{c}$$

The relativistic Doppler effect has been used to measure the shifts in frequency of light emitted by other moving galaxies. The observation of "red shift" suggests that galaxies are receding from us and that the Universe is still expanding.

69. **INTERPRET**   We shall use the relativistic momentum, and Newton's second law ($F = dp/dt$), to find an equation for the force on a particle in terms of its acceleration. We limit ourselves to acceleration parallel to the velocity.

DEVELOP   Equation 33.7 gives the relativistic momentum as $p = mu/\sqrt{1-u^2/c^2}$, and Newton's second law is $F = dp/dt$. Thus, we shall differentiate the relativistic momentum to find an expression for the force.

EVALUATE   Differentiating the momentum gives

$$F = \frac{dp}{dt} = \frac{dp}{du}\frac{du}{dt} = \left[\frac{m}{\sqrt{1-\frac{u^2}{c^2}}} + \frac{mu^2}{c^2\left(1-\frac{u^2}{c^2}\right)^{3/2}}\right]\frac{du}{dt}$$

$$= \frac{m}{\sqrt{1-\frac{u^2}{c^2}}}\left[1 + \frac{u^2}{c^2\left(1-\frac{u^2}{c^2}\right)}\right]\frac{du}{dt} = \gamma m\left[1 + \left(\frac{\gamma u}{c}\right)^2\right]a$$

ASSESS   This is quite a bit more complicated than $F = ma$, but for $u \ll c$, $\gamma = 1$ and the term in square brackets reduces to unity, which gives $F = ma$, as expected.

71. **INTERPRET**   You're trying to understand a television tube by the relativistic effects on the electrons that travel through it.

DEVELOP   You're given the length of the tube in the electrons' reference frame, which is shorter than what you measure in the rest-frame due to length contraction: $\Delta x' = \Delta x\sqrt{1-v^2/c^2}$. From this you can determine the electrons' speed, as well as the kinetic energy: $K = (\gamma-1)mc^2$ (Equation 33.8), where the relativistic factor is $\gamma = 1/\sqrt{1-v^2/c^2}$. To achieve this kinetic energy, you'll need a voltage difference given by $K = e\Delta V$.

EVALUATE   Solving for the speed from the provided lengths gives

$$v = c\sqrt{1-\left(\frac{\Delta x'}{\Delta x}\right)^2} = c\sqrt{1-\left(\frac{57\text{ cm}}{60\text{ cm}}\right)^2} = 0.312c \approx 0.31c$$

The required voltage difference is

$$\Delta V = \frac{1}{e}(1-\gamma)mc^2 = \frac{1}{e}\left(1 - \frac{1}{\sqrt{1-0.312^2}}\right)(511\text{ keV}) = 27\text{ kV}$$

ASSESS   The voltage difference is typical for a television tube. Transformers (recall Chapter 28) are needed to increase the voltage from the wall outlet to several tens of kilovolts.

73. **INTERPRET**   You consider possible relativistic effects on a high-speed interstellar voyage.

DEVELOP   From the Earth's perspective, the voyage will be completed in a time of $\Delta t = \Delta x/v$, where Δx is the distance to Proxima Centauri as measured from Earth. On the spacecraft, this time will be dilated: $\Delta t' = \Delta t\sqrt{1-v^2/c^2}$.

EVALUATE   By journey's end, you will have aged by

$$\Delta t' = \frac{\Delta x}{v}\sqrt{1-v^2/c^2} = \frac{4\text{ ly}}{0.8c}\sqrt{1-0.8^2} = 3\text{ y}$$

The answer is (a).

ASSESS   From Earth's perspective, you will have aged 5 years.

75. **INTERPRET**   You consider possible relativistic effects on a high-speed interstellar voyage.

DEVELOP   The distance from Earth to the star will be length contracted in your reference frame: $\Delta x' = \Delta x\sqrt{1-v^2/c^2}$.

EVALUATE   Plugging in the values, you find

$$\Delta x' = (4\text{ ly})\sqrt{1-0.8^2} = 2.4\text{ ly}$$

The answer is (a).

ASSESS   This agrees with the result from Problem 33.73, since this is the distance your ship will have traveled in 3 years time.

PARTICLES AND WAVES

EXERCISES

Section 34.2 Blackbody Radiation

15. **INTERPRET**   This is a problem about blackbody radiation. We want to explore the connection between temperature and the radiated power.

    DEVELOP   From the Stefan-Boltzmann law (Equation 34.1), $P = \sigma A T^4$, we see that the total radiated power, or luminosity, of a blackbody is proportional to T^4.

    EVALUATE   Doubling the absolute temperature increases the luminosity by a factor of $2^4 = 16$.

    ASSESS   A blackbody is a perfect absorber of electromagnetic radiation. As the temperature of the blackbody increases, its radiated power also goes up.

17. **INTERPRET**   We are given the temperature of a blackbody (i.e., the Earth) and asked to find the wavelengths that correspond to peak radiance and median radiance.

    DEVELOP   The wavelength at which a blackbody at a given temperature radiates the maximum power is given by Wien's displacement law (Equation 34.2a):

    $$\lambda_{peak} T = 2.898 \text{ mm} \cdot \text{K}$$

    Similarly, the median wavelength, below and above which half the power is radiated, is given by Equation 34.2b:

    $$\lambda_{median} T = 4.11 \text{ mm} \cdot \text{K}$$

    EVALUATE   Using the above formulas, we obtain

    $$\lambda_{peak} = \frac{2.898 \text{ mm} \cdot \text{K}}{288 \text{ K}} = 10.1 \text{ } \mu\text{m}$$

    $$\lambda_{median} = \frac{4.11 \text{ mm} \cdot \text{K}}{288 \text{ K}} = 14.3 \text{ } \mu\text{m}$$

    ASSESS   The wavelengths are in the infrared. Note that $\lambda_{median} > \lambda_{peak}$.

19. **INTERPRET**   We are to find the wavelength for the peak radiance of solar blackbody radiation, and the median wavelength. In both cases, we'll use the per-unit-wavelength basis; Equations 34.2a and 34.2b.

    DEVELOP   Wien's law (Equation 34.2a) gives us the peak wavelength: $\lambda_{peak} T = 2.898 \text{ mm} \cdot \text{K}$. The median wavelength is given by Equation 34.2b: $\lambda_{median} T = 4.11 \text{ mm} \cdot \text{K}$. The temperature of the Sun is $T = 5800$ K, so we can use these equations to solve for the respective wavelengths.

    EVALUATE   Inserting the temperature gives

    $$\text{(a) } \lambda_{peak} = \frac{2.898 \text{ mm} \cdot \text{K}}{5800 \text{ K}} = 5.000 \times 10^{-4} \text{ mm} = 500.0 \text{ nm}$$

    $$\text{(b) } \lambda_{median} = \frac{4.11 \text{ mm} \cdot \text{K}}{5800 \text{ K}} = 7.086 \times 10^{-4} \text{ mm} = 708.6 \text{ nm}$$

    ASSESS   The peak wavelength is near the center of the visible spectrum (green) and the median wavelength is just beyond the visible in the near-infrared region.

Section 34.3 Photons

21. **INTERPRET**   We're asked to express the range of human eye sensitivity in terms of photon energies.

DEVELOP   The photon energy is given by Equation 34.6: $E = hf$, or in terms of wavelength: $E = hc / \lambda$.

EVALUATE   The limits of human eye sensitivity are

$$E_{min} = \frac{hc}{\lambda_{max}} = \frac{1240 \text{ eV} \cdot \text{nm}}{700 \text{ nm}} = 1.8 \text{ eV}$$

$$E_{max} = \frac{hc}{\lambda_{min}} = \frac{1240 \text{ eV} \cdot \text{nm}}{400 \text{ nm}} = 3.1 \text{ eV}$$

ASSESS   We've used the common shorthand of $hc = 1240 \text{ eV} \cdot \text{nm}$.

23. **INTERPRET**   The problem asks for a comparison of the power output by a red laser and a blue laser. The lasers emit photons at the same rate, but the photon energy of each laser is different.

DEVELOP   Using $\lambda = c/f$ and Equation 34.6, $E = hf$, the ratio of the photon energies is

$$\frac{E_{blue}}{E_{red}} = \frac{f_{blue}}{f_{red}} = \frac{\lambda_{red}}{\lambda_{blue}}$$

EVALUATE   Using the above equation, the ratio of the energies is

$$\frac{E_{blue}}{E_{red}} = \frac{\lambda_{red}}{\lambda_{blue}} = \frac{650 \text{ nm}}{450 \text{ nm}} = 1.44$$

Since the lasers emit photons at the same rate, this is also the ratio of their power outputs. Thus the power of the blue laser is 1.44 times that of the red laser.

ASSESS   Blue lasers, with shorter wavelength (higher frequency), are more energetic than red lasers.

Section 34.4 Atomic Spectra and the Bohr Atom

25. **INTERPRET**   This problem is about the energy levels of a hydrogen atom using the Bohr model. We are interested in the wavelengths of the first three lines in the Lyman series.

DEVELOP   The wavelength can be calculated using Equation 34.9:

$$\frac{1}{\lambda} = R_H \left(\frac{1}{n_2^2} - \frac{1}{n_1^2} \right)$$

where $R_H = 1.097 \times 10^7 \text{ m}^{-1}$ is the Rydberg constant and $n_2 = 1$ for the Lyman series.

EVALUATE   The first three lines correspond to $n_1 = 2, 3,$ and $4,$ and the wavelengths are, respectively,

$$\lambda = \frac{1}{R_H} \frac{n_1^2 n_2^2}{n_1^2 - n_2^2} = \frac{1}{R_H} \frac{n_1^2}{n_1^2 - 1} = 122 \text{ nm, } 103 \text{ nm, and } 97.2 \text{ nm}$$

Note that $R_H^{-1} = (0.01097)^{-1} \text{ nm} = 91.2 \text{ nm},$ which is the Lyman series limit.

ASSESS   The wavelengths are less than 400 nm. Therefore, the Lyman spectral lines are in the ultraviolet regime.

27. **INTERPRET**   This problem is about the ionization energy of a hydrogen atom in its ground state. We want to find the wavelength that corresponds to a photon carrying this much energy.

DEVELOP   The energy of the ground state of hydrogen is given by Equation 34.12b (with $n = 1$): $E_1 = -13.6 \text{ eV}.$ Therefore, the ionization energy is $E_I = |E_1| = 13.6 \text{ eV}$ (the subscript "I" is for ionization). For a photon whose wavelength is λ, the energy it carries is (Equation 34.6) $E = hf = hc/\lambda$.

EVALUATE   A photon with energy $E_I = 13.6 \text{ eV}$ has wavelength

$$\lambda = \frac{hc}{E_I} = \frac{1240 \text{ eV} \cdot \text{nm}}{13.6 \text{ eV}} = 91.2 \text{ nm}$$

ASSESS   This is the same as the Lyman series limit (Equation 34.9 with $n_2 = 1$ and $n_1 = \infty$) $R_H^{-1} = hc/13.6 \text{ eV},$ and lies in the ultraviolet.

Section 34.5 Matter Waves

29. **INTERPRET** In this problem, we are asked to find the de Broglie wavelength of the Earth orbiting the Sun and an electron moving at the given speed.

    DEVELOP For nonrelativistic momentum, Equation 34.14 becomes

    $$\lambda = \frac{h}{p} = \frac{h}{mv}$$

    EVALUATE (a) Using the orbital speed given and the Earth's mass from Appendix E gives

    $$\lambda = \frac{h}{mv} = \frac{6.626 \times 10^{-34} \text{ J} \cdot \text{s}}{\left(5.97 \times 10^{24} \text{ kg}\right)\left(30 \text{ km/s}\right)} = 3.7 \times 10^{-63} \text{ m}$$

    (b) For the given electron,

    $$\lambda = \frac{h}{mv} = \frac{6.626 \times 10^{-34} \text{ J} \cdot \text{s}}{\left(9.11 \times 10^{-32} \text{ kg}\right)\left(10 \times 10^{3} \text{ m/s}\right)} = 73 \text{ nm}$$

    ASSESS The Earth's de Broglie wavelength is much smaller than the smallest physically meaningful distance.

31. **INTERPRET** The problem asks what relative speed must an electron have in order to have the same de Broglie wavelength as a proton.

    DEVELOP Since $\lambda = h/p$ (Equation 34.14), the same de Broglie wavelength means the same momentum.

    EVALUATE At non-relativistic speeds, the equal momenta implies $m_p v_p = m_e v_e$, or

    $$\frac{v_e}{v_p} = \frac{m_p}{m_e} = \frac{1.672 \times 10^{-27} \text{kg}}{9.109 \times 10^{-31} \text{kg}} = 1836$$

    This says that the electron will need to be moving 1836 times faster than the proton.

    ASSESS The problem would be more complicated if the momenta were relativistic: $p = \gamma m v$ (Equation 33.7).

Section 34.6 The Uncertainty Principle

33. **INTERPRET** We want to find the minimum uncertainty in the velocity of a proton, given the uncertainty in its position.

    DEVELOP To find Δv, use the uncertainty principle, $\Delta x \Delta p \geq \hbar$ (Equation 34.15) with $\Delta p = m\Delta v$ and $\Delta x = 1$ fm.

    EVALUATE The above equation gives

    $$\Delta v = \frac{\Delta p}{m} \geq \frac{\hbar}{m\Delta x} = \frac{\left(197.3 \text{ MeV} \cdot \text{fm}/c\right)}{\left(938 \text{ MeV}\right)\left(1 \text{ fm}\right)} = 0.21c = 6 \times 10^{7} \text{ m/s}$$

    ASSESS The quantity $\Delta p = \hbar/\Delta x = 197.3$ MeV/c is barely small enough compared to $mc = 938$ MeV/c to justify using the nonrelativistic relation $p = mv$, but this is good enough for the purpose of approximation.

35. **INTERPRET** In this problem, we want to find the uncertainty in the position of a proton given the uncertainty in its velocity.

    DEVELOP To find Δx, we use the uncertainty principle, $\Delta x \Delta p \geq \hbar$ (Equation 34.15), where $\Delta p = m\Delta v$. We take the uncertainty in velocity to be the full range of variation given; that is, $\Delta v = 0.25 \text{ m/s} - \left(-0.25 \text{ m/s}\right) = 0.50 \text{ m/s}$.

    EVALUATE The position uncertainty of the proton is

    $$\Delta x \geq \frac{\hbar}{m \Delta v} = \frac{1.055 \times 10^{-34} \text{ J} \cdot \text{s}}{\left(1.67 \times 10^{-27} \text{ kg}\right)\left(0.50 \text{ m/s}\right)} = 130 \text{ nm}$$

    ASSESS The smaller the uncertainty Δv in velocity, the greater the uncertainty Δx in position.

37.   **INTERPRET**   The neutron is confined in the uranium nucleus with Δx equal to the diameter of the nucleus. We are to find the minimum energy of the neutron using the uncertainty principle.

DEVELOP   Using the same reasoning as given in Example 34.6, for a neutron $(mc^2 = 940 \text{ MeV})$ confined to a uranium nucleus $(\Delta x \approx 15 \text{ fm})$, the uncertainty principle requires that

$$K = \frac{p^2}{2m} \geq \frac{1}{2m}\left(\frac{\hbar}{2\Delta x}\right)^2$$

EVALUATE   From the above equation, we find the minimum kinetic energy to be

$$K_{min} = \frac{1}{2m}\left(\frac{\hbar}{2\Delta x}\right)^2 = \frac{1}{2mc^2}\left(\frac{\hbar c}{2\Delta x}\right)^2 = \frac{1}{2(940 \text{ MeV})}\left(\frac{197.3 \text{ MeV} \cdot \text{fm}}{2(15 \text{ fm})}\right)^2 = 23 \text{ keV}$$

ASSESS   This is smaller than the 5 MeV estimated for the nucleon in Example 34.6 by a factor of $15^2 = 225$ because Δx is 15 times larger. Most estimates of nuclear energies for single-particle states, based on the uncertainty principle, give values of the order of 1 MeV, consistent with experimental measurements.

PROBLEMS

39.   **INTERPRET**   We are given the temperature of the Sun, which we shall treat as a blackbody, and asked to compare its radiance at two different wavelengths.

DEVELOP   The radiance of a blackbody is given by Equation 34.4:

$$R(\lambda, T) = \frac{2\pi hc^2}{\lambda^5\left(e^{hc/\lambda kT} - 1\right)}$$

This equation allows us to compare the radiance at two different wavelengths.

EVALUATE   From the above equation (also see Example 34.1), the ratio of the blackbody radiances for the two given wavelengths is

$$\frac{R(\lambda_2, T)}{R(\lambda_1, T)} = \left(\frac{\lambda_1}{\lambda_2}\right)^5\left(\frac{e^{hc/\lambda_1 kT} - 1}{e^{hc/\lambda_2 kT} - 1}\right) = \left(\frac{5}{2}\right)^5\left(\frac{146.9}{2.66 \times 10^5}\right) = 5.4 \times 10^{-2}$$

where $\lambda_1 = 500 \text{ nm}$, $\lambda_2 = 200 \text{ nm}$, $T = 5800 \text{ K}$, and $hc/k = 1.449 \times 10^{-2} \text{ m} \cdot \text{K}$.

ASSESS   The characteristic radiance as a function of wavelength is shown in Figure 34.2. For a given wavelength, the radiance increases with temperature.

41.   **INTERPRET**   This problem is about blackbody radiation. We are given the wavelengths that correspond to peak radiance and asked to find the temperature of the blackbody. We also want to compare the radiance at two different wavelengths.

DEVELOP   The wavelength at which a blackbody at a given temperature radiates the maximum power is given by Wien's displacement law (Equation 34.2a): $\lambda_{peak} T = 2.898 \text{ mm} \cdot \text{K}$. For part **(b)**, to compare the radiance at two different wavelengths, we use Equation 34.4:

$$R(\lambda, T) = \frac{2\pi hc^2}{\lambda^5\left(e^{hc/\lambda kT} - 1\right)}$$

EVALUATE   **(a)** Equation 34.2a gives

$$T = \frac{2.898 \text{ mm} \cdot \text{K}}{\lambda_{peak}} = \frac{2.898 \text{ mm} \cdot \text{K}}{660 \text{ nm}} = 4.39 \times 10^3 \text{ K}$$

(b) With $hc/(kT) = 3.30 \ \mu\text{m}$, the ratio of the radiances is

$$\frac{R(\lambda = 400 \text{ nm})}{R(\lambda = 700 \text{ nm})} = \left(\frac{700}{400}\right)^5\left(\frac{e^{3.30/0.7} - 1}{e^{3.30/0.4} - 1}\right) = 0.474$$

ASSESS   The wavelengths considered are in the visible spectrum. Note that the characteristic radiance as a function of wavelength is shown in Figure 34.2. For a given wavelength, the radiance increases with temperature.

43.   **INTERPRET**   We are given the power output at various frequencies and asked to find the rate of photon emission.

DEVELOP   The rate dN/dt of photon emission is the electromagnetic power output divided by the photon energy:

$$\frac{dN}{dt} = \frac{P}{E} = \frac{P}{hf} = \frac{P\lambda}{hc}$$

where we have used Equation 34.6 $E = hf$.

EVALUATE   **(a)** For the antenna, the rate is

$$\frac{dN}{dt} = \frac{P}{hf} = \frac{1.0 \text{ kW}}{(6.626\times10^{-34} \text{ J} \cdot \text{s})(89.5 \text{ MHz})} = 1.7\times10^{28} \text{ s}^{-1}$$

(b) For the laser, we have

$$\frac{dN}{dt} = \frac{P\lambda}{hc} = \frac{(1.0 \text{ mW})(633 \text{ nm})}{(6.626\times10^{-34} \text{ J} \cdot \text{s})(3.00\times10^{8} \text{ m/s})} = 3.2\times10^{15} \text{ s}^{-1}$$

(c) Similarly, for the X-ray machine, the rate is

$$\frac{dN}{dt} = \frac{P\lambda}{hc} = \frac{(2.5 \text{ kW})(0.10 \text{ nm})}{(6.626\times10^{-34} \text{ J} \cdot \text{s})(3.00\times10^{8} \text{ m/s})} = 1.3\times10^{18} \text{ s}^{-1}$$

ASSESS   For a general device at a given power output, the rate of photon production decreases with the energy of the photon; the more energetic the photons, the smaller the rate of production because each photon carries more energy.

45.   **INTERPRET**   This problem is about the photoelectric effect. We want to find the cutoff frequency and the maximum energy of electrons ejected by shining light with the given frequency on copper.

DEVELOP   At the cutoff frequency, $K_{max} = 0$ and the photon energy equals the work function, $\phi = hf_{cutoff}$ (see Equation 34.7), which we can find in Table 34.1. For part (b), apply Equation 34.7 to find the maximum kinetic energy possible for the given frequency f

$$K_{max} = hf - \phi$$

EVALUATE   **(a)** The work function of copper is $\phi_{Cu} = 4.65 \text{ eV}$. Therefore, the cutoff frequency is

$$f_{cutoff} = \frac{\phi_{Cu}}{h} = \frac{4.65 \text{ eV}}{4.136\times10^{-15} \text{ eV} \cdot \text{s}} = 1.12\times10^{15} \text{ Hz}$$

(b) The maximum kinetic energy of the ejected electrons is

$$K_{max} = hf - \phi = (4.136\times10^{-15} \text{ eV} \cdot \text{s})(1.8\times10^{15} \text{ Hz}) - 4.65 \text{ eV} = 2.79 \text{ eV}$$

ASSESS   Upon illuminating copper with photons at 7.44 eV, it takes 4.65 eV to overcome the work function of copper, leaving the electrons with 2.79 eV of kinetic energy.

47.   **INTERPRET**   We are asked to explain why plants are green using the absorption peaks in the chlorophyll molecule.

DEVELOP   We can convert the wavelength peaks into energy peaks using $E = hc/\lambda$, and the shorthand $hc = 1240 \text{ eV} \cdot \text{nm}$.

EVALUATE   **(a)** The energy peaks in chlorophyll's absorption spectrum are at

$$E_1 = \frac{1240 \text{ eV} \cdot \text{nm}}{430 \text{ nm}} = 2.9 \text{ eV}$$

$$E_2 = \frac{1240 \text{ eV} \cdot \text{nm}}{662 \text{ nm}} = 1.9 \text{ eV}$$

(b) These absorption peaks correspond to blue and red wavelengths, near the limits of the human visible range. The light that is not absorbed is reflected, and this is what we humans observe. Since the reflected light is primarily in the green region of the visible spectrum between blue and red, we perceive plants to be green.

ASSESS   Plants also reflect a lot of infrared light with wavelengths longer than 700 nm.

49.   **INTERPRET**   This problem is about the photoelectric effect. We are given the maximum speed of electrons ejected from potassium and asked to find the wavelength of the light that ejected the electrons.

DEVELOP   The maximum speed of the ejected electrons is related to the wavelength of the light by Einstein's photoelectric effect equation (Equation 34.7):

$$K_{max} = \frac{1}{2}mv_{max}^2 = hf - \phi = \frac{hc}{\lambda} - \phi$$

We shall use this equation to find λ.

EVALUATE   The maximum kinetic energy of the electron is

$$K_{max} = \frac{1}{2}mv_{max}^2 = \frac{1}{2}\left(mc^2\right)\left(\frac{v_{max}}{c}\right)^2 = \frac{1}{2}(511\,keV)\left(\frac{4.2\times10^5\ m/s}{3.00\times10^8\,m/s}\right) = 0.501\,eV$$

From Equation 34.7 and using Table 34.1 to find the work function ϕ, we find the wavelength to be

$$\lambda = \frac{hc}{K_{max}+\phi} = \frac{1240\ eV\cdot nm}{0.501\ eV + 2.30\ eV} = 440\ nm$$

to two significant figures.

ASSESS   Strictly speaking, the result should be reported as 44×10^1 nm to make the significant figures more obvious. The cutoff wavelength of potassium is

$$\lambda_{cutoff} = hc/\beta = (1240\ eV\cdot nm)/(2.30\ eV) = 539\ nm$$

For the photoelectric effect to take place, we require $\lambda \le \lambda_{cutoff}$.

51.   **INTERPRET**   This problem is about Compton scattering of a photon with an electron. We are interested in the wavelength of the scattered photon and the kinetic energy of the electron.

DEVELOP   The Compton shift of wavelength is given by Equation 34.8:

$$\Delta\lambda = \frac{h}{mc}(1-\cos\theta) = \lambda_C(1-\cos\theta)$$

where $\lambda_C = h/(mc) = 2.43$ pm is the Compton wavelength of the electron. By conservation of energy, the kinetic energy of the scattered electron is equal to the energy lost by the photon.

EVALUATE   **(a)** From Equation 34.8, the wavelength of the scattered photon is

$$\lambda' = \lambda + \Delta\lambda = 150\ pm + 2.43\ pm\left[1-\cos(135°)\right] = 150\ pm + 4.15\ pm = 154\ pm$$

(b) The kinetic energy of the scattered electron is

$$K = E - E' = \frac{hc}{\lambda} - \frac{hc}{\lambda'} = \frac{hc\Delta\lambda}{\lambda\lambda'} = \frac{(1240\ eV\cdot nm)(4.15\ pm)}{(150\ pm)(154\ pm)} = 222\ eV$$

ASSESS   For X rays, the wavelength is in the range 0.01–10 nm, so the detection of the Compton shift in X rays is difficult.

53.   **INTERPRET**   This problem involves the photoelectric effect. We are given enough information to find the work function of the material, and are asked if this material will emit electrons when illuminated with radiation at a longer wavelength and, if so, what will be the maximum electron energy.

DEVELOP   To determine whether or not the photoelectric effect can occur, we shall first find the work function from Einstein's photoelectric effect equation (Equation 34.7). Using the data for the blue light, this gives the work function of the photocathode material as

$$\phi = E - K_{max} = \frac{hc}{\lambda} - K_{max} = \frac{1240\ eV\cdot nm}{430\ nm} - 0.85\ eV = 2.03\ eV$$

EVALUATE   The energy of a photon of the red light is only

$$\frac{hc}{\lambda} = \frac{1240\ eV\cdot nm}{633\ nm} = 1.96\ eV$$

and therefore is insufficient to eject photoelectrons.

ASSESS   A wavelength of 633 nm is greater than the cutoff wavelength of $\lambda_{cutoff} = hc/\phi = 610$ nm for the photocathode material.

55. **INTERPRET** This problem is about the wavelength and energy of the photon emitted when a Rydberg hydrogen atom undergoes a transition.

DEVELOP The wavelength of the photon can be calculated using Equation 34.9:

$$\frac{1}{\lambda} = R_H \left(\frac{1}{n_2^2} - \frac{1}{n_1^2} \right)$$

where $R_H = 1.097 \times 10^7$ m^{-1} is the Rydberg constant. Once we know the wavelength, the energy of the photon may be found using $E = hf = hc/\lambda$.

EVALUATE (a) The above equation gives

$$\lambda = \frac{1}{R_H} \frac{n_1^2 n_2^2}{n_1^2 - n_2^2} = \left(\frac{1}{1.097 \times 10^7 \text{ m}^{-1}} \right) \frac{(179)^2 (180)^2}{(180)^2 - (179)^2} = 26.4 \text{ cm}$$

(b) The energy of the emitted photon is

$$E = \frac{hc}{\lambda} = \frac{1.24 \times 10^{-6} \text{ eV} \cdot \text{m}}{0.264 \text{ m}} = 4.70 \text{ } \mu\text{eV}$$

ASSESS The long wavelength corresponds to the radio region of the electromagnetic spectrum.

57. **INTERPRET** The hydrogen atom undergoing a downward transition emits a photon. We are interested in the original state of the atom, given the energy of the photon and the quantum number of the final state of the atom.

DEVELOP The wavelength of the photon can be calculated using Equation 34.9:

$$\frac{1}{\lambda} = R_H \left(\frac{1}{n_2^2} - \frac{1}{n_1^2} \right)$$

where $R_H = 1.097 \times 10^7$ m^{-1} is the Rydberg constant. Once we know the wavelength, the energy of the photon may be found using Equation 34.6:

$$E = hf = \frac{hc}{\lambda} = hcR_H \left(\frac{1}{n_2^2} - \frac{1}{n_1^2} \right)$$

EVALUATE Solving for n_1 gives

$$n_1 = \left[n_2^{-2} - E/(hcR_H) \right]^{-1/2} = [(225)^{-2} - (9.32 \text{ } \mu\text{eV}/(13.6 \text{ eV})]^{-1/2} = 229$$

Note that $hcR_H = 13.6$ eV is the ionization energy.

ASSESS The wavelength of the emitted photon is 0.133 m, which falls into the radio wave spectrum.

59. **INTERPRET** This problem involves the Bohr model of the atom. We are to find the ionization energy of a hydrogen atom in its first excited state.

DEVELOP Using Equation 34.12b, the energy of the first excited state ($n = 2$) is

$$E_2 = \frac{-13.6 \text{ eV}}{2^2} = -3.40 \text{ eV}$$

whereas an ionized atom (with zero electron kinetic energy) has energy zero.

EVALUATE Thus, we must supply an energy E_I such that

$$0 = E_I + (-3.40 \text{ eV})$$
$$E_I = 3.40 \text{ eV}$$

ASSESS The ionization energy here is only ¼ of the case where the atom is in the ground state. The higher the value of n, the smaller the ionization energy because the electron is less tightly bound to the nucleus.

61. **INTERPRET** He$^+$ is a hydrogen-like atom with a nuclear charge $+2e$. We are to apply the Bohr model to this system to find the ground-state electron radius and the energy difference between the $n = 2$ and $n = 1$ state.

    DEVELOP Modifying the treatment of the Bohr atom in the text (see derivation of Equations 34.11 and 34.12) for singly ionized helium (He$^+$) by replacing the nuclear charge with $2e$, one gets $r_n = -2ke^2/(2E_n)$ and $E_n = -(2ke^2)^2 m/(2\hbar^2 n^2)$. Thus,

    $$E_n = -4\frac{ke^2}{2n^2 a_0} = -2^2\left(\frac{13.6 \text{ eV}}{n^2}\right)$$

    and

    $$r_n = \frac{n^2 a_0}{2}$$

    EVALUATE (a) The radius of the ground state of He$^+$ is

    $$r_1 = \frac{a_0}{2} = \frac{0.0529 \text{ nm}}{2} = 0.0265 \text{ nm}$$

    (b) The energy released in the transition from $n = 2$ to $n = 1$ is

    $$\Delta E = 4(13.6 \text{ eV})\left(1 - \frac{1}{4}\right) = 40.8 \text{ eV}$$

    Note that there is also a small change in m for helium, different from the correction in hydrogen, for the motion of the nucleus.

    ASSESS In general, replacing the nuclear charge with Ze gives results for any one-electron Bohr atom, where Z is the number of protons for the atom (i.e., the atomic number).

63. **INTERPRET** The resolution of a microscope depends on the wavelength used. A smaller de Broglie wavelength will improve the resolution of an electron microscope. We are to find the minimum electron speed that will make its de Broglie wavelength less than 450 nm.

    DEVELOP The resolution of the electron microscope is better than the optical microscope with 450-nm light if the de Broglie wavelength λ of the electrons is less than 450 nm. Thus,

    $$\lambda = \frac{h}{p} < 450 \text{ nm}$$

    Since $p = mv$ (for nonrelativistic electrons), the above condition allows us to obtain the minimum electron speed.

    EVALUATE The above inequality gives

    $$v > \frac{h}{m(450 \text{ nm})} = \frac{6.626 \times 10^{-34} \text{ J} \cdot \text{s}}{(9.11 \times 10^{-31} \text{ kg})(450 \text{ nm})} = 1.62 \text{ km/s}$$

    So the minimum speed is 1.62 km/s.

    ASSESS Since $450 \text{ nm} \gg \lambda_c = h/mc \doteq 2.43 \text{ pm}$ (Compton wavelength of the electron), our use of the nonrelativistic momentum was justified. The electron microscope can provide resolutions down to about 1 nm and magnifications of 10^6.

65. **INTERPRET** This problem involves the uncertainty principle. We want to find the minimum velocity of an electron based on its uncertainty in position.

    DEVELOP Using the uncertainty principle given in Equation 34.15 with $\Delta x = 20$ nm (the width of the well), we have

    $$\Delta p \geq \frac{\hbar}{\Delta x} = \frac{197.3 \text{ eV} \cdot \text{nm}/c}{20 \text{ nm}} = 9.87 \text{ eV}/c$$

    This is small compared to $mc = 511 \text{ keV}/c$, so nonrelativistic formulas are sufficient (see Example 34.6). Therefore, $\Delta v = \Delta p/m = 2v$.

    EVALUATE The above conditions lead to

    $$v = \frac{\Delta p}{2m} \geq \frac{9.87 \text{ eV}/c}{2(511 \text{ keV}/c^2)} = (9.65 \times 10^{-6})(3.00 \times 10^8 \text{ m/s}) = 2.9 \text{ km/s}$$

Thus, the minimum speed is $v_{min} = 2.9$ km/s.

ASSESS   Quantum wells have important applications in the field of semiconductor fabrication. Note that the result is given to two significant digits, as warranted by the data.

67.   **INTERPRET**   This problem involves energy-time uncertainty. We are interested in the minimum measurement time needed to measure the energy with the desired precision.

DEVELOP   The electron is nonrelativistic ($v/c = 1/300 \ll 1$), so we can use the nonrelativistic expression $K = mv^2/2$ for kinetic energy. The desired uncertainty in the kinetic energy is

$$\Delta E = 2(0.01\%) \times \left(\frac{1}{2}mv^2\right) = 10^{-4} mc^2 \left(\frac{v}{c}\right)^2 = 10^{-4}(511 \text{ keV})\left(\frac{1}{300}\right)^2 = 5.68 \times 10^{-4} \text{ eV}$$

The minimum time can then be calculated using Equation 34.16, $\Delta E \Delta t \geq \hbar$.

EVALUATE   An energy measurement of this precision requires a time

$$\Delta t \geq \frac{\hbar}{\Delta E} = \frac{6.582 \times 10^{-16} \text{ eV} \cdot \text{s}}{5.68 \times 10^{-4} \text{ eV}} = 1 \text{ ps}$$

to a single significant figure.

ASSESS   The energy-time uncertainty principle implies that the minimum measurement time must necessarily go up in order to achieve a greater accuracy in energy measurement.

69.   **INTERPRET**   In this problem, we want to show that if a photon's wavelength is equal to a particle's Compton wavelength, then the photon's energy is equal to the particle's rest energy.

DEVELOP   From Equation 34.8, we see that the Compton wavelength of a particle is $\lambda_C = h/mc$. The rest energy of a particle is $E = mc^2$ (see discussion preceding Equation 33.9). From Equation 34.6, we see that the energy of a photon is $E = hf = hc/\lambda$.

EVALUATE   When the wavelength of the photon is $\lambda = \lambda_C$, its energy is

$$E = \frac{hc}{\lambda_C} = \frac{hc}{h/(mc)} = mc^2$$

which is the same as the particle's rest energy.

ASSESS   This result is not surprising because photons have zero rest mass, so their energy is completely kinetic. Have you ever seen a photon that is not moving at the speed of light?

71.   **INTERPRET**   This problem involves Compton scattering of a photon off an electron that is initially at rest (zero kinetic energy). We are to find an expression for the initial photon energy given the final total energy (kinetic energy plus rest energy) of the electron.

DEVELOP   For Compton scattering at 90°, Equation 34.8 reduces to $\lambda = \lambda_0 + \lambda_C$. In terms of the photon energy (Equation 34.6) $E = hf = hc/\lambda$ and the electron's Compton wavelength [Equation 34.8, $\lambda_C = hc/(m_e c^2)$], this can be written as

$$\frac{1}{E} = \frac{1}{E_0} + \frac{1}{m_e c^2}$$

or

$$E = \frac{E_0 m_e c^2}{E_0 + m_e c^2}$$

The recoil electron's kinetic energy is

$$K_e = (\gamma - 1)m_e c^2 = E_0 - E = E_0 - \frac{E_0 m_e c^2}{E_0 + m_e c^2} = \frac{E_0^2}{E_0 + m_e c^2}$$

This is a quadratic equation in E_0, namely $E_0^2 - (\gamma - 1)mc^2 (E_0 + m_e c^2) = 0$. The positive solution corresponds to the initial photon energy that we seek.

EVALUATE   The positive solution for E_0 is

$$E_0 = \frac{1}{2}\left[(\gamma - 1)m_e c^2 + \sqrt{(\gamma - 1)^2 m_e^2 c^4 + 4(\gamma - 1)m^2 c^4}\right] = \frac{1}{2}m_e c^2 \left[(\gamma - 1) + \sqrt{(\gamma - 1)(\gamma + 3)}\right]$$

ASSESS   With some algebra, the kinetic energy of the recoiled electron can be written as

$$K_e = (\gamma-1)m_e c^2 = \frac{E_0^2}{E_0 + m_e c^2} = \frac{1}{2}m_e c^2 \frac{\left[(\gamma-1)+\sqrt{(\gamma-1)(\gamma+3)}\right]^2}{(\gamma+1)+\sqrt{(\gamma-1)(\gamma+3)}}$$

In the nonrelativistic limit where $\gamma \approx 1 + \frac{1}{2}v^2/c^2$, the above expression reduces to the expected result $K_e \approx \frac{1}{2}m_e v^2$.

73.   **INTERPRET**   We shall use conservation of energy and conservation of momentum (with relativistic expressions for both energy and momentum) to derive equations related to Compton's equation, and then derive the equation for the Compton shift.

DEVELOP   The energy of a photon is $E = hf = hc/\lambda$ (Equation 34.6) and for a particle it is $E = \gamma mc^2$ (Equation 33.9). The relativistic momentum is $\vec{p} = \gamma m\vec{u}$ for the electron and $\vec{p} = \hat{i}\,h/\lambda_0$ for the photon, where $\hat{i}$ is the direction of the photon's motion. We will use conservation of energy to obtain one of the desired equations, and conservation of momentum in two dimensions to obtain the other two equations.

EVALUATE   The initial energy is the energy of the photon plus the rest energy of the electron:

$$E_i = hc/\lambda_0 + mc^2.$$

The final energy is the energy of the new photon plus the relativistic energy of the moving electron:

$$E_f = hc/\lambda + \gamma mc^2$$

Equating these two energies (by conservation of energy) gives us the first of the three desired equations:

$$\frac{hc}{\lambda_0} + mc^2 = \frac{hc}{\lambda} + \lambda mc^2.$$

The next two equations come from the initial momentum, $\vec{p}_i = \hat{i}\,h/\lambda_0$ and the components of final momentum $p_{fx} = \frac{h}{\lambda}\cos\theta + \gamma mu\cos\phi$ and $p_{fy} = 0 = \gamma mu\sin\phi - \frac{h}{\lambda}\sin\theta$. By conservation of momentum, we can equate the initial and final momentum in each direction, which leads to

$$p_{ix} = p_{fx} \quad \Rightarrow \quad \frac{h}{\lambda_0} = \frac{h}{\lambda}\cos\theta + \gamma mu\cos\phi$$

$$p_{iy} = p_{fy} = 0 \quad \Rightarrow \quad 0 = \frac{h}{\lambda}\sin\theta - \gamma mu\sin\phi$$

These are the second two of the desired relationships we were to derive. Solving these three equations for $\lambda_0 - \lambda$ directly is a lengthy algebraic process. An easier approach is to start with the momentum in vector form and use the law of cosines:

$$\vec{p}_\gamma = \vec{p}_{\gamma'} + \vec{p}_{e'} \to p_{e'}^2 = p_\gamma^2 + p_{\gamma'}^2 - 2p_\gamma p_{\gamma'}\cos\theta$$

$$p_{e'}^2 = \left(\frac{h}{\lambda_0}\right)^2 + \left(\frac{h}{\lambda}\right)^2 - 2\frac{h}{\lambda_0}\frac{h}{\lambda}\cos\theta$$

We now use conservation of energy in the form $\frac{hc}{\lambda_0} + mc^2 = \frac{hc}{\lambda} + \sqrt{m^2c^4 + p_{e'}^2 c^2}$, and solve for $p_{e'}^2$ to obtain

$$p_{e'}^2 = \frac{\left(\frac{hc}{\lambda_0} - \frac{hc}{\lambda} + mc^2\right)^2 - m^2c^4}{c^2}$$

We equate the two equations for $p_{e'}^2$ to obtain

$$\frac{\left(\frac{hc}{\lambda_0}-\frac{hc}{\lambda}+mc^2\right)^2-m^2c^4}{c^2}=\left(\frac{h}{\lambda_0}\right)^2+\left(\frac{h}{\lambda}\right)^2-2\frac{h}{\lambda_0}\frac{h}{\lambda}\cos\theta$$

$$\left(\frac{hc}{\lambda_0}\right)^2+\left(\frac{hc}{\lambda}\right)^2-\frac{2c^3hm}{\lambda}+\frac{2c^3hm}{\lambda_0}-\frac{2c^2h^2}{\lambda\lambda_0}=\left(\frac{hc}{\lambda_0}\right)^2+\left(\frac{hc}{\lambda}\right)^2-2\frac{hc}{\lambda_0}\frac{hc}{\lambda}\cos\theta$$

$$\frac{mc}{\lambda_0}-\frac{mc}{\lambda}=\frac{h}{\lambda\lambda_0}(1-\cos\theta)$$

$$\lambda-\lambda_0=\frac{h}{mc}(1-\cos\theta)$$

ASSESS   We have derived the equation for the Compton shift, using conservation of energy and momentum.

75.   **INTERPRET**   We are to integrate the radiance equation (Equation 34.3, Planck's law) over all wavelengths and show that the resulting total power radiated per unit area is equivalent to the Stefan-Boltzmann law (Equation 34.1).

DEVELOP   The Stefan-Boltzmann law gives the power per area as $P/A=\sigma T^4$, and the radiance equation is

$$R(\lambda,T)=\frac{2\pi hc^2}{\lambda^5\left(e^{hc/(\lambda kT)}-1\right)}$$

Substituting $hc/(\lambda kT)$ by the integration variable x gives

$$R(x,T)=\frac{2\pi k^5T^5}{h^4c^3}\left(\frac{x^5}{e^x-1}\right)$$

The differentials dx and $d\lambda$ are related by

$$d\lambda=-\frac{hc}{x^2kT}dx$$

We will integrate $R(x, T)d\lambda$ over all λ.

EVALUATE   Performing the integration gives

$$\frac{P}{A}=\int_0^\infty R(\lambda,T)d\lambda=\frac{2\pi k^5T^5}{h^4c^3}\int_0^\infty\left(\frac{x^5}{e^x-1}\right)\left(-\frac{hc}{x^2kT}\right)dx=-\frac{2\pi k^4T^4}{h^3c^2}\int_0^\infty\left(\frac{x^3}{e^x-1}\right)dx$$

$$=\frac{2k^4\pi^5}{15c^2h^3}T^4$$

ASSESS   This is equivalent to the Stefan-Boltzmann law, with $\sigma=2k^4\pi^5/(15h^3c^2)$.

77.   **INTERPRET**   We use conservation of momentum to find the recoil angle of the electron in Compton scattering.
DEVELOP   The momentum conservation equations from problem 34.73 are

$$\frac{h}{\lambda_0}=\frac{h}{\lambda}\cos\theta+\gamma mu\cos\phi$$

$$0=\frac{h}{\lambda}\sin\theta-\gamma mu\sin\phi$$

The Compton scattering equation is $\lambda-\lambda_0=\frac{h}{mc}(1-\cos\theta)$. We will solve the momentum equations for $\sin\phi$ and $\cos\phi$, and then take the ratio to see if we can get the desired equation.
EVALUATE   First solving for $\sin\phi$ and $\cos\phi$, we get

$$\sin\phi=\frac{1}{\gamma mu}\left(\frac{h}{\lambda}\sin\theta\right)$$

$$\cos\phi=\frac{1}{\gamma mu}\left(\frac{h}{\lambda_0}-\frac{h}{\lambda}\cos\theta\right)$$

Taking the ratio and multiplying the numerator and denominator by $\lambda\lambda_0$, we arrive at the desired result:

$$\tan\phi = \frac{\lambda_0 \sin\theta}{\lambda - \lambda_0 \cos\theta}$$

ASSESS   To remove the dependence on the final wavelength, λ, we use the Compton shift equation:

$$\tan\phi = \frac{\lambda_0 \sin\theta}{\lambda - \lambda_0 + \lambda_0(1-\cos\theta)} = \frac{\sin\theta}{(1+\lambda_C/\lambda_0)(1-\cos\theta)}$$

where $\lambda_C = h/mc$ is the Compton wavelength. This shows that as you decrease λ_0, i.e., as you go towards higher energy photons, the recoil angle, ϕ, of the electron decreases (for a given recoil angle, θ, for the photon).

79.   **INTERPRET**   We investigate the relation between particle lifetimes and uncertainty in their rest energy measurements.

DEVELOP   From Equation 34.16, the uncertainties in the energy and time are constrained by $\Delta E \Delta t \geq \hbar$. Therefore, the lifetime of a given particle will be inversely proportional to the uncertainty in its rest energy: $\tau \propto 1/\Delta E$.

EVALUATE   The shortest lifetime will correspond to the curve with the largest uncertainty in its rest energy. In the graph, this is particle C.

The answer is (c).

ASSESS   The distribution width shown in the graph is called the natural line width and is denoted by Γ. It is called "natural" to signify that this uncertainty is inherent to the particle and does not, like other uncertainties, come simply from the imperfect instruments used to collect the data.

81.   **INTERPRET**   We investigate the relation between particle lifetimes and uncertainty in their rest energy measurements.

DEVELOP   The inverse relation between energy uncertainty and lifetime is $\Delta E \sim \hbar/\tau$.

EVALUATE   A longer lifetime leads to a narrower range in the energy measurement. By Einstein's mass-energy equivalence, this corresponds to a narrower range in the mass, as well.

The answer is (d).

ASSESS   Some particles, like the proton and the electron, appear to have infinite lifetimes, so we'd expect the uncertainty in their mass to be near to zero.

QUANTUM MECHANICS

EXERCISES

Section 35.2 The Schrödinger Equation

11. **INTERPRET** We are given the wave function of a particle and are to deduce from this the most probable position of the particle and the position(s) where the probability of finding the particle is 50%.

    DEVELOP The quantity $\psi^2(x)$ represents the probability of finding the particle at the position x. Therefore, the particle is most likely to be found at the position where the probability density $\psi^2(x)$ is a maximum.

    EVALUATE (a) The maximum of $\psi^2(x) = A^2 e^{-2x^2/a^2}$ occurs where $d\left[\psi^2(x)\right]/dx = 0$ and $d^2\left[\psi^2(x)\right]/dx^2 < 0$. Evaluating the first derivative gives

    $$\frac{d}{dx}\psi^2(x) = -\frac{4A^2 x}{a^2} e^{-2x^2/a^2} = 0$$

    $$x = 0$$

    Evaluating the second derivative at $x = 0$ gives

    $$\left.\frac{d^2}{dx^2}\psi^2(x)\right|_{x=0} = \left[-\frac{4A^2}{a^2} e^{-2x^2/a^2}\left(1 - \frac{4x}{a^2}\right)\right]_{x=0} = -\frac{4A^2}{a^2} < 0$$

    which shows that the extremum at $x = 0$ is indeed a maximum. Thus, most probable place to find the particle is at $x = 0$.

    (b) The probability density $\psi^2(x)$ falls to half its maximum value when $\psi^2(x) = \psi^2(0)/2$. Solving this equation gives $x = \pm a\sqrt{\ln 2/2} = \pm 0.589a$.

    ASSESS The probability distribution is shown below. Note that $\psi^2(x)/A^2 = e^{-2x^2/a^2}$ peaks at $x = 0$ and appears to be halved at $x/a \sim \pm 0.6$.

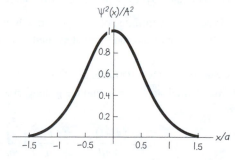

Section 35.3 Particles and Potentials

13. **INTERPRET** We are to find the principal quantum number of a particle in an infinite square well given the particle's energy relative to the ground state.

    DEVELOP The energy levels for a particle in an infinite square well are (Equation 35.5) $E_n = E_0 n^2$, where $E_0 = h^2/(8mL^2)$ is the ground state energy. Thus, we must find the quantum number n such that $E_n = 25E_0$.

EVALUATE Solving the equation gives

$$E_0 n^2 = 25 E_0$$
$$n = 5$$

ASSESS The energy levels go as n squared.

15. **INTERPRET** We have an electron confined in an infinite potential well, and are to find its ground-state energy.

DEVELOP The energy levels for an infinite square potential well are given by Equation 35.5:

$$E_n = \frac{n^2 \pi^2 \hbar^2}{2mL^2} = \frac{n^2 h^2}{8mL^2}$$

The ground-state energy corresponds to $n = 1$.

EVALUATE From the above equation, we find the ground-state energy to be

$$E_1 = \frac{h^2}{8mL^2} = \frac{(hc)^2}{8(mc^2)L^2} = \frac{(1240 \text{ eV} \cdot \text{nm})^2}{8(511 \text{ keV})(10 \text{ nm})^2} = 3.8 \times 10^{-3} \text{ eV} = 6.0 \times 10^{-22} \text{ J}$$

ASSESS A nonzero ground-state energy is a common feature of quantum systems. Note that the energy levels are quantized and proportional to n^2.

17. **INTERPRET** The problem asks us to compute the lowest energy states of a quantum wire.

DEVELOP The energy levels for an infinite square well potential is given in Equation 35.5:

$$E = \frac{n^2 h^2}{8mL^2}$$

To simplify the calculation, we will multiply the top and bottom of the fraction by c^2, so that we can use the rest energy of the electron, $mc^2 = 511$ keV, and the shorthand $hc = 1240$ eV $\cdot$ nm.

EVALUATE (a) Plugging in the tube diameter for the well width, we get for the $n = 1$ ground state

$$E_1 = \frac{n^2 (hc)^2}{8mc^2 L^2} = \frac{1^2 (1240 \text{ eV} \cdot \text{nm})^2}{8(511 \text{ keV})(1.0 \text{ nm})^2} = 0.376 \text{ eV} \approx 0.38 \text{ eV}$$

(b) The first excited state has $n = 2$ and its energy is 4 times the ground state:

$$E_2 = n^2 E_1 = 2^2 (0.376 \text{ eV}) = 1.5 \text{ eV}$$

ASSESS The answers seem reasonable. In the lowest energy states, electrons in the nanotube will have roughly the same energy as electrons accelerated by a 1 V potential difference.

19. **INTERPRET** This problem involves and infinite potential well in which an unknown particle is trapped. Given the difference in energy between the ground state and the first excited state, we are to determine if the particle is an electron or a proton.

DEVELOP Use the result of Schrödinger's equation applied to an infinite potential well (Equation 35.5). The energy difference ΔE between the first excited state ($n = 2$) and the ground-state ($n = 1$) of a one-dimensional infinite square well is

$$\Delta E = \frac{(4-1)h^2}{8mL^2} = \frac{3}{8} \left(\frac{hc}{L} \right)^2 \left(\frac{1}{mc^2} \right)$$

EVALUATE Using given values, we find that

$$mc^2 = \frac{3}{8} \left(\frac{hc}{L} \right)^2 \frac{1}{\Delta E} = \frac{3}{8} \left(\frac{1240 \text{ eV} \cdot \text{nm}}{1.0 \text{ nm}} \right)^2 \frac{1}{1.13 \text{ eV}} = 510 \text{ keV}$$

which is very close to the electron's rest energy.

ASSESS The particle must be an electron.

21. **INTERPRET** We are to find the ground-state energy of an alpha particle (i.e., He^{2+}) trapped in the given infinite quantum well.

DEVELOP Use the result of Schrödinger's equation applied to an infinite potential well (Equation 35.5). For the ground state, $n = 1$. From Appendix C, we find that $1\ u = 931.5\ MeV/c^2$.

EVALUATE With $n = 1$, the lowest alpha-particle energy in a one-dimensional infinite square well of width $L = 15$ fm is

$$E = \frac{h^2}{8mL^2} = \left(\frac{hc}{L}\right)^2 \frac{1}{8mc^2} = \left(\frac{hc}{L}\right)^2 \frac{1}{8(4u)c^2}$$

$$= \left(\frac{1240\ MeV \cdot fm}{15\ fm}\right)^2 \left(\frac{1}{8(4 \times 931.5\ MeV)}\right) = 0.2\ MeV$$

ASSESS This is a rather high energy, which is not surprising given the small size of the potential well. Note that the Bohr radius is $a_0 = 52.9$ pm gives the typical atomic size, which includes the electrons. Thus, our alpha particle is confined to a space much less than the size of an atom.

23. **INTERPRET** We are to find the energy of the lowest state for a particle in a harmonic oscillator potential.

DEVELOP Apply Equation 35.7. The lowest state is the ground state and has quantum number $n = 0$ for a harmonic oscillator.

EVALUATE With $n = 0$, Equation 35.7 gives the ground-state energy as

$$E_0 = \tfrac{1}{2}\hbar\omega = \tfrac{1}{2}\left(6.582 \times 10^{-16}\ eV \cdot s\right)\left(1.0 \times 10^{17}\ s^{-1}\right) = 33\ eV$$

ASSESS Note that the energy of a particle in a harmonic oscillator potential is independent of mass.

25. **INTERPRET** We are to find the energy separation between states in a harmonic oscillator given the ground-state energy.

DEVELOP Equation 35.7 gives the energy levels in a harmonic oscillator: $E_n = \left(n + \tfrac{1}{2}\right)\hbar\omega$, so the separation between levels is $\hbar\omega$. Given that $E_0 = 4.0$ eV for $n = 0$, we can solve for $\hbar\omega$.

EVALUATE The energy spacing $\hbar\omega$ between adjacent levels is

$$E_0 = 4.0\ eV = \tfrac{1}{2}\hbar\omega$$

$$\hbar\omega = 8.0\ eV$$

ASSESS Unlike the infinite square well, the energy levels for the harmonic oscillator are linear in n.

Section 35.4 Quantum Mechanics in Three Dimensions

27. **INTERPRET** We are to consider a particle three-dimensionally confined in a cubic potential well. If the length of all sides of the box is doubled, how is the particle's ground-state energy affected?

DEVELOP From Equation 35.8, we see that the energy of a particle in a cubical box is inversely proportional to the square of the length of the sides of the box. In the ground state, $n_x = n_y = n_z = 1$, so the ground-state energy is

$$E_0 = \frac{h^2}{8mL^2}$$

If we double the length, $L \rightarrow 2L$, so we can recalculate the new ground-state energy.

EVALUATE Replacing L by 2L, we find that the ground state energy becomes

$$E_0' = \frac{h^2}{8m(2L)^2} = \frac{1}{4}\left(\frac{h^2}{8mL^2}\right) = \frac{1}{4}E_0$$

ASSESS The ground-state energy is inversely proportional to L^2, so doubling L reduces the ground-state energy by a factor of $L^2 = 4$.

29.   **INTERPRET**   This problem is similar to the previous problem, except that we are now given energy of the photon emitted by an electron confined to a cubic quantum potential well and are asked for the size of the cube.

DEVELOP   From the solution to the previous problem, we know that the energy difference between the energy of the ground-state and of the first excited state is

$$\Delta E = \frac{3h^2}{8mL^2}$$

Set this equal to the photon energy (Equation 34.5) $E_\gamma = hf = hc/\lambda$ to find the cubic box size L.

EVALUATE   Inserting the given quantities and solving for L gives

$$\frac{hc}{\lambda} = \frac{3h^2}{8mL^2} \quad \Rightarrow \quad L = \sqrt{\frac{3\lambda hc}{8mc^2}} = \sqrt{\frac{3(1240 \text{ eV} \cdot \text{nm})(950 \text{ nm})}{8(511 \text{ keV})}} = 930 \text{ pm}$$

where we have used $m_e = 511 \text{ MeV}/c^2$.

ASSESS   This cubic potential well is about the same size as in the previous problem, but the corresponding energy difference is much less because the electron's mass is much less than that of the proton.

PROBLEMS

31.   **INTERPRET**   For this problem, we are to show that if two wave functions are solutions of the Schrödinger equation, then their linear combination must also be a solution.

DEVELOP   The time-independent one-dimensional Schrödinger equation is given by Equation 35.1:

$$-\frac{\hbar^2}{2m}\frac{d^2\psi(x)}{dx^2} + U(x)\psi(x) = E\psi(x)$$

We want to show that for any constants a and b, $a\psi_1 + b\psi_2$ is a solution if ψ_1 and ψ_2 are.

EVALUATE   Substituting $\psi = a\psi_1 + b\psi_2$ into the Schrödinger equation gives

$$\left[-\frac{\hbar^2}{2m}\frac{d^2}{dx^2} + U(x)\right](a\psi_1 + b\psi_2) = a\left[-\frac{\hbar^2}{2m}\frac{d^2\psi_1}{dx^2} + U(x)\psi_1\right] + b\left[-\frac{\hbar^2}{2m}\frac{d^2\psi_2}{dx^2} + U(x)\psi_2\right]$$

$$= aE\psi_1 + bE\psi_2 = E(a\psi_1 + b\psi_2) = E\psi$$

ASSESS   The Schrödinger equation is a linear differential equation. Therefore, the result follows directly from the superposition principle.

33.   **INTERPRET**   We are to find the energy and wavelength of the photon emitted as an electron trapped in an infinite square well makes a transition to the adjacent energy level.

DEVELOP   The energy levels for an infinite square potential well are given by Equation 35.5:

$$E_n = \frac{n^2 h^2}{8mL^2}$$

Thus, the energy of the photon emitted when the electron drops from n_i to $n_f < n_i$ is

$$E_\gamma = \Delta E = \left(n_i^2 - n_f^2\right)\frac{h^2}{8mL^2}$$

From Equation 34.6, the wavelength of the photon is $\lambda = hc/E_\gamma$.

EVALUATE   **(a)** Substituting the values given, we find

$$E_\gamma = \left(n_i^2 - n_f^2\right)\frac{h^2}{8mL^2} = \left(n_i^2 - n_f^2\right)\frac{(hc)^2}{8(mc^2)L^2} = \left(7^2 - 6^2\right)\frac{(1240 \text{ eV} \cdot \text{nm})^2}{8(511 \text{ keV})(1.5 \text{ nm})^2} = 2.2 \text{ eV}$$

(b) The wavelength of the photon is

$$\lambda = \frac{hc}{E_\gamma} = \frac{8mc^2 L^2}{\left(n_i^2 - n_f^2\right)hc} = 570 \text{ nm}$$

to two significant figures.

ASSESS   The wavelength is in the visible region of the electromagnetic spectrum.

35.   **INTERPRET**   An electron trapped in the given one-dimensional potential well must absorb a photon in order to make a transition from the ground state to an excited state. We want to know the maximum wavelength associated with this transition.

DEVELOP   Equation 35.5 describes the allowed energy states for a particle trapped in a one-dimensional potential well. The allowed quantum numbers n are $n = 1, 2, 3, \ldots$, so the smallest transition energy is to the first excited state (i.e., $n = 1$ to $n = 2$). The energy difference may be found from inserting these quantum numbers into Equation 35.5: so

$$\Delta E = E_{n=2} - E_{n=1}$$

$$= \left(2^2 - 1^2\right)\frac{h^2}{8mL^2} = \frac{3(hc)^2}{8(mc^2)L^2} = \frac{3\left(1240 \text{ eV} \cdot \text{nm}\right)^2}{8\left(511 \text{ keV}\right)\left(4.4 \text{ nm}\right)^2} = 0.0583 \text{ eV}$$

The wavelength that corresponds to this energy is (see Equation 34.6) $E_\gamma = hf = hc/\lambda$. Because of the inverse relationship between wavelength and energy, the smallest energy corresponds to the largest wavelength. Therefore, the maximum wavelength that can cause a transition is

$$E_\gamma = \Delta E \quad \Rightarrow \quad \lambda_{\max} = \frac{hc}{\Delta E}$$

EVALUATE   Thus, the maximum wavelength that can be absorbed is

$$\lambda_{\max} = \frac{hc}{\Delta E} = \frac{1240 \text{ eV} \cdot \text{nm}}{0.0583 \text{ eV}} = 21 \text{ μm}$$

ASSESS   The wavelength corresponds to the infrared region of the electromagnetic spectrum.

37.   **INTERPRET**   There are various ways for the electron which is initially in the $n = 4$ state to make a transition to the ground state. We want to find the wavelengths associated with all possible spectral lines in this process.

DEVELOP   The energy levels for a one-dimensional infinite square potential well are given by Equation 35.5:

$$E_n = \frac{n^2 h^2}{8mL^2}$$

Thus, the energy of the photon emitted when the electron drops from n_i to $n_f < n_i$ is

$$E_\gamma = \Delta E = \left(n_i^2 - n_f^2\right)\frac{h^2}{8mL^2}$$

From Equation 34.6, we know that the corresponding photon wavelengths are

$$\lambda = \frac{hc}{E_\gamma} = \frac{8\left(mc^2\right)L^2}{\left(n_i^2 - n_f^2\right)hc} = \frac{8\left(511 \text{ keV}\right)\left(0.1 \text{ nm}\right)^2}{\left(n_i^2 - n_f^2\right)\left(1240 \text{ eV} \cdot \text{nm}\right)} = \frac{33.0 \text{ nm}}{n_i^2 - n_f^2}$$

EVALUATE   Starting from the $n = 4$ state, the possible transitions to the ground state are $4 \to 1, 4 \to 2, 4 \to 3,$ $3 \to 1, 3 \to 2,$ and $2 \to 1,$ with corresponding wavelengths

$$\lambda_{4 \to 1} = \frac{33.0 \text{ nm}}{4^2 - 1^2} = 2.2 \text{ nm}, \quad \lambda_{4 \to 2} = \frac{33.0 \text{ nm}}{4^2 - 2^2} = 2.8 \text{ nm}$$

$$\lambda_{4 \to 3} = \frac{33.0 \text{ nm}}{4^2 - 3^2} = 4.7 \text{ nm}, \quad \lambda_{3 \to 1} = \frac{33.0 \text{ nm}}{3^2 - 1^2} = 4.1 \text{ nm}$$

$$\lambda_{3 \to 2} = \frac{33.0 \text{ nm}}{3^2 - 2^2} = 6.6 \text{ nm}, \quad \lambda_{2 \to 1} = \frac{33.0 \text{ nm}}{2^2 - 1^2} = 11 \text{ nm}$$

ASSESS   Only photons of these discrete wavelengths will be emitted during the transition from $n = 4$ to $n = 1$. The wavelengths correspond to the ultraviolet region of the electromagnetic spectrum.

39.   **INTERPRET**   We are given a potential that is the same as in Fig. 35.5, except that the origin of coordinates is at the center of the well. We want to find expressions for the normalized wave function for even and odd quantum numbers and the corresponding energy levels.

DEVELOP The wave function for this well can be found by using Equation 35.6 (which is already normalized), but replacing x by $x' + \frac{1}{2}L$, where $-\frac{1}{2}L \le x' \le \frac{1}{2}L$. The normalized wave function thus takes the form

$$\psi_n(x') = \sqrt{\frac{2}{L}} \sin\frac{n\pi}{L}\left(x' + \frac{L}{2}\right) = \sqrt{\frac{2}{L}}\left(\sin\frac{n\pi x'}{L}\cos\frac{n\pi}{2} + \cos\frac{n\pi x'}{L}\sin\frac{n\pi}{2}\right)$$

We need to distinguish between even and odd values of n.

EVALUATE (a) If n is odd, then $\cos\frac{1}{2}n\pi = 0$ and $\sin\frac{1}{2}n\pi = \pm1$, so the wave function is

$$\psi_{n\text{-odd}}(x') = \sqrt{\frac{2}{L}}\cos\left(\frac{n\pi x'}{L}\right)$$

The probability density $\psi_n^2(x)$ is unaffected by the overall sign of $\psi_n(x)$, so we choose the sign to be positive. If n is even, then $\cos\frac{1}{2}n\pi = \pm1$ and $\sin\frac{1}{2}n\pi = 0$, and the wave function is

$$\psi_{n\text{-even}}(x') = \sqrt{\frac{2}{L}}\sin\left(\frac{n\pi x'}{L}\right)$$

(b) The energy levels are the same, $E_n = n^2h^2/8mL^2$ regardless of how the potential is parameterized.

ASSESS These results can be confirmed by direct solution of the Schrödinger equation. For $-\frac{1}{2}L \le x \le \frac{1}{2}L$, Equation 35.4 has two solutions: $A\sin(kx)$ or $A\cos(kx)$. In order for ψ to vanish at $x = \pm\frac{1}{2}L$, one must use the cosine solution for odd quantum numbers [$\cos(\pm kL/2)$] vanishes for kL equal to odd multiples of π and the sine solution for even quantum numbers [$\sin(\pm kL/2)$] vanishes for kL equal to even multiples of π. Since the average of $\sin^2$ or $\cos^2$ over an integer number of half-cycles is ½, the normalization constant is $\sqrt{2/L}$ for either wave function (or use integrals in Appendix A). Note that these wave functions have even or odd parity about the center of the potential well (see Section 39.2).

41. **INTERPRET** We're given the energy of the photon emitted in a transition between adjacent quantum states and asked to find the width of the one-dimensional infinite potential well.

DEVELOP The energy levels for a one-dimensional infinite square potential well are given by Equation 35.5:

$$E_n = \frac{n^2h^2}{8mL^2}$$

Thus, the energy of the photon emitted when the electron drops from n_i to $n_f < n_i$ is

$$\Delta E = \left(n_i^2 - n_f^2\right)\frac{h^2}{8mL^2}$$

EVALUATE From the above equation, the energy difference between $n_i = 2$ and $n_f = 1$ is $\Delta E = 3h^2/(8mL^2)$, so the width of the potential well is

$$L = \frac{\sqrt{3}hc}{\sqrt{8mc^2\Delta E}} = \frac{\sqrt{3}(1240\text{ eV}\cdot\text{nm})}{\sqrt{8(511\text{ keV})(1.96\text{ eV})}} = 0.759\text{ nm}$$

ASSESS The result is consistent with the typical quantum well width (about the size of an atom).

43. **INTERPRET** The problem asks if quantum mechanics should be considered for macromolecules trapped inside a biological cell.

DEVELOP We will treat the biological cell like a one-dimensional square well, so that the energy levels are given by Equation 35.5: $E = n^2h^2/8mL^2$.

EVALUATE The energy difference between the ground state and the first excited state is

$$\Delta E = \frac{\left(2^2 - 1^2\right)\left(6.63\times10^{-34}\text{ J}\cdot\text{s}\right)^2}{8(250,000\text{ u})(10\text{ μm})^2}\left[\frac{1\text{ u}}{1.661\times10^{-27}\text{ kg}}\right] = 4.0\times10^{-36}\text{ J} = 2.5\times10^{-17}\text{ eV}$$

This is so much smaller than the energy of biochemical reactions (1 eV) that quantization is not relevant.

ASSESS Quantization is rarely considered in biology because the objects of interest are too large. One exception is photosynthesis, where the conversion of sunlight into chemical energy appears to exhibit some quantum effects.

45. **INTERPRET**   This problem concerns a particle in a one-dimensional infinite square potential well. We are asked to find the probability that the particle will be found within a given ranged centered at two different points in the square well.

DEVELOP   As in Problem 42, the probability of finding the particle between x_1 and x_2 (where $0 \leq x_1 < x_2 \leq L$) is

$$P = \frac{2}{L} \int_{x_1}^{x_2} \sin^2\left(\frac{\pi x}{L}\right) dx = \frac{(x_2 - x_1)}{L} - \frac{1}{2\pi}\left[\sin\left(\frac{2\pi x_2}{L}\right) + \sin\left(\frac{2\pi x_1}{L}\right)\right]$$

We shall evaluate this probability function for the two ranges given.

EVALUATE   (a) The probability P of finding the particle between $x_1 = 0.500L - 0.075L = 0.425L$ and $x_2 = 0.500L + 0.075L = 0.575L$ is

$$P = \frac{0.575L - 0.425L}{L} - \frac{1}{2\pi}\left\{\sin\left[2\pi(0.575)\right] - \sin\left[2\pi(0.5425)\right]\right\} = 0.29$$

(b) The probability P of finding the particle between $x_1 = 0.250L - 0.075L = 0.425L$ and $x_2 = 0.250L + 0.075L = 0.575L$ is

$$P = \frac{0.175L - 0.325L}{L} - \frac{1}{2\pi}\left\{\sin\left[2\pi(0.325)\right] - \sin\left[2\pi(0.175)\right]\right\} = 0.15$$

Assess   The probability is greater for the particle to be found in the center of the well than near the edges, which is reasonable in view of the probability distribution function (see Figure 35.7).

47. **INTERPRET**   We are to show that the Schrödinger equation has nonzero solutions in classically forbidden regions where $E < U$.

DEVELOP   We are given solutions of the form $\psi(x) = Ae^{\pm\sqrt{2m(U-E)}x/\hbar}$, so all we need to do is substitute this wavefunction into the time-independent Schrödinger equation

$$\left(-\frac{\hbar^2}{2m}\frac{d^2}{dx^2} + U\right)\psi = E\psi$$

(see Section 35.3) and see if it fits.

EVALUATE   Inserting the trial solution into the Schrödinger equation gives

$$\frac{d^2}{dx^2}\psi = A\frac{2m(U-E)}{\hbar^2}e^{\pm\sqrt{2m(U-E)}x/\hbar}$$

Substitute this into the time-independent Schrödinger equation:

$$-\frac{\hbar^2}{2m}\left[A\frac{2m(U-E)}{\hbar^2}e^{\pm\sqrt{2m(U-E)}x/\hbar}\right] + UAe^{\pm\sqrt{2m(U-E)}x/\hbar} = EAe^{\pm\sqrt{2m(U-E)}x/\hbar}$$

$$-(U-E) + U = E$$

$$E = E$$

Thus, we have shown that the proposed solution is a valid solution for the time-independent Schrödinger equation.

ASSESS   This form of $\psi(x)$ is a solution to the time-independent Schrödinger equation. Note that this form of solution is not "wave like": it exponentially decays (or increases) instead of oscillating.

49. **INTERPRET**   We are to verify that the given wave function is a solution to the three-dimensional Schrödinger equation and that the derived energy levels match those given by Equation 35.8.

DEVELOP   For the given wave function $\psi(x, y, z)$ the second partial derivatives are

$$\frac{\partial^2 \psi}{\partial x^2} = -\left(\frac{n_x \pi}{L}\right)^2 \psi, \quad \frac{\partial^2 \psi}{\partial y^2} = -\left(\frac{n_y \pi}{L}\right)^2 \psi \quad \text{and} \quad \frac{\partial^2 \psi}{\partial z^2} = -\left(\frac{n_z \pi}{L}\right)^2 \psi$$

Substituting into the Schrödinger equation, with a potential for a cubical box ($U = 0$ inside and $U = \infty$ outside), we find

$$-\frac{\hbar^2}{2m}\left(\frac{\partial^2 \psi}{\partial x^2}+\frac{\partial^2 \psi}{\partial y^2}+\frac{\partial^2 \psi}{\partial z^2}\right)=-\frac{1}{2m}\left(\frac{h}{2\pi}\right)^2\left[-\left(\frac{n_x \pi}{L}\right)^2-\left(\frac{n_y \pi}{L}\right)^2-\left(\frac{n_z \pi}{L}\right)^2\right]\psi$$

$$=\frac{h^2}{8mL^2}\left(n_x^2+n_y^2+n_z^2\right)\psi=E\psi$$

EVALUATE   The above derivation **(a)** demonstrates that $\psi(x,y,z)$ is a solution, and **(b)** shows that the energy levels match those given in Equation 35.8.

ASSESS   This wave function also satisfies the boundary conditions appropriate for confinement ($\psi = 0$ at $x = 0$ or L, $y = 0$ or L, and $z = 0$ or L, which are points where $U = \infty$), since n_x, n_y and n_z are integers.

51.   **INTERPRET**   When transitioning to lower states, electrons emit photons to release energy. We are interested in all the visible photon wavelengths associated with all possible electronic transitions in the given quantum well.

   DEVELOP   The energy levels for an infinite square potential well are given by Equation 35.5:

$$E_n = \frac{n^2 h^2}{8mL^2}$$

Thus, the energy of the photon emitted when the electron drops from initial state n_i to final state $n_f < n_i$ is

$$E_\gamma = \Delta E = \left(n_i^2 - n_f^2\right)\frac{h^2}{8mL^2}$$

The possible photon wavelengths are

$$\lambda = \frac{hc}{E_\gamma}=\frac{8mc^2 L^2}{\left(n_i^2 - n_f^2\right)hc}=\frac{8(511\text{ keV})(1.2\text{ nm})^2}{\left(n_i^2 - n_f^2\right)(1240\text{ eV}\cdot\text{nm})}=\frac{4.75\ \mu\text{m}}{n_i^2 - n_f^2}$$

   EVALUATE   **(a)** Visible photons fall within the wavelength range $0.4\ \mu\text{m} < \lambda < 0.7\ \mu\text{m}$, or

$$4.75/0.7 = 6.78 < n_i^2 - n_f^2 < 11.9 = 4.75/0.4$$

There are four transitions satisfying this condition: $3 \to 1$, $4 \to 3$, $5 \to 4$, and $6 \to 5$.

   ASSESS   Since energy is quantized, only photons with wavelengths that satisfy the above condition will be emitted during the transitions.

53.   **INTERPRET**   You're asked to develop the Schrödinger equation and the ground state wave function for the simple harmonic oscillator.

   DEVELOP   You start with the time-independent Schrödinger equation:

$$-\frac{\hbar^2}{2m}\frac{d^2\psi(x)}{dx^2}+U(x)\psi(x)=E\psi(x)$$

   EVALUATE   **(a)** Plugging in the potential energy of a harmonic oscillator, you arrive at

$$-\frac{\hbar^2}{2m}\frac{d^2\psi(x)}{dx^2}+\tfrac{1}{2}m\omega^2 x^2\psi(x)=E\psi(x)$$

(b) For $n = 0$, the energy is $E = \tfrac{1}{2}\hbar\omega$, and the Schrödinger equation reduces to:

$$\frac{d^2\psi(x)}{dx^2}=\left(\frac{m^2\omega^2}{\hbar^2}x^2 - \frac{m\omega}{\hbar}\right)\psi(x)=\alpha^2\left(\alpha^2 - 1\right)\psi(x)$$

where $\alpha^2 = m\omega/\hbar$. If $\psi(x)=A_0 e^{-\alpha^2 x^2/2}$, then the first derivative is $\frac{d\psi(x)}{dx}=A_0\left(-\alpha^2 x\right)e^{-\alpha^2 x^2/2}$, and the second derivative is

$$\frac{d^2\psi(x)}{dx^2}=-\alpha^2 A_0\left[e^{-\alpha^2 x^2/2}+x\left(-\alpha^2 x\right)e^{-\alpha^2 x^2/2}\right]=\alpha^2\left(\alpha^2 - 1\right)\psi(x)$$

This proves that $\psi(x)=A_0 e^{-\alpha^2 x^2/2}$ is a solution to the ground state.

(c) To find the normalization constant, we use the normalization condition (Equation 35.3):

$$\int_{-\infty}^{+\infty} \psi^2 \, dx = \int_{-\infty}^{+\infty} A_0^2 e^{-\alpha^2 x^2} \, dx = 1$$

The integral is a Gaussian. One can find the value in an integral table, but we will give a short derivation here. We first define $I = \int_{-\infty}^{+\infty} e^{-\alpha^2 x^2} \, dx$ and then square both sides. We combine the right-hand side into a single exponent and then change to polar coordinates, which puts the integral into a more solvable form:

$$I^2 = \int_{-\infty}^{+\infty} e^{-\alpha^2 x^2} \, dx \cdot \int_{-\infty}^{+\infty} e^{-\alpha^2 y^2} \, dy = \int_{-\infty}^{+\infty} \int_{-\infty}^{+\infty} e^{-\alpha^2 \left(x^2 + y^2 \right)} dx dy$$

$$= \int_0^{2\pi} \int_0^{+\infty} e^{-\alpha^2 r^2} r \, dr \, d\theta = 2\pi \left[\frac{-1}{2\alpha^2} e^{-\alpha^2 r^2} \right]_0^{+\infty} = \frac{\pi}{\alpha^2}$$

Therefore, $I = \sqrt{\pi} / \alpha$, and the normalization constant must be $A_0 = \left(\alpha^2 / \pi \right)^{1/4}$. In summary, the normalized ground state wave function of the simple harmonic oscillator is

$$\psi(x) = \left(\frac{\alpha^2}{\pi} \right)^{1/4} e^{-\alpha^2 x^2 / 2}$$

Assess The probability density for the ground state is a Gaussian, or "bell curve," as depicted in Figure 35.11a.

55. **Interpret** We evaluate the energy levels of a quantum dot.

Develop A quantum dot is basically a three-dimensional square well. If we assume the qdot is a cube, the energy levels are given by Equation 35.8:

$$E = \frac{h^2}{8mL^2} \left(n_x^2 + n_y^2 + n_z^2 \right)$$

Evaluate The ground state has $n_x = n_y = n_z = 1$, whereas the first excited state has one of the n's equal to 2. The energy difference between the two states is

$$\Delta E = \frac{h^2}{8mL^2} \left[\left(2^2 + 1^2 + 1^2 \right) - \left(1^2 + 1^2 + 1^2 \right) \right] = \frac{3h^2}{8mL^2}$$

If the qdot decreases in size, the energy difference increases. The photon emitted when the qdot drops to its ground state will, therefore, have a smaller wavelength, since

$$\lambda = \frac{hc}{\Delta E} \propto L^2$$

The answer is (b).

Assess The advantage of qdots is that they are like tunable atoms. You can essentially choose the wavelength at which it absorbs or emits by simply adjusting its size.

57. **Interpret** We evaluate the energy levels of a quantum dot.

Develop The first excited state of a cubically symmetric qdot has energy of $E = \frac{h^2}{8mL^2} \left(n_x^2 + n_y^2 + n_z^2 \right)$, where one of the three quantum numbers equals 2, while the other two equal 1.

Evaluate There are three states that have the energy $E = \frac{6h^2}{8mL^2}$, i.e., n_x, n_y, n_z can equal 2,1,1 or 1,2,1 or 1,1,2. See Figure 35.17. We say this state is three-fold degenerate.

The answer is (c).

Assess Degeneracy often depends on there being some sort of symmetry. In this case, it is the symmetry of the cube. If the qdot's three sides were not equal, then the first excited state would nondegenerate.

36

ATOMIC PHYSICS

Note: For the problems in this chapter, useful numerical values of Planck's constant, in SI and atomic units, are: $h = 6.626 \times 10^{-34}$ J·s $= 4.136 \times 10^{-15}$ eV·s $= 1240$ eV·nm/c, and $\hbar = h/2\pi = 1.055 \times 10^{-34}$ J·s $= 6.582 \times 10^{-16}$ eV·s $= 197.3$ MeV·fm/c.

Useful constants and combinations, in SI and atomic units, are:

$$h = 6.626 \times 10^{-34} \text{ J·s} = 4.136 \times 10^{-15} \text{ eV·s} = 1240 \text{ eV·nm}/c$$

$$\hbar = h/2\pi = 1.055 \times 10^{-34} \text{ J·s} = 6.582 \times 10^{-16} \text{ eV·s} = 197.3 \text{ MeV·fm}/c$$

$$c = 2.998 \times 10^{8} \text{ m/s}, ke^2 = 1.440 \text{ eV·nm},$$

$$1u = 1.661 \times 10^{-27} \text{ kg} = 931.5 \text{ MeV}/c^2$$

$$k_B = 1.381 \times 10^{-23} \text{ J/K} = 8.617 \times 10^{-5} \text{ eV/K}$$

EXERCISES

Section 36.1 The Hydrogen Atom

15. **INTERPRET** Our system consists of a group of hydrogen atoms in an excited state characterized by the quantum number n. The minimum energy needed to ionize these atoms is 1.5 eV, and we are to find the quantum number n of this system.

DEVELOP The hydrogen energy levels are given by Equation 36.6:

$$E_n = \frac{E_1}{n^2} = \frac{-13.6 \text{ eV}}{n^2}$$

where n is the principal quantum number. The ionization energy for this state is the difference between the vacuum energy (i.e., $E = 0$) and the energy of the state, so

$$1.5 \text{ eV} = -E_n = \frac{13.6 \text{ eV}}{n^2}$$

EVALUATE Solving for the principle quantum number n yields

$$n = \sqrt{\frac{13.6 \text{ eV}}{1.5 \text{ eV}}} = 3$$

ASSESS The result makes sense since principal quantum must be a positive integer.

17. **INTERPRET** This problem requires us to enumerate the first few allowed values of the orbital angular momentum of a hydrogen atom.

DEVELOP The magnitude of orbital angular momentum is given by Equation 36.9:

$$L = \sqrt{l(l+1)}\hbar, \quad l = 0,1,2,\ldots$$

EVALUATE Successive values of $\sqrt{l(l+1)}$, starting with $l = 3$, are:

$\sqrt{3 \times 4} = \sqrt{12}, \sqrt{4 \times 5} = \sqrt{20}, \sqrt{5 \times 6} = \sqrt{30}, \sqrt{6 \times 7} = \sqrt{42}, \sqrt{7 \times 8} = 56$ .... Evidently, (d) is an erroneous possibility.

ASSESS Since orbital angular momentum is quantized, only certain discrete values are allowed.

19.   **INTERPRET**   We're given the magnitude of the orbital angular momentum of an electron and asked to find its corresponding orbital quantum number.

DEVELOP   The magnitude of orbital angular momentum is given by Equation 36.9:

$$L = \sqrt{l(l+1)}\hbar, \quad l = 0,1,2,\ldots$$

where l is the orbital quantum number.

EVALUATE   From the above equation, we have

$$L = \sqrt{30}\hbar = \sqrt{l(l+1)}\hbar$$
$$l(l+1) = 30$$
$$l = 5$$

ASSESS   In this case, the orbital quantum number l is easily found by inspection. In general, l is a nonnegative integer solution of the quadratic formula.

21.   **INTERPRET**   This problem is the inverse of the preceding problem. We are now given the energy and orbital angular momentum of an electron and asked to give its label.

DEVELOP   We need to find both principal and orbital quantum numbers for this state. The principle quantum number n for hydrogen energy levels are related to the energy of the state by Equation 36.6:

$$E_n = \frac{E_1}{n^2} = \frac{-13.6\,\text{eV}}{n^2}$$

Similarly, the magnitude of orbital angular momentum L is related to the orbital quantum number l by Equation 36.9:

$$L = \sqrt{l(l+1)}\hbar, \quad l = 0,1,2,\ldots$$

EVALUATE   From Equation 36.6, we find $n = \sqrt{13.6/1.51} = 3$, and from Equation 36.9, we have $l(l+1) = 6 = 2(3)$. Thus, $n = 3$ and $l = 2$, so this is a $3d$ state.

ASSESS   Our result makes sense since orbital quantum number l can only take on values between 0 and $n-1$.

Section 36.2 Electron Spin

23.   **INTERPRET**   We are given the spin quantum number and asked to find the magnitude of spin angular momentum for the theoretical particle called the graviton.

DEVELOP   The magnitude of spin angular momentum is given by Equation 36.11:

$$S = \sqrt{s(s+1)}\hbar$$

EVALUATE   With $s = 2$, the above equation gives

$$S = \sqrt{2\times3}\hbar = \sqrt{6}\hbar = (2\pi)^{-1}\sqrt{6}\left(6.626\times10^{-34}\,\text{J}\cdot\text{s}\right) = 2.58\times10^{-34}\,\text{J}\cdot\text{s}$$

for the graviton.

ASSESS   Spin angular momentum is quantized, just like the orbital angular momentum, so only a discrete set of values is allowed.

25.   **INTERPRET**   We have a hydrogen atom in the $3D$ state, and we want to know all the possible values for the quantum number j.

DEVELOP   The total angular momentum of the hydrogen atom is (see Equation 36.15) $\vec{J} = \vec{L} + \vec{S}$. For the $3D$ state, $l = 2$ and $s = \frac{1}{2}$ (hydrogen has one electron). Apply Equation 36.17a, $j = l \pm \frac{1}{2}$, to find all the j values.

EVALUATE   The possible j values are $j = l - \frac{1}{2} = \frac{3}{2}$ and $j = l + \frac{1}{2} = \frac{5}{2}$.

ASSESS   Thus, there are two possible values of j for this state, which correspond to total angular momentum

$$J = \sqrt{j(j+1)} = \begin{cases} \sqrt{\frac{3}{2}\left(\frac{3}{2}+1\right)}\hbar = \sqrt{\frac{15}{4}}\hbar \\ \sqrt{\frac{5}{2}\left(\frac{5}{2}+1\right)}\hbar = \sqrt{\frac{35}{4}}\hbar \end{cases}$$

Section 36.3 The Exclusion Principle

27. **INTERPRET**   We are to find the electronic highest energy possible given several electrons in a harmonic oscillator potential. We will assume that the electrons are in the lowest states possible and use the exclusion principle.

  DEVELOP   The exclusion principle states that no more than one electron may be in any given state. Electrons have spin $\frac{1}{2}$, so two electrons may exist in each energy state n; one with spin $+\frac{1}{2}$ (spin up) and one with spin $-\frac{1}{2}$ (spin down). We will find the energy of the highest-energy electron by "filling" the energy levels with pairs of electrons, and finding the lowest energy level available to the 21st electron. The energy levels are given by (Equation 35.7) $E_n = \left(n+\frac{1}{2}\right)\hbar\omega$.

  EVALUATE   The first 20 electrons fill levels 0–9, so the 21st electron must be in level $n = 10$. The corresponding energy is

  $$E = \left(10+\tfrac{1}{2}\right)\hbar\omega = \frac{21\hbar\omega}{2}$$

  ASSESS   Electrons are fermions, so they follow the exclusion principle. If the particles in question were bosons, they could all be in the lowest energy state because bosons are not subject to the exclusion principle.

Section 36.4 Multielectron Atoms and the Periodic Table

29. **INTERPRET**   We are asked for the symbolic label for a neutral scandium atom, such as those in given in Table 36.2.

  DEVELOP   Scandium ($Z = 21$) is the first of the transition metals and has one $3d$ electron in addition to the configuration of the preceding element, calcium ($Z = 20$; see explanation following Table 36.2).

  EVALUATE   Thus, $1s^2 2s^2 2p^6 3s^2 3p^6 4s^2 3d^1$ is the full electronic configuration for scandium.

  ASSESS   As a useful check, we note that the sum of all the superscripts is 21.

Section 36.5 Transitions and Atomic Spectra

31. **INTERPRET**   We are to find the relationship between the wavelength and energy of a photon, with wavelength expressed in nm and energy in eV.

  DEVELOP   Using Equation 34.6, the energy of a photon can be written as

  $$E = hf = \frac{hc}{\lambda}$$

  Writing Planck's constant and the speed of light in the appropriate units leads to the given expression.

  EVALUATE   Writing $h = 6.626\times10^{-34}$ J·s $= 4.136\times10^{-15}$ eV·s and $c = 3\times10^{17}$ nm/s, we obtain

  $$\lambda = \frac{hc}{E} = \frac{\left(4.136\times10^{-15}\ \text{eV·s}\right)\left(3\times10^{17}\ \text{nm/s}\right)}{E} = \frac{1240\ \text{eV·nm}}{E}$$

  ASSESS   The units eV and nm are more appropriate for photons. The above formula allows us to calculate the photon energy (in eV) easily, once its wavelength (in nm) is given.

33. **INTERPRET**   In this problem we are asked about the energy splitting between two $4p$ states of sodium, given the wavelengths of the two spectral lines produced by transition from these states to a lower level that is not split.

  DEVELOP   Because the $3s$ level is not split, the fine structure splitting of the $4p$ levels is equal to the difference in the photon energies. We can therefore apply the result of Problem 36.31 to find the energy of each transition, and take the difference to obtain the energy splitting.

  EVALUATE   The energy difference is

  $$\Delta E = hc\left(\frac{1}{\lambda_1} - \frac{1}{\lambda_2}\right) = (1240\ \text{eV·nm})\left(\frac{1}{330.2\ \text{nm}} - \frac{1}{330.3\ \text{nm}}\right) = 1.137\times10^{-3}\ \text{eV}$$

  ASSESS   The two states correspond to $4p_{3/2}$ and $4p_{1/2}$, with $4p_{3/2}$ having a slightly higher energy compared to $4p_{1/2}$.

PROBLEMS

35. **INTERPRET**   We are given the energy and orbital angular momentum of an electron and asked to find the principal quantum number and orbital quantum number.

DEVELOP   The hydrogen energy levels are given by Equation 36.6:

$$E_n = \frac{E_1}{n^2} = \frac{-13.6\,\text{eV}}{n^2}$$

where n is the principal quantum number. Similarly, the magnitude of orbital angular momentum is given by Equation 36.9:

$$L = \sqrt{l(l+1)}\hbar, \quad l = 0,1,2,\dots$$

EVALUATE   From Equation 36.6, we get $n = \sqrt{13.6/0.85} = 4,$ and from Equation 36.9, we have $l(l+1) = 12 = 3(4).$ Thus, $n = 4,$ and $l = 3.$

ASSESS   This is a $4f$ state. Our result makes sense since orbital quantum number l can only take on values between 0 and $n - 1$.

37. **INTERPRET**   We are to find the orbital quantum number l for the Moon, assuming its orbital angular momentum is quantized (which is not possible to prove with current technology). We would expect that for a macroscopic object such as this, the number would be rather high—so high that the difference between angular momentum states is smaller than anything we could measure.

DEVELOP   We know that the angular momentum is quantized by $L = \sqrt{l(l+1)}\hbar.$ The classical angular momentum of the Moon is $L = mvr = mr^2\omega,$ where r = 384,000 km, $m = 7.35 \times 10^{22}$ kg , and ω is the angular velocity of the moon, $\omega = \frac{2\pi\,\text{radians}}{28\,\text{days}} = 2.60 \times 10^{-6}$ rad/s. We will equate these values of L, and solve for the orbital quantum number l.

EVALUATE   The orbital angular momentum quantum number is

$$\sqrt{l(l+1)}\hbar = mr^2\omega$$

$$l^2 + l - \frac{m^2 r^4 \omega^2}{\hbar^2} = 0$$

The constant $m^2 r^4 \omega^2 / \hbar^2$ has the value

$$\frac{m^2 r^4 \omega^2}{\hbar^2} = \frac{4\pi^2 \left(7.35 \times 10^{22}\,\text{kg}\right)^2 \left(3.84 \times 10^8\,\text{m}\right)^4 \left(2.60 \times 10^{-6}\,\text{s}^{-1}\right)^2}{\left(6.626 \times 10^{-34}\,\text{J} \cdot \text{s}\right)^2} = 7.14 \times 10^{136}$$

so the orbital quantum number is

$$l = \frac{-1 \pm \sqrt{1 + 4\left(7.14 \times 10^{136}\right)}}{2} = 2.67 \times 10^{68}$$

ASSESS   This is such a large value of l that there is no way that we could measure the difference between angular momentum states, and the angular momentum appears to be a continuous function as classical theory predicts.

39. **INTERPRET**   This problem involves finding the possible orientations of the angular momentum vector in the $l = 2$ state.

DEVELOP   For $l = 2$, the magnetic orbital angular momentum quantum number has $2l + 1 = 5$ values; namely $m_l = 0, \pm 1, \pm 2$. The corresponding angles that the angular momentum $\vec{L}$ makes with the z axis are given by

$$\theta = \cos^{-1}\left(\frac{L_z}{L}\right) = \cos^{-1}\left(\frac{m_l}{\sqrt{l(l+1)}}\right) = \cos^{-1}\left(\frac{m_l}{\sqrt{6}}\right)$$

EVALUATE   Substituting the possible values for m_l, we find the angles to be 90.0°, 65.9°, 114°, 35.3°C, or 145° for $m_l = 0, \pm 1, \pm 2$, respectively.

ASSESS   The angular momentum is quantized not only in magnitude but also in direction. Note that the "angle" has meaning only in the vector model for angular momentum.

41. **INTERPRET**   We are given the state of the electron in a hydrogen atom and are to find the possible values that may be taken by L_z when measurements are performed.

    DEVELOP   An f state has $l = 3$, so $m_l = \pm1, \pm2$, and ±3.

    EVALUATE   Since $L_z = m_l\hbar$ (Equation 36.10), when a measurement is performed, the possible results are (in units of $\hbar$) $m_l = \pm1, \pm2$, and ±3.

    ASSESS   These are the possible components of the orbital angular momentum in units of $\hbar$. Note that L_z can only take on certain discrete values. This is a consequence of space quantization in quantum mechanics.

43. **INTERPRET**   This problem is about the radial probability distribution of the electron in the hydrogen ground state. We want to show that the probability density has a maximum at $r = a_0$, where a_0 is the Bohr radius.

    DEVELOP   Using Equations 36.3 and 36.5, the radial probability in the ground state can be written as

    $$P(r) = 4\pi r^2 \psi^2 = 4\pi r^2 A^2 e^{-2r/a_0}$$

    EVALUATE   We differentiate $P(r)$ with respect to r to get

    $$\frac{dP(r)}{dr} = 4\pi A^2 e^{-2r/a_0}\left(2r - 2r^2/a_0\right) = 0$$

    The result shows that $r = a_0$ is a maximum of $P(r)$. Physically, this means that the electron is most likely to be found at $r = a_0$.

    ASSESS   The radial probability density function is plotted in Fig. 36.4. From the figure, we see that $P(r)$ is indeed a maximum at $r = a_0$.

45. **INTERPRET**   We're asked to give the energy and angular momentum of a hydrogen atom in an excited state.

    DEVELOP   The hydrogen atom is in the state $4F_{5/2}$. The first number, 4, gives the quantum number $n = 4$, the letter F corresponds to $l = 3$, and the subscript is for $j = 5/2$. $\lambda_2 = \lambda_1/n_2$.

    EVALUATE   (a) Equation 36.6 gives the energy levels of hydrogen:

    $$E_4 = \frac{E_1}{n^2} = \frac{E_1}{16}$$

    (b) The orbital angular momentum is delineated in Equation 36.9:

    $$L = \sqrt{l(l+1)}\hbar = \sqrt{3(4)}\hbar = \sqrt{12}\hbar$$

    (c) The atom's total angular momentum has magnitude given by Equation 36.16:

    $$J = \sqrt{j(j+1)}\hbar = \sqrt{\tfrac{5}{2}\left(\tfrac{7}{2}\right)}\hbar = \tfrac{1}{2}\sqrt{35}\hbar$$

    ASSESS   Since $j - l = -\tfrac{1}{2}$, we can surmise that the spin $\left(\vec{S}\right)$ and orbital $\left(\vec{L}\right)$ angular momenta are pointing in roughly opposite directions (see Figure 36.10).

47. **INTERPRET**   We have eight particles in the ground state of a harmonic oscillator potential, and we want to know the energy of the system when the particles are electrons and when they are spin-1 particles.

    DEVELOP   The energy levels of the one-dimensional harmonic oscillator are $E_n = \left(n + \tfrac{1}{2}\right)\hbar\omega$. Each level can be occupied by, at most, two electrons, as required by the Pauli exclusion principle for spin-1/2 fermions. The exclusion principle does not apply to spin-1 particles because they are bosons, so the number of them in one level is unlimited.

    EVALUATE   (a) For electrons, the lowest energy state is that with 2 electrons in the $n=0, 1, 2$ and 3 levels. The energy in this case is

    $$E_{\text{tot}} = 2E_0 + 2E_1 + 2E_2 + 2E_3 = \left[2\left(\tfrac{1}{2}\right) + 2\left(\tfrac{3}{2}\right) + 2\left(\tfrac{5}{2}\right) + 2\left(\tfrac{7}{2}\right)\right]\hbar\omega = 16\hbar\omega$$

    (b) For spin-1 particles, the lowest energy state is that with 8 electrons in the $n=0$ level:

    $$E_{\text{tot}} = 8E_0 = 8\left(\tfrac{1}{2}\hbar\omega\right) = 4\hbar\omega$$

    ASSESS   The ground state of the boson system is less energetic than that of the fermion system. This is expected, since fermions are often forced to higher energy levels due to the exclusion principle.

49. **INTERPRET**   We want to find the electronic configuration of copper, whose atomic number is 29.

    DEVELOP   Generally electrons fill in order of increasing shell number n, and within each shell with increasing l $(s, p, d, f, \ldots)$. In the case of copper which is a transition element, we actually run into an exception.

EVALUATE   As mentioned in the text following Table 36.2, the electronic configuration of copper is $1s^2 2s^2 2p^6 3s^2 3p^6 4s^1 3d^{10}$, instead of $\ldots 4s^2 3d^9$. The energy of the closed $3d$ subshell (with total angular momentum zero) is low enough (in spin-orbit, orbit-orbit, and spin-spin interactions) to compensate for the $4s - 3d$ difference.

ASSESS   As a useful check, we note that the sum of all the superscripts is 29.

51.   INTERPRET   We are given the wavelength (or energy) of each photon produced by the laser and its overall power output. From this we are to calculate the number of electronic transitions that occur within the laser gain medium each second.

DEVELOP   From Equation 34.6 we find that the energy released in the emission of a photon of wavelength 30 μm is

$$E = hf = \frac{hc}{\lambda} = \frac{\left(6.626 \times 10^{-34} \text{ J} \cdot \text{s}\right)\left(3.00 \times 10^8 \text{ m/s}\right)}{3.0 \times 10^{-5} \text{ m}} = 6.626 \times 10^{-21} \text{ J}$$

The power output is $P = d(NE)/dt = E(dN/dt)$, where dN/dt is the rate of emission of photons. By conservation of energy (assuming no photons are absorbed in the gain medium), this rate must be the same as the electronic transition rate.

EVALUATE   Thus, the electronic transition rate is

$$\frac{dN}{dt} = \frac{P}{E} = \frac{2.0 \times 10^{-3} \text{ J/s}}{6.626 \times 10^{-21} \text{ J}} = 3.02 \times 10^{17} \text{ s}^{-1}$$

transitions per second, so the number of transitions that occur in 1.0 seconds is

$$N = \left(1.0 \text{ s}\right)\left(3.02 \times 10^{17} \text{ s}^{-1}\right) = 3.0 \times 10^{17}$$

ASSESS   This is a far-infrared laser since it emits photons in the far-infrared. The result is given to two significant figures to reflect the precision of the data.

53.   INTERPRET   This problem involves the radial probability distribution of an electron in the ground state of a hydrogen atom. We want to know the probability that the electron in the hydrogen ground state is found in in the range $r = a_0 \pm 0.1 a_0$, where a_0 is the Bohr radius.

DEVELOP   Using Equations 36.3 and 36.5, the radial probability distribution of the ground state can be written as

$$P(r) = 4\pi^2 \psi^2 = 4\pi r^2 A^2 e^{-2r/a_0}$$

where $A^2 = 1/(\pi a_0^3)$. The probability can be found by integrating $P(r)$ from $0.9 a_0$ to $1.1 a_0$.

EVALUATE   Integrating over the range, we find the probability to be

$$P = 4\pi \int_{0.9a_0}^{1.1a_0} \psi^2(r) r^2 dr = 4\pi \left(\frac{1}{\pi a_0^3}\right) \left\{ \frac{r^2 e^{-2r/a_0}}{-2/a_0} - \frac{2}{-2/a_0}\left[\frac{e^{-2r/a_0}}{(-2/a_0)^2}\left(-\frac{2r}{a_0} - 1\right)\right] \right\}_{0.9a_0}^{1.1a_0}$$

$$= 2\left\{ (0.9)^2 e^{-1.8} - (1.1)^2 e^{-2.2} + \frac{1}{2}\left[-(3.2)e^{-2.2} + (2.8)e^{-1.8}\right] \right\} = 0.1$$

ASSESS   In Problem 43, we have shown that the radial probability density function (plotted in Fig. 36.4) has a maximum at $r = a_0$. Our result indicates that the probability of finding the electron within $r = a_0 \pm 0.1 a_0$ is about 10%.

55.   INTERPRET   We want to find the maximum orbital angular momentum we can add to a hydrogen $6d$ electron without changing the principle quantum number.

DEVELOP   The orbital quantum number l ranges from $l = 0$ to $l = n - 1$ (see discussion preceding Equation 36.9). If the electron stays in an $n = 6$ state, then the maximum value possible for the orbital quantum number is $l = 5$.

EVALUATE   The angular momentum is $L = \sqrt{l(l+1)}\hbar$, so the difference between the $6h$ ($l = 5$) and $6d$ ($l = 2$) states is $\sqrt{5(5+1)}\hbar - \sqrt{2(2+1)}\hbar = 3\hbar$. The new state corresponds to the $6h$ state.

ASSESS   The transition is $6d \rightarrow 6h$.

57.   INTERPRET   Given the state of a hydrogen atom, we are to find the probability that its electron will be found beyond a single Bohr radii and beyond 10 Bohr radii.

DEVELOP   Apply the result of Example 36.1, with the lower limit of integration set to a_0 and $10a_0$ for parts (a) and (b), respectively. Introduce the dimensionless variable $x = r/a_0$ to express the radial probability density for the hydrogen 2s wave function as

$$P(r)dr = 4\pi r^2 \psi_{2s}^2 dr$$
$$= \frac{4\pi r^2 (2 - r/a_0)^2}{16(2\pi a_0^3)} e^{-r/a_0} a_0 d\left(\frac{r}{a_0}\right)$$
$$= \frac{1}{8} x^2 (2 - x)^2 e^{-x} dx = P(x)dx$$

The probability that the electron in a hydrogen atom 2s state lies beyond the dimensionless distance x_0 is

$$P(x > x_0) = \int_{x_0}^{\infty} P(x)dx = \frac{1}{8}\int_{x_0}^{\infty} x^2 \left(4 - 4x + x^2\right) e^{-x} dx$$
$$= \frac{e^{-x_0}}{2}\left(x_0^2 + 2x_0 + 2\right) - \frac{e^{-x_0}}{2}\left(x_0^3 + 3x_0^2 + 6x_0 + 6\right) + \frac{1}{8}e^{-x_0}\left(x_0^4 + 4x_0^3 + 12x_0^2 + 24x_0 + 24\right)$$
$$= \frac{1}{8}e^{-x_0}\left(x_0^4 + 4x_0^2 + 8x_0 + 8\right)$$

EVALUATE   (a) For $x_0 = r/a_0 = 1$, $P(x > 1) = (21/8)e^{-1} = 0.966$.
(b) For $x_0 = 10$, $P(x > 10) = (10,488/8)e^{-10} = 5.95 \times 10^{-2}$.

ASSESS   As expected, the probability of finding the electron beyond 10 Bohr radii from the nucleus is much, much less than that of finding the electron beyond a single Bohr radius.

59.   INTERPRET   We are to find the most likely radial position for an electron in a hydrogen atom.

DEVELOP   Equations 36.5 gives the radial probability distribution function in terms of an arbitrary wave function ψ. For the 2s wave function given by Equation 3.7, this takes the form

$$P(r)dr = 4\pi r^2 \psi_{2s}^2 dr$$
$$= \frac{4\pi r^2 (2 - r/a_0)^2}{16(2\pi a_0^3)} e^{-r/a_0} a_0 d\left(\frac{r}{a_0}\right)$$
$$= \frac{1}{8} x^2 (2 - x)^2 e^{-x} dx = P(x)dx$$

where we have introduced the dimensionless variable $x = r/a_0$.

EVALUATE   The maximum probability can be found numerically, or by differentiation. We seek the maximum of the function $x^2(2 - x)^2 e^{-x}$, where $x = r/a_0$. Setting the derivative equal to zero, we find

$$0 = -x^2\left(2 - x^2\right)e^{-x} + 2x(2 - x)(2 - 2x)e^{-x} = x(2 - x)\left(x^2 - 6x + 4\right)e^{-x}.$$

The roots $x = 0$ and $x = 2$ are minima, and $x = 3 \pm \sqrt{5}$ are maxima. Direct computation shows that $P(x)$ is greatest for $x = 3 + \sqrt{5}$, hence $r_{max} = 5.24a_0$.

ASSESS   The secondary maximum at $x = 0.764$ is about 27% as great. $P(x)$ is sketched below.

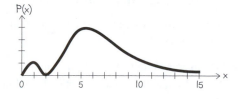

61.   INTERPRET   We are to verify Equation 36.8 for the energy levels of a general single-electron atom.

DEVELOP   For part (a), substitute Ze for the nuclear charge in the stead of ke^2 with Zke^2 in Equations 36.1, 36.2, and 36.4 and use

$$-\frac{\hbar^2}{2ma_0^2} + \frac{\hbar^2}{mra_0} - \frac{ke^2}{r} = E_1$$

so a_0 is replaced by a_0/Z for the radius of the ground state.

EVALUATE   (a) When this radius is inserted into Equation 36.6, the result is Equation 36.8.

(b) The ionization energy of a one-electron atom is the magnitude of Equation 36.8 with $n = 1$ (see paragraph following Example 34.3). Numerical values for the given nuclei are:

Z	2	8	26	82	92
Z^2 (13.6 eV)	54.4 eV	870 eV	9.19 keV	91.4 keV	115 keV

ASSESS   As expected, the ionization energy increases for larger nuclei because they have more charge holding the electron tighter in its orbit.

63.   INTERPRET   This problem is about transitions allowed by the given selection rule between different states of an infinite square-well potential.

DEVELOP   The energy levels for an electron in this one-dimensional infinite square well (Equation 35.5) are

$$E_n = \frac{n^2 h^2}{8mL^2} = \frac{n^2 (hc/L)^2}{8mc^2} = \frac{n^2 (1240 \text{ eV} \cdot \text{nm})/(0.2 \text{ nm})^2}{8(511 \text{ keV})} = n^2 (9.40 \text{ eV})$$

The energies associated with these transitions are

$$\Delta E_{i \to f} = \left(n_i^2 - n_f^2\right)(9.40 \text{ eV})$$

EVALUATE   (a) The allowed transitions (Δn odd) from the $n = 4$ level are $4 \to 1$, $4 \to 3$, $3 \to 2$, and $2 \to 1$, as shown below:

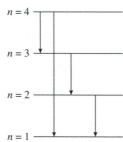

(b) The photon energies associated with these transitions are

$$\Delta E_{4 \to 1} = \left(4^2 - 1^2\right)(9.40 \text{ eV}) = 15(9.40 \text{ eV}) = 141 \text{ eV}$$

$$\Delta E_{4 \to 3} = \left(4^2 - 3^2\right)(9.40 \text{ eV}) = 7(9.40 \text{ eV}) = 65.8 \text{ eV}$$

$$\Delta E_{3 \to 2} = \left(3^2 - 2^2\right)(9.40 \text{ eV}) = 5(9.40 \text{ eV}) = 47.0 \text{ eV}$$

$$\Delta E_{2 \to 1} = \left(2^2 - 1^2\right)(9.40 \text{ eV}) = 3(9.40 \text{ eV}) = 28.2 \text{ eV}$$

ASSESS   The wavelengths of the photons (using the equation given in Exercise 31) range from 8.8 nm to 44 nm and they are in the far ultraviolet region.

65.   INTERPRET   We are to find the average radius of an electron in the ground state of hydrogen by using the probability density from Equation 36.5 with the wave function from Equation 36.3 and Example 36.1.

DEVELOP   Given a probability density $P(r)$, the average value of r is

$$\langle r \rangle = \int_0^\infty P(r)\, dr$$

We will use the square of the wave function given in Equation 36.3 and normalized in Example 36.1:

$$P(r) = 4\pi r^2 \psi^2$$

with

$$\psi(r) = \frac{1}{\sqrt{\pi a_0^3}} e^{-r/a_0}$$

EVALUATE   The average radial distance $\langle r \rangle$ of the electron from the nucleus is

$$P(r) = 4\pi r^2 \left(\frac{1}{\pi a_0^3} e^{-2r/a_0} \right) = \frac{4r^2}{a_0^3} e^{-2r/a_0}$$

$$\langle r \rangle = \int_0^\infty P(r) r\, dr = \frac{4}{a_0^3} r^3 e^{-2r/a_0} = \frac{3}{2} a_0$$

ASSESS   The probability density is not symmetric, so the *average* value of r is (in this case) greater than the *most probable* value of r.

67. **INTERPRET**   We are to find the g-factor for a classical electron orbit and for electron spin. The g-factor is defined as the ratio of the magnetic moment in units of μ_B to the angular momentum in units of $\hbar$.

DEVELOP   We first find the classical orbital g-factor, in which we find the orbital magnetic moment by treating the electron orbit as a current loop. The magnetic moment of a current loop is $\mu = IA$, where the current is $I = -e/T$, the area is $A = \pi r^2 = \pi a_0^2$, and the period can be found from the speed $v = n\hbar/(mr)$ of the electron. We will express this magnetic moment in terms of the Bohr magneton $\mu_B = e\hbar/(2m)$ and divide by the angular momentum $mvr = n\hbar$. For the second part of the problem, we use $\mu = -e/(mS)$ where $S = \frac{1}{2}\hbar$, and divide by the spin angular momentum $\frac{1}{2}\hbar$.

EVALUATE   **(a)** The classical electron speed is $v = n\hbar/(mr)$ so the orbital period is $T = 2\pi a_0/v = 2\pi m a_0^2/(n\hbar)$. The current in this orbital loop is then $I = -e/T = -ne\hbar/(2\pi m a_0^2)$. The magnetic moment of a loop is

$$\mu = IA = -\frac{ne\hbar}{2\pi m a_0^2}\pi a_0^2 = -n\left(\frac{e\hbar}{2m}\right) = -n\mu_B$$

The classical orbital angular momentum is $L = mvr = n\hbar$, so the g-factor is

$$g = \frac{\mu/\mu_B}{L/\hbar} = 1$$

(b) $\mu = -e/(mS)$ but $S = \frac{1}{2}\hbar$, so $\mu = -e\hbar/(2m) = -\mu_B$. The angular momentum is $S = \pm\frac{1}{2}\hbar$, so g = 2.

ASSESS   We have shown what was required.

69. **INTERPRET**   We consider characteristic X-rays that are emitted when an electron from an outer shell drops to an unoccupied inner shell of an atom.

DEVELOP   The wavelength of light is related to the energy by $\lambda = hc/E$.

EVALUATE   Plugging in the energy, the wavelength is

$$\lambda = \frac{hc}{E} = \frac{(1240\text{ eV}\cdot\text{nm})}{17.4\text{ keV}} = 70\text{ pm} \sim 100\text{ pm}$$

The answer is (b).

ASSESS   This is roughly 1 Angstrom, which is the characteristic size of an atom.

71. **INTERPRET**   We consider characteristic X-rays that are emitted when an electron from an outer shell drops to an unoccupied inner shell of an atom.

DEVELOP   Again, we don't have a generic formula for the energy levels of an atom, but we can assume that to a rough approximation the energy will increase with the square of the atomic number, as in Equation 36.8:
$E_n \approx -Z^2/n^2$.

EVALUATE We're told that element B has twice the atomic number as element A $\left(Z_B = 2Z_A\right)$, so the ratio of their $K\alpha$ X-ray energies will be

$$\frac{E_B}{E_A} \approx \frac{-Z_B^2\left(1-\frac{1}{4}\right)}{-Z_A^2\left(1-\frac{1}{4}\right)} = \left(\frac{2Z_A}{Z_A}\right)^2 = 4$$

The answer is (d).

ASSESS As we did in the previous problem, we will use the graph in Figure 36.20a to check whether our assumptions are reasonable. The peaks in the plot represent the energy of certain characteristic X-rays for lead (Pb) and arsenic (As), which have atomic numbers of 82 and 33, respectively. The $K\alpha$ X-ray of As has roughly the same energy as the $L\alpha$ X-ray of Pb. If we consider both of these atoms to be roughly hydrogen-like, then the ratio of their transition energies should

$$\frac{E_{PbL\alpha}}{E_{AsK\alpha}} \approx \left(\frac{Z_{Pb}}{Z_{As}}\right)^2 \left[\frac{-\frac{1}{9}+\frac{1}{4}}{-\frac{1}{4}+1}\right] = \left(\frac{82}{33}\right)^2 \frac{5}{27} = 1.1$$

This roughly agrees with the data presented in the figure, so we are justified in assuming that the energy levels for these atoms roughly obey the relation $E_n \approx -Z^2/n^2$.

MOLECULES AND SOLIDS

EXERCISES

Section 37.2 Molecular Energy Levels

17. **INTERPRET** We are looking for the wavelength of the photons that would cause a transition in an oxygen molecule to the first excited rotational energy state above the ground state.

    DEVELOP Using Equation 37.2, the difference in energy between the $l = 1$ and $l = 0$ states is

    $$\Delta E_{rot} = \frac{\hbar^2}{2I}1(1+1) - 0 = \frac{\hbar^2}{I}$$

    The photon wavelength corresponding to this transition is $\lambda = hc/\Delta E_{rot}$.

    EVALUATE Substituting the value of I given in the problem, we get

    $$\lambda = \frac{hc}{\Delta E} = \frac{hc}{\hbar^2/I} = \frac{2\pi cI}{\hbar} = \frac{2\pi(3.00\times 10^8\,\text{m/s})(1.95\times 10^{-46}\,\text{kg}\cdot\text{m}^2)}{1.055\times 10^{-34}\,\text{J}\cdot\text{s}} = 3.48\,\text{mm}$$

    ASSESS Transition between adjacent rotational energy levels requires absorbing a photon in the microwave region (frequency $f \sim 10^{11}$ Hz).

19. **INTERPRET** The gas molecules must absorb a photon in order to make a transition to the excited rotational state. We are given the wavelength of the photon, and asked to find the rotational inertia of the molecule.

    DEVELOP Using Equation 37.2, the difference in energy between the $l = 1$ and $l = 0$ states is

    $$\Delta E_{l\to 0} = \frac{\hbar^2}{2I}l(l+1) - 0 = \frac{\hbar^2}{I}$$

    The energy of the absorbed photon must equate to this difference in energy: $\Delta E_{rot} = hf = hc/\lambda$ Thus,

    $$\frac{\hbar^2}{I} = \Delta E_{l\to 0} = \frac{hc}{\lambda} = \frac{1240\,\text{eV}\cdot\text{nm}}{1.68\,\text{cm}} = 7.38\times 10^{-5}\,\text{eV} = 1.18\times 10^{-23}\,\text{J}$$

    We can find I from the above equation.

    EVALUATE The rotational inertia is

    $$I = \frac{\hbar^2}{\Delta E_{l\to 0}} = \frac{(1.055\times 10^{-34}\,\text{J}\cdot\text{s})^2}{1.18\times 10^{-23}\,\text{J}} = 9.41\times 10^{-46}\,\text{kg}\cdot\text{m}^2$$

    ASSESS The value of I is reasonable for a molecule (compare with previous problem). We can estimate the bond length of the molecule using $I = mR^2$. With $m \sim 10^{-26}$ kg, we get $R = 0.3$ nm, which is also a reasonable value.

21. **INTERPRET** This problem is about the vibrational motion of the N_2 molecule. We are given the energy spacing between adjacent vibrational energy levels of the molecule and are asked to find the corresponding classical frequency of vibration.

    DEVELOP The quantized vibrational energy levels are given by Equation 37.3:

    $$E_{vib} = (n+1/2)\hbar\omega$$

    Therefore, the energy difference between the adjacent levels is $\Delta E_{vib} = \hbar\omega = hf$.

    EVALUATE The classical vibrational frequency is

    $$f = \frac{\Delta E_{vib}}{h} = \frac{0.293\,\text{eV}}{4.136\times 10^{-15}\,\text{eV}\cdot\text{s}} = 7.08\times 10^{13}\,\text{Hz}$$

ASSESS   The frequency associated with vibrational motion is $f_{vib} = 10^{13}$ Hz, which is higher than that associated with rotation, $f_{rot} = 10^{11}$ Hz. This means that a more energetic photon must be absorbed by the molecule in order to excite the vibrational modes.

Section 37.3 Solids

23.   INTERPRET   This problem is an exercise in unit conversion. We are to express the ionic cohesive energy of NaCl in units of kcal/mol.

DEVELOP   To convert eV to kcal/mol, we use the following conversion factors (see Appendix C):

$$1\,eV = 1.602 \times 10^{-19}\,J$$
$$1\,kcal = 4184\,J$$
$$1\,mol = 6.022 \times 10^{23}$$

EVALUATE   Using the above conversion factors, we find:

$$(7.84\,eV)\overbrace{\left(\frac{1.602 \times 10^{-19}\,J}{eV}\right)}^{=1}\overbrace{\left(\frac{1\,kcal}{4184\,J}\right)}^{=1}\overbrace{\left(\frac{6.022 \times 10^{23}}{mol}\right)}^{=1} = 181\,kcal/mol$$

ASSESS   The result means that it takes 181 kcal to break one mole of NaCl into its constituent ions.

25.   INTERPRET   We are to find the wavelength of light emitted by electrons that make transitions across the band gap of Gallium phosphide.

DEVELOP   Gallium phosphide is a semiconductor with a band-gap energy of 2.26 eV (see Table 37.1). From Equation 34.6, $E = hf$, we know that a photon of energy corresponding to the band gap would have a wavelength of $E = hf = hc/\lambda$, where E is the band-gap energy.

EVALUATE   Solving the expression above for the wavelength gives

$$\lambda = \frac{hc}{E} = \frac{1240\,eV \cdot nm}{2.26\,eV} = 549\,nm$$

ASSESS   This light is green in color.

27.   INTERPRET   From the list of materials given in Table 37.1, we want to know which would emit the longest wavelength if used in a light-emitting diode.

DEVELOP   The wavelength emitted depends on the energy gap according to $\lambda = hc/E$. Thus, the maximum wavelength for the materials in Table 37.1 corresponds to the smallest energy gap.

EVALUATE   The material with the smallest energy gap is InAs, which has an energy gap of $E = 0.350$ eV. The wavelength of this emission is

$$\lambda = \frac{hc}{E} = \frac{1240\,eV \cdot nm}{0.350\,eV} = 3.54\,\mu m$$

ASSESS   The wavelength is in the infrared portion of the electromagnetic spectrum.

PROBLEMS

29.   INTERPRET   We have a molecule that emits a photon to make a transition to a state with lower rotational energy. We want to know the energy of the photon emitted for the $l = 1 \rightarrow l = 0$ transition.

DEVELOP   Using Equation 37.2, the difference in energy between the $l = 2$ and the $l = 1$ state is

$$\Delta E_{2 \rightarrow 1} = \frac{\hbar^2}{2I}\left[2(2+1) - 1(1+1)\right] = \frac{2\hbar^2}{I}$$

This means that $\hbar^2/I = \Delta E_{2 \rightarrow 1}/2$. For the $l = 1 \rightarrow l = 0$ transition,

$$\Delta E_{1 \rightarrow 0} = \frac{\hbar^2}{2I}\left[1(1+1) - 0\right] = \frac{\hbar^2}{I} = \frac{1}{2}\Delta E_{2 \rightarrow 1}$$

EVALUATE   Since $\Delta E_{2 \rightarrow 1} = 2.50$ meV, we have $\Delta E_{1 \rightarrow 0} = (2.50\,meV)/2 = 1.25$ meV.

ASSESS   The result shows that the energy levels are not evenly spaced. In general, the energy of a photon emitted in a transition between rotational levels with $|\Delta l| = 1$ is $\Delta E_{l \rightarrow (l-1)} = l\hbar^2/I$.

31.   **INTERPRET**   A molecule must absorb a photon to make a transition to a higher energy rotational state. We want to find an expression for the energy of the photon associated with the transition $l-1 \rightarrow l$.

     DEVELOP   The quantized rotational energy of a molecule is given by Equation 37.2:

$$E_{\text{rot}} = \frac{\hbar^2}{2I} l(l+1)$$

By conservation of energy, we know that the energy of the photon is equal to the energy difference between levels l and $l-1$.

     EVALUATE   Thus, the photon energy may be expressed as

$$E = \Delta E_{l \rightarrow l-1} = E_{\text{rot}} = \frac{\hbar^2}{2I} \left[l(l+1) - (l-1)(l-1+1) \right]$$

$$= \frac{l\hbar^2}{I}$$

     ASSESS   The energy difference between two adjacent rotational levels is proportional to the upper l-value.

33.   **INTERPRET**   In this problem we are asked to find the energy separation between the atomic rotational levels in the O_2 molecule.

     DEVELOP   The separation of the rotational spectral lines in energy is $\Delta E = \hbar^2 / I$ (see Example 37.1), or $\hbar^2 / I = 0.356$ meV for O_2. In a diatomic molecule, with equal-mass atoms and atomic separation R, each atom rotates about the center of mass at a distance of $R/2$, so $I = 2m_0 (R/2)^2 = 2(16.0\text{u})(R/2)^2 = 8.00uR^2$, where the mass of an oxygen atom is taken to be $m_O = 16.0\text{u}$.

     EVALUATE   The above expressions can be simplified to give

$$R^2 = \frac{I}{8.00\text{u}} = \frac{\hbar^2}{(8.00 \text{ u})(0.356 \text{ meV})}$$

     or

$$R = \frac{\hbar c}{\sqrt{(8.00 \text{ uc}^2)(0.356 \text{ meV})}} = \frac{197.3 \text{ eV} \cdot \text{nm}}{\sqrt{(8.00 \times 931.5 \text{ MeV})(0.356 \text{ meV})}} = 0.121 \text{ nm}$$

     ASSESS   Our result is in reasonable agreement with the experimental value of 0.146 nm.

35.   **INTERPRET**   This problem is about the vibrational energy levels of the HCl molecule. We are given its classical vibration frequency and are to find its ground-state energy and the energy spacing between its vibrational levels.

     DEVELOP   The quantized vibrational energy levels are given by Equation 37.3:

$$E_{\text{vib}} = (n+1/2)\hbar\omega = (n+1/2)hf$$

where we have used $\omega = 2\pi f$. In the ground state energy $n = 0$, and the energy spacing between the adjacent levels is

$$\Delta E_{\text{vib}} = E_{\text{vib}}^{(n)} - E_{\text{vib}}^{(n-1)}$$

     EVALUATE   **(a)** The vibrational ground-state energy is

$$E_{\text{vib},0} = \frac{1}{2} hf = \frac{1}{2}(4.136 \times 10^{-15} \text{ eV} \cdot \text{s})(8.66 \times 10^{13} \text{ Hz}) = 0.179 \text{ eV}$$

(b) The photon energy for allowed transitions ($\Delta n = 1$) is

$$\Delta E_{\text{vib}} = E_{\text{vib}}^{(n)} - E_{\text{vib}}^{(n-1)} = \left(n + \frac{1}{2}\right)hf - \left[(n-1) + \frac{1}{2}\right]hf = hf = 2E_{\text{vib},0} = 0.358 \text{ eV}$$

     ASSESS   Since $f = 10^{14}$ Hz is in the infrared region of the spectrum, the study of molecular vibrations typically involves infrared spectroscopy.

37.   **INTERPRET**   The problem concerns one of the dominant absorption wavelengths of carbon dioxide molecules.

     DEVELOP   The relation between transition energy and photon wavelength is given by $\Delta E = hc / \lambda$.

     EVALUATE   Solving for the wavelength and using a shorthand expression for the quantity hc,

$$\lambda = \frac{hc}{\Delta E} = \frac{1240 \text{ eV} \cdot \text{nm}}{82.96 \text{ meV}} = 14.95 \text{ } \mu\text{m}$$

     ASSESS   The wavelength is longer than the limit of the visible region (700 nm), as we'd expect for infrared light.

39.   INTERPRET   We are given the distance between the atoms in diatomic hydrogen (i.e., H_2) and are asked to find the energy spacing between the rotational ground state and the first rotational excited state.

DEVELOP   From Equation 10.12, the rotational inertia of an H_2 molecule is

$$I = 2m_H (R/2)^2 = \tfrac{1}{2}m_H R^2$$

where $R = 74$ pm and $R/2$ is the distance from each atom to the center-of-mass, which is the point about which they rotate. From Equation 37.2, the energy difference between the rotational ground state and first excited state ($l = 0$ and $l = 1$, respectively) is

$$\Delta E_{rot}^{1\to0} = 0 - \frac{\hbar^2}{2I}1(1+1) = -\frac{\hbar^2}{I}$$

By conservation of energy, this is the magnitude of the energy that must be carried away by the photon emitted by this transition.

EVALUATE   The energy of the photon released in the $l = 1$ to $l = 0$ transition is thus

$$E = \left|\Delta E_{rot}^{1\to0}\right| = \frac{\hbar^2}{I} = \frac{2\hbar^2}{m_H R^2} = \frac{2(hc)^2}{4\pi^2 m_H c^2 R^2} = \frac{2(197.3\text{ eV}\cdot\text{nm})^2}{4\pi^2(938\text{ MeV})(74\text{ pm})^2} = 15.2\text{ meV}$$

ASSESS   The wavelength of this photon is ~81 μm, which is in the far infrared portion of the electromagnetic spectrum. Note that we have used the proton's mass as the mass for the hydrogen atom, which is warranted because it is an accurate value for the precision of this problem.

41.   INTERPRET   This problem involves both rotational and vibrational transitions of a KCl molecule. We are given the classical vibration frequency and the rotational inertia and are asked to find the energy difference for a given transition.

DEVELOP   The quantized vibrational energy levels are given by Equation 37.3:

$$E_{vib} = (n+1/2)\hbar\omega$$

Therefore, the energy difference between adjacent levels is $\Delta E_{vib} = \hbar\omega = hf$. Similarly, using Equation 37.2, the difference in energy between the $l = 2$ and $l = 3$ states is

$$\Delta E_{rot}^{3\to2} = \frac{\hbar^2}{2I}2(2+1) - \frac{\hbar^2}{2I}3(3+1) = -\frac{3\hbar^2}{I}$$

Thus, the difference in energy between these vibrational-rotational levels is

$$\Delta E = \Delta E_{vib} + \Delta E_{rot} = hf - \frac{3\hbar^2}{I}$$

$$= (4.136\times10^{-15}\text{ eV}\cdot\text{s})(8.40\times10^{12}\text{ Hz}) - \frac{3(4.136\times10^{-15}\text{ eV}\cdot\text{s})^2(1.602\times10^{-19}\text{ J/eV})}{4\pi^2(2.43\times10^{-45}\text{ J}\cdot\text{s}^2)}$$

$$= 34.7\text{ meV}$$

EVALUATE   The energy corresponds to a photon wavelength of $\lambda = hc/\Delta E = 35.8$ μm .

ASSESS   Since $\Delta E > 0$, the final state has higher energy than the initial state, so the transition involves absorbing a photon in the infrared.

43.   INTERPRET   This problem is about the ionic cohesive energy of KCl. We are asked to solve for the constant n in Equation 37.4.

DEVELOP   As shown in Example 37.3, the constant n (not to be confused with the quantum number n used above) is given by

$$n = \left(1 + \frac{U_0 r_0}{\alpha k e^2}\right)^{-1}$$

Since the crystal structures of KCl and NaCl are the same, $\alpha = 1.748$.

EVALUATE Substituting the values given, we get

$$n = \left[1 + \frac{(-7.21 \text{ eV})(0.315 \text{ nm})}{(1.748)(1.44 \text{ eV} \cdot \text{nm})} \right]^{-1} = 10.2$$

where we used a convenient value of ke^2 in atomic units:

$$ke^2 = \left(9.0 \times 10^9 \text{ N} \cdot \text{m}^2/\text{C}^2\right)\left(1.60 \times 10^{-19} \text{ C}\right)^2 \frac{1 \text{ eV}}{1.602 \times 10^{-19} \text{ J}} = 1.44 \text{ eV} \cdot \text{nm}$$

ASSESS The result can be compared with $n = 8.22$ for NaCl (see Example 37.3). The large value of the exponent implies that KCl is strongly resistant to compression.

45. **INTERPRET** For this problem, we are to find the ionic cohesive energy of LiCl, which has the same structure as NaCl.

DEVELOP As calculated in Example 37.3, the ionic cohesive energy for NaCl is

$$U_0 = -\frac{\alpha k e^2}{r_0}\left(1 - \frac{1}{n}\right)$$

EVALUATE With $n = 7$ and $r_0 = 0.257$ nm, the cohesive energy for LiCl is

$$U_0 = -\frac{\alpha k e^2}{r_0}\left(1 - \frac{1}{n}\right) = -\frac{(1.748)(1.44 \text{ eV} \cdot \text{nm})}{0.257 \text{ nm}}\left(1 - \frac{1}{7}\right) = -8.40 \text{ eV}$$

ASSESS The result can be compared with $U_0 = -7.84$ eV for NaCl (see Example 37.3). The large value of the exponent implies that LiCl is strongly resistant to compression.

47. **INTERPRET** We are to integrate the density of states for a metal over all occupied states to find the number of conduction electrons per unit volume.

DEVELOP Equation 37.5 tells us that the density of states is

$$g(E) = \left(\frac{2^{7/2} \pi m^{3/2}}{h^3}\right)\sqrt{E}$$

We will integrate this density from $E = 0$ to $E = E_F$ to find the electron number density.

EVALUATE Performing the integration gives

$$n = \int_0^{E_F} g(E)\,dE = \frac{2^{7/2} \pi m^{3/2}}{h^3}\int_0^{E_F} E^{1/2}\,dE = \frac{2^{7/2} \pi m^{3/2}}{h^3}\left[\frac{2}{3}E^{3/2}\right]_0^{E_F}$$
$$= \frac{2^{9/2} \pi m^{3/2}}{3h^3}E_F^{3/2}$$

ASSESS We have shown what was required.

49. **INTERPRET** We shall use the results of Problem 47 to calculate the Fermi energy for calcium, given the number density.

DEVELOP The number density of conduction electrons for calcium is $n = 4.6 \times 10^{28}$ m^{-3}. We use this, and the electron mass $m = 9.109 \times 10^{-31}$ kg, to solve $n = E_F^{3/2}\, 2^{9/2} \pi m^{3/2}/(3h^3)$ for E_F.

EVALUATE Solving for the Fermi energy gives

$$E_F = \left(\frac{3nh^3}{2^{9/2}\pi m^{3/2}}\right)^{2/3} = 4.68 \text{ eV}$$

ASSESS This is a reasonable value for the Fermi energy in an alkali metal.

51. **INTERPRET** We are to find the Fermi temperature for silver, given its Fermi energy.

DEVELOP From the problem statement, we see that the Fermi temperature can be expressed mathematically as

$$T_F = E_F/k_B$$

EVALUATE   With $E_F = 5.48$ eV, we find the Fermi temperature of silver to be

$$T_F = \frac{E_F}{k_B} = \frac{5.48 \text{ eV}}{8.617 \times 10^{-5} \text{ eV} \cdot \text{K}^{-1}} = 6.36 \times 10^4 \text{ K}$$

This is about 212 times the room temperature (300 K).

ASSESS   The characteristic Fermi temperature is on the order of 10^5 K for a metal, and the Fermi energy is about 1 to 10 eV, much higher than the thermal energy at typical temperatures (0.025 eV at room temperatures).

53.   **INTERPRET**   We want to know whether zinc selenide would make a good photovoltaic cell. We shall do this by treating the Sun as a blackbody and finding the median wavelength of solar emission then comparing that wavelength with the band-gap energy for ZnSe.

DEVELOP   We approximate the Sun as a 5800-K blackbody, and use $\lambda_{median} T = 4.11$ mm·K to find the median wavelength. Next, we convert that wavelength to an energy using Equation 34.6, $E = hf = hc/\lambda$ and compare the energy of the median photon with the band-gap energy of ZnSe, $E_g = 3.6$ eV.

EVALUATE $\lambda_{median} = \frac{4.11 \text{ mm·K}}{5800 \text{ K}} = 709$ nm, so the energy of the median photon is

$$E = \frac{hc}{\lambda_{median}} = \frac{1240 \text{ eV} \cdot \text{nm}}{709 \text{ nm}} = 2.80 \times 10^{-19} \text{ J} = 1.75 \text{ eV}$$

ASSESS   The energy of this photon is less than the energy of the band gap for ZnSe, so ZnSe would not make a good photovoltaic cell.

55.   **INTERPRET**   In this problem, we're asked about the current required in a solenoid to achieve the critical magnetic field strength of the given superconductor.

DEVELOP   The magnetic field inside a long thin solenoid is given by Equation 26.21, $B = \mu_0 nI$, where $n = 5000/(0.75 \text{ m})$ is the number of turns per unit length in the solenoid.

EVALUATE   Substituting the values given, we find the current is

$$I = \frac{B}{\mu_0 n} = \frac{15 \text{ T}}{(4\pi \times 10^{-7} \text{ T} \cdot \text{m/A})(5000/0.75 \text{ m})} = 1.8 \text{ kA}$$

ASSESS   The current required is enormous! Note that niobium titanium is a Type-II superconductor with a critical temperature of about 10 K.

57.   **INTERPRET**   We are asked to find the emission wavelength of a fluorescent protein given the energy separation of its final transition.

DEVELOP   The relation between transition energy and photon wavelength is given by $\Delta E = hc/\lambda$.

EVALUATE   We're told that the protein absorbs light and then goes through an intermediate transition to an excited state that is 2.44 eV above the ground state. When the protein relaxes to its ground state, it will emit a photon with wavelength

$$\lambda = \frac{hc}{\Delta E} = \frac{1240 \text{ eV} \cdot \text{nm}}{2.44 \text{ eV}} = 508 \text{ nm}$$

ASSESS   This green wavelength is longer than the 395-nm wavelength that the protein absorbs, which implies that the emitted photon has less energy than the absorbed photon. The "missing energy" was expelled when the protein dropped from its initial excited state to the second excited state in the intermediate transition. This is typically how fluorescence works: a molecule absorbs a short wavelength and is excited to a high energy level; it then drops down to a slightly lower energy level before emitting a photon of longer wavelength.

59.   **INTERPRET**   You want to find the bond length of a diatomic molecule using the spacing between rotational energy levels inferred from spectral lines. You will use a more accurate model of the moment of inertia.

DEVELOP   In Example 37.1, the energy spacing between adjacent pairs of rotational energy levels was given for the HCl molecule. Using the quantized rotational energy levels in Equation 37.2: $E = \hbar^2 l(l+1)/2I$, an expression for the rotational inertia was found in terms of the energy spacing:

$$I = \frac{\hbar^2}{\Delta(\Delta E)} = \frac{\left(\frac{1}{2\pi}6.63\times10^{-34}\,\text{J}\cdot\text{s}\right)^2}{2.63\,\text{meV}} = 2.65\times10^{-47}\,\text{kg}\cdot\text{m}^2$$

If the HCl molecule is approximated by the hydrogen atom rotating around a fixed chlorine atom, then $I = m_H R^2$, and one can solve for the internuclear separation: $R \approx 0.126$ nm. You now want to redo the calculation without assuming the chlorine atom is fixed.

EVALUATE   For a general diatomic molecule, the center of mass is located along the line between the two atoms. Let r_1 and r_2 be the distance between the center of mass and the masses m_1 and m_2, respectively. If you choose the origin at the center of mass, then the center of mass equation gives

$$r_{cm} = 0 = \frac{\sum m_i r_i}{\sum m_i} = \frac{m_1 r_1 - m_2 r_2}{m_1 + m_2} \quad \rightarrow \quad m_1 r_1 = m_2 r_2$$

Combining this with the fact that $r_1 + r_2 = R$, you find $r_1 = m_2 R/(m_1 + m_2)$ and $r_2 = m_1 R/(m_1 + m_2)$. Plugging these values into the rotational inertia equation:

$$I = \sum m_i r_i^2 = m_1 \left(\frac{m_2 R}{m_1 + m_2}\right)^2 + m_2 \left(\frac{m_1 R}{m_1 + m_2}\right)^2 = \frac{m_1 m_2 R^2}{m_1 + m_2}$$

With this expression in hand, the internuclear separation in HCl is

$$R = \sqrt{\frac{(m_H + m_{Cl})I}{m_H m_{Cl}}} = \sqrt{\frac{(1.008\,\text{u} + 35.45\,\text{u})(2.65\times10^{-47}\,\text{kg}\cdot\text{m}^2)}{(1.008\,\text{u})(35.45\,\text{u})\left[1.661\times10^{-27}\,\frac{\text{kg}}{\text{u}}\right]}} = 0.128\,\text{nm}$$

Notice that if we neglect the mass of the hydrogen atom (i.e., assume $m_H \ll m_{Cl}$), this equation would reduce to the same solution in Example 37.1, $R = \sqrt{I/m_H}$.

ASSESS   Neglecting the mass of the hydrogen atom only introduces an error of 1.5% in the calculation of internuclear separation.

61. **INTERPRET**   We are to calculate the Madelung constant for a hypothetical one-dimensional crystal that consists of alternating positive and negative ions.

DEVELOP   The Madelung constant expresses the magnitude of the electrostatic potential energy of an ion in the crystal (see discussion preceding Equation 37.4). For any ion in the one-dimensional chain shown in Figure 37.24, there are two oppositely charged ions at distances of r_0, two similarly charged ions at distances of $2r_0$, two opposite ions at $3r_0$, etc. We can find the Madelung constant by summing these terms and using the series expansion of $\ln(1 + x)$.

EVALUATE   Thus, the electrostatic potential energy of any ion is

$$U = -\frac{2ke^2}{r_0} + \frac{2ke^2}{2r_0} - \frac{2ke^2}{3r_0} + \cdots = -\frac{2ke^2}{r_0}\left(1 - \frac{1}{2} + \frac{1}{3} - \cdots\right) = -\alpha\frac{ke^2}{r_0}$$

The Taylor expansion of $\ln(1 + x)$ is

$$\ln(1+x) = \ln(1) + x\frac{d}{dx}\ln(1+x) + \frac{x}{2!}\frac{d^2}{dx^2}\ln(1+x) + \frac{x}{3!}\frac{d^3}{dx^3}\ln(1+x) + \cdots$$

$$= 0 + \frac{x}{1+x} + \left(\frac{x}{2!}\right)\frac{-1}{(1+x)^2} + \left(\frac{x}{3!}\right)\frac{2}{(1+x)^3} + \cdots$$

When evaluated at $x = 1$, this gives

$$\ln(2) = 1 - \frac{1}{2} + \frac{1}{3} + \cdots$$

Comparing this with the series expression above for U shows that $\alpha = 2\ln 2$ for this "crystal."

ASSESS   The convergence of this series, which needs special consideration for $x = 1$, is discussed in many first-year calculus textbooks.

63.   **INTERPRET**   We are asked to derive the density of states equation for a metallic conductor.

DEVELOP   Equation 35.8 denotes the energy levels in a three dimensional cube of side L: $E = \frac{h^2}{8mL^2}\left(n_x^2 + n_y^2 + n_z^2\right)$, where n_x, n_y and n_z are integers greater than or equal to zero. Each set of n's corresponds to a unique energy level that can be occupied by at most two electrons, according to the exclusion principle. So there are two electron states for each possible combination of n_x, n_y and n_z.

EVALUATE   We need to count all the electron states that have energy less than some given energy E, so let's define an arbitrary vector $\vec{n} = \left(n_x, n_y, n_z\right)$. Counting electron states will be the same as counting the number of vectors $\vec{n}$ that have magnitude less than some radius r, where r is defined by the energy limit:

$$\left|\vec{n}\right| = \sqrt{n_x^2 + n_y^2 + n_z^2} \le r = \sqrt{\frac{8mL^2 E}{h^2}}$$

The number of vectors $\vec{n}$ satisfying this inequality with integer components is just the volume inside a sphere of radius r, see the figure below. We don't consider negative values for the n's, so the relevant volume is that of just one octant of the sphere: $V = \frac{1}{8} \cdot \frac{4\pi}{3} r^3$. Recalling that there are 2 electron states for each set of n's, the number of states with energy less than or equal to E is

$$N(E) = 2V = \frac{\pi}{3} r^3 = \frac{\pi}{3}\left(\frac{8mL^2 E}{h^2}\right)^{3/2} = \frac{2^{9/2}\pi m^{3/2}L^3}{3h^3} E^{3/2}$$

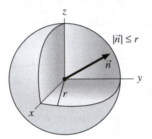

(b) Differentiating the above result, we find the density of states:

$$g(E) = \frac{d}{dE} N(E) = \frac{2^{7/2}\pi m^{3/2}L^3}{h^3} E^{1/2}$$

ASSESS   The density of states is the number of states with energy equal to E, as opposed to $N(E)$, which is the number of states with energy less than or equal to E. Notice that $N(E)$ is the integral of $g(E)$:

$$N(E) = \int_0^E g(E')dE'$$

From this, it is clear why $g(E)$ is the derivative of $N(E)$.

65.   **INTERPRET**   You need to specify the maximum current in a solenoid in order to avoid exposing the superconductors in your device to the critical magnetic field.

DEVELOP   Recall Equation 26.21, $B = \mu_0 nI$, for the magnetic field of a solenoid with n turns of wire per unit length. You want the maximum that the current can be in order to keep the magnetic field below half the critical field, i.e., $B < \frac{1}{2}B_{crit}$.

EVALUATE   Solving for the maximum current allowed gives

$$I_{max} = \frac{\frac{1}{2}B_{crit}}{\mu_0 n} = \frac{\frac{1}{2}(12\ \text{T})}{\left(4\pi \times 10^{-7}\ \frac{\text{T·m}}{\text{A}}\right)\left(75\ \text{m}^{-1}\right)} = 64\ \text{kA}$$

ASSESS   A magnetic field of 6 T is fairly high, so it makes sense that the maximum allowed current is also fairly high.

67.   **INTERPRET**   We examine the physics behind photovoltaic cells that use semiconductors to convert sunlight into electricity.

DEVELOP   In the previous problem, we showed how one can calculate the power emitted per unit area of the Sun for energies greater than the band gap energy of silicon:

$$P\left(E > E_{Si}\right) \approx \frac{2\pi}{h^3 c^2} \int_{E_{Si}}^{\infty} E^3 e^{-E/kT} dE$$

We can find the number of photons emitted by dividing out one factor of the energy from the integral:

$$N\left(E > E_{Si}\right) \approx \frac{2\pi}{h^3 c^2} \int_{E_{Si}}^{\infty} E^2 e^{-E/kT} dE$$

EVALUATE   We could evaluate the integral for $N\left(E > E_{Si}\right)$, but it's not necessary. It should be clear that dividing out one factor of E shifts more of the "weight" of the integral to smaller energies below E_{Si}. Therefore, the percentage of solar photons that silicon can absorb is less than the percentage of solar energy that silicon can absorb.

The answer is (a).

ASSESS   A simple way to think about this is that there are less photons with wavelengths smaller than the median wavelength (700 nm for the Sun) than there are with larger wavelengths, even though the sum of each set of photons has the same amount of energy. This is because photons at the short-wavelength-end of the spectrum carry more energy per photon.

69.   **INTERPRET**   We examine the physics behind photovoltaic cells that use semiconductors to convert sunlight into electricity.

DEVELOP   Light that is not absorbed by a semiconductor passes through it and therefore can be used by another semiconductor below it. In a multi-layer cell, the first *PN* junction from the top should absorb a small fraction of the solar spectrum, so that there is plenty of light left for the next *PN* junction below. This second junction should only absorb a small fraction of the remainder, so there is still plenty left for the third junction, and so on and so forth. See the figure below, showing a representation of the solar spectrum with the absorption limits of three staggered *PN* junctions.

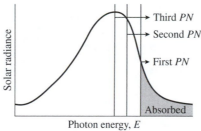

EVALUATE   A *PN* junction will absorb the light above its band gap energy. Looking at the figure, it's clear then that the first junction (closest to the top) should have the largest energy gap.

The answer is (a).

ASSESS   Since the solar spectrum has a median wavelength of 700 nm, the largest band gap should be at a shorter wavelength (higher energy) in the visible or ultraviolet. This shows that answer (c) is false. Similarly, the smallest band gap should be at infrared wavelengths longer than 700 nm, which contradicts answer (d).

NUCLEAR PHYSICS

EXERCISES

Section 38.1 Elements, Isotopes, and Nuclear Structure

13. **INTERPRET** This problem involves writing the conventional symbols for three isotopes of radon.

    DEVELOP The conventional symbol for a nucleus X is $_Z^A X$, where A is the mass number and Z is the atomic number.

    EVALUATE With the number of protons ($Z = 86$ for all radon isotopes) and neutrons ($N = A - Z$) given, the mass numbers A of the three isotopes are, respectively,

    $$A = Z + N = \begin{cases} 86 + 125 = 211 \\ 86 + 134 = 220 \\ 86 + 136 = 222 \end{cases}$$

    Therefore, the nuclear symbols are $_{86}^{211}\text{Ra}$, $_{86}^{220}\text{Ra}$, and $_{86}^{222}\text{Ra}$.

    ASSESS Isotopes of a given element have the same number Z of protons but different number of neutrons (and thus a different value for A).

15. **INTERPRET** We are given the symbol for two nuclei and asked to compare the number of nucleons and charges between them.

    DEVELOP The comparison can be made by noting that the conventional symbol for a nucleus X is $_Z^A X$ where A is the mass number and Z is the atomic number.

    EVALUATE (a) The mass number (number of nucleons) is $A = 35$ for both.

    (b) The charge Ze of a potassium nucleus ($Z = 19$) is two electronic charge units greater than that for a chlorine nucleus, which has $Z = 17$.

    ASSESS Equality in mass number A does not imply equality in atomic number Z. Two nuclei have the same Z only when they are isotopes.

17. **INTERPRET** We are to deduce the size of the fission products of $_{92}^{235}\text{U}$, and find the radius of the daughter nuclei, given that they are of the same size.

    DEVELOP The nuclear radius can be estimated using Equation 38.1:

    $$R = R_0 A^{1/3} = (1.2 \text{ fm}) A^{1/3}$$

    Given that the daughter nuclei are of the same size, the will each have half the number of nucleons of the parent nucleus, which has $A = 235$. Thus, the daughter nuclei will have $A = 235/2 = 117$ or 118.

    EVALUATE Inserting $A = 117$ or 118 into the expression above for the radius gives

    $$R = (1.2 \text{ fm}) A^{1/3} = (1.2 \text{ fm})(117)^{1/3} \approx 5.9 \text{ fm}$$

    ASSESS Equation 38.1 is a good approximation for R since nucleons are packed tightly into the nucleus. The tight packing also suggests that all nuclei have roughly the same density.

Section 38.2 Radioactivity

19. **INTERPRET**   In this problem we are asked to write down all possible beta-decay processes for $_{29}^{64}\text{Cu}$.

DEVELOP   As detailed by Equations 38.5a, 38.5b, and 38.5c, beta decay in $_{29}^{64}\text{Cu}$ can involve positron-neutrino or electron-anti-neutrino emission, or electron capture. In each reaction, charge and mass number must be conserved.

EVALUATE   The reactions are

$$_{29}^{64}\text{Cu} \rightarrow \, _{30}^{64}\text{Zn} + \beta^- + \bar{\nu} \quad (40\%)$$

$$_{29}^{64}\text{Cu} \rightarrow \, _{28}^{64}\text{Ni} + \beta^+ + \nu \quad (19\%)$$

$$_{29}^{64}\text{Cu} + e^- \rightarrow \, _{28}^{64}\text{Ni} + \nu \, (41\%)$$

ASSESS   In each decay mode, charge and mass number are conserved.

21. **INTERPRET**   This problem is an exercise in conversion between Ci and Bq.

DEVELOP   Section 38.2 gives the relation between Ci and Bq: $1 \text{ Ci} = 3.7 \times 10^{10}$ Bq.

EVALUATE   The activity of the milk sample in SI units is

$$\left(450 \text{ pCi/L}\right)\left(3.7 \times 10^{10} \text{ Bq/Ci}\right) = 16.7 \text{ Bq/L}$$

ASSESS   The more modern unit is the Bq.

23. **INTERPRET**   This is a problem about the decay of ^{90}Sr. We are to find the time it will take for the activity of $_{90}\text{Sr}$ to decay to the 99% and 99.9% of its original activity.

DEVELOP   From Table 38.1, we find the half-life of ^{90}Sr to be 29 years. Solving for the elapsed time t in Equation 38.3b, $N = N_0 2^{-t/t_{1/2}}$, the time elapsed since the bomb tests is

$$t = t_{1/2} \frac{\ln\left(N_0/N\right)}{\ln 2}$$

The fraction N/N_0 of ^{90}Sr that remains is one minus the fraction that decays.

EVALUATE   **(a)** For 99% of the radioactive contaminant to decay, or 1% to remain ($N/N_0 = 1.00 - 0.99 = 0.01$) the time required is

$$t = t_{1/2} \frac{\ln\left(N_0/N\right)}{\ln 2} = \left(29 \text{ y}\right)\frac{\ln\left(1/0.01\right)}{\ln 2} = 193 \text{ y}$$

(b) Similarly, for 99.9% of the radioactive contaminant to decay, or 0.1% to remain, the time required is

$$t = \left(29 \text{ y}\right)\frac{\ln\left(1/0.001\right)}{\ln 2} = 289 \text{ y}$$

ASSESS   Since $\left(1/2\right)^7 = 0.0078 < 0.01 < \left(1/2\right)^6 = 0.0156$, for 1% of contaminant to remain, $t/t_{1/2}$ must be between 6 and 7, so our result of $193/29 = 6.66$ is reasonable. Similarly reasoning also shows that our result for **(b)** is about right.

Section 38.3 Binding Energy and Nucleosynthesis

25. **INTERPRET**   We're given the binding energy per nucleon for $_{28}^{60}\text{Ni}$, and asked to find its atomic mass.

DEVELOP   Using Equation 38.7, the binding energy of a nucleus can be written as

$$E_b = Z m_p c^2 + \left(A - Z\right) m_n c^2 - m_N c^2$$

where m_b, m_n, and m_N are the masses of the protons, neutrons, and the nucleus, respectively. The total nuclear binding energy of $_{28}^{60}\text{Ni}$ is the number of nucleons ($A = 60$) times the given binding energy per nucleon, or

$$E_b = 60\left(8.8 \text{ MeV}\right)\left(\frac{1 \text{ u} \cdot c^2}{931.5 \text{ MeV}}\right) = 0.567 \text{ u} \cdot c^2$$

The result can be used to solve for the atomic mass of $_{28}^{60}\text{Ni}$.

EVALUATE   If we express Equation 38.7 in terms of atomic masses, by adding $Z = 28$ electron rest energies ($m_e c^2$) to both sides, and neglect atomic binding energies, we obtain:

$$M\left({}^{60}_{28}\text{Ni}\right) = 28M\left({}^{1}_{1}\text{H}\right) + (60-28)m_n - E_b/c^2 = 28(1.00783\text{ u}) + 32(1.00867\text{ u}) - 0.567\text{ u}$$
$$= 59.93\text{ u}$$

where we have retained only two figures after the decimal point because the binding energy is known with that precision.

ASSESS The actual binding energy of ${}^{60}\text{Ni}$ is so close to 8.8 MeV/nucleon that the accuracy of the atomic mass just calculated is better than one might expect from data given to two significant figures.

27. **INTERPRET** In this problem we are asked to find the binding energy per nucleon for ${}^{7}_{3}\text{Li}$ given its nuclear mass.

    DEVELOP Using Equation 38.7, the binding energy per nucleon of a nucleus can be written as

    $$\frac{E_b}{A} = \frac{1}{A}\left[Zm_p c^2 + (A-Z)m_n c^2 - m_N c^2\right]$$

    where m_p, m_n, and m_N are the masses of the protons, neutrons, and the nucleus, respectively.

    EVALUATE Substituting the values given, we find

    $$\frac{E_b}{A} = \frac{1}{7}\left[3(1.00728\text{ u})c^2 + (7-3)(1.00867\text{ u})c^2 - 7.01435\text{ u}\cdot c^2\right]\left(931.5\text{ MeV}\cdot\text{u}^{-1}\cdot c^{-2}\right)$$
    $$= 5.612\text{ MeV/nucleon}$$

    ASSESS The binding energy per nucleon as a function of A is plotted in Fig. 38.9. For very light nuclides such as ${}^{7}_{3}\text{Li}$, the energy is low because the nuclear force is not yet saturated for so few nucleons.

Section 38.4 Nuclear Fission

29. **INTERPRET** This problem is about the number of neutrons released in the fission of ${}^{235}\text{U}$. We are given the daughter nuclei produced in neutron-induced fission of ${}^{235}\text{U}$ and are asked to find the number of neutrons released.

    DEVELOP This reaction is described by the formula

    $$1{}^{1}_{0}\text{n} + {}^{235}_{92}\text{U} \rightarrow {}^{139}_{53}\text{I} + {}^{95}_{39}\text{Y} + X{}^{1}_{0}\text{n}$$

    Thus, the initial number of neutrons is $1 + (235 - 92) = 144$. The final number of neutrons, which must be the same because charge is conserved and the number of protons does not change, is $144 = (139 - 53) + (95 - 39) + X$.

    EVALUATE Solving the equation above for the number of neutrons X gives $X = 144 - (139 - 53) - (95 - 39) = 2$.

    ASSESS As expected, the fission process produces two middle-weight products with notably unequal masses. The number of neutrons released is in accordance with conservation of the number of nucleons.

31. **INTERPRET** We are given the power output in a fission reactor and energy released per fission event and are asked to find the fission rate.

    DEVELOP We assume that all of the energy released in fissions goes into thermal power. Thus,

    $$P = \frac{dE}{dt} = \frac{d(NE_1)}{dt} = E_1 \frac{dN}{dt}$$

    where $E_1 = 200$ MeV is the energy released per fission event and $P = 3.2$ GW.

    EVALUATE The fission rate dN/dt is

    $$\frac{dN}{dt} = \frac{P}{E_1} = \frac{3.2\text{ GW}}{(200\text{ MeV/fission})(1.602\times10^{-19}\text{ J/eV})} = 1.0\times10^{20}\text{ fissions/s}$$

    ASSESS A very high fission rate is required to achieve this power output.

Section 38.5 Nuclear Fusion

33. **INTERPRET** This problem is about the Lawson criterion for D-T fusion. We are interested in the density of the nuclei required to meet this criteria for the given confinement time.

    DEVELOP The Lawson criterion for D-T fusion is $n\tau > 10^{20}$ s/m^3 (Equation 38.11). This condition allows us to solve for n.

EVALUATE   For $\tau = 0.5$ s, a particle density of

$$n > \frac{10^{20} \text{ s/m}^3}{0.5 \text{ s}} = 2 \times 10^{20} \text{ m}^{-3}$$

would be required.

ASSESS   This is a very high density. From the Lawson criterion, we see that in order to achieve self-sustaining fusion, we must have a high nuclei density and confine them long enough such that the product $n\tau$ exceeds 10^{20} s/m^3. So far, the approach has not yet produced a sustained energy yield.

35.   INTERPRET   Using the Lawson criterion, we are to find the required confinement time for D-T fusion, given the plasma density.

DEVELOP   The D-T Lawson criterion is $n\tau > 10^{22}$ s/m^3 (Equation 38.10), and the number density given in the problem is $n = 10^{19}$ m^{-3}. We will solve for the confinement time τ.

EVALUATE   The confinement time must satisfy

$$\tau > \frac{10^{22} \text{ s} \cdot \text{m}^{-3}}{10^{19} \text{ m}^{-3}} \frac{10^{22} \text{ s·m}^{-3}}{10^{19} \text{ m}^{-3}} = 10^3 \text{ s}.$$

ASSESS   This confinement time is inconveniently long, but creating a greater particle density so as to decrease the time is also difficult.

PROBLEMS

37.   INTERPRET   We are to find the energy required to flip the spin state of a proton, which acts like a magnetic dipole, in the Earth's magnetic field.

DEVELOP   As discussed in Example 38.1, a proton acts like a magnetic dipole whose component along the magnetic field is $\mu_p = 1.41 \times 10^{-26}$ J/T. The energy needed to flip the spin (Equation 26.16) is

$$\Delta U = \mu_p B - \left(-\mu_p B \right) = 2\mu_p B$$

EVALUATE   With $B = 30$ μT we get

$$\Delta U = 2\mu_p B = 2(30 \text{ }\mu\text{T})(1.41 \times 10^{-26} \text{ J/T}) = 8.5 \times 10^{-31} \text{ J} = 5.3 \times 10^{-12} \text{ eV}$$

ASSESS   The frequency of a photon with this energy, 1.28 kHz, is in the audible range!

39.   INTERPRET   In this problem, we are asked to find the binding energy per nucleon for $^{56}_{26}$Fe.

DEVELOP   Using Equation 38.7, the binding energy of a nucleus can be written as

$$E_b = Zm_p c^2 + (A - Z)m_n c^2 - m_N c^2$$

where m_p, m_n, and m_N are the masses of the proton, neutron, and the nucleus, respectively.

EVALUATE   Since the nuclear mass is given, the above equation gives directly

$$E_b\left(^{56}_{26}\text{Fe}\right) = \left[26(1.00728 \text{ u}) + 30(1.00867 \text{ u}) - 55.9206 \text{ u} \right]c^2 = \left(0.5288 \text{ u} \cdot c^2\right)\left(931.5 \text{ MeV} \cdot \text{u}^{-1} \cdot c^{-2}\right)$$
$$= 492.6 \text{ MeV}$$

Therefore, the binding energy per nucleon is

$$\frac{E_b}{A} = \frac{492.5 \text{ MeV}}{56} = 8.796 \text{ MeV/nucleon}$$

The result is in good agreement with the curve given in Fig. 38.9.

ASSESS   This result is very close to the peak of the curve in Fig. 38.9. Nuclei with mass numbers around $A = 60$ are most tightly bound.

41.   INTERPRET   This problem is about the age of the Earth in half-lives of the isotopes specified.

DEVELOP   From Table 38.1, the half-lives are 5730 y for ^{14}C, 4.46×10^9 y for ^{238}U, and 1.25×10^9 y for ^{40}K. Starting with $N_0 = 10^6$, the number of atoms remaining after 4.5 billion years can be found with Equation 38.3b, $N = N_0 2^{-t/t_{1/2}}$.

EVALUATE   For C-14, the number of half-lives since the Earth formed is so large, $t/t_{1/2} = 7.85 \times 10^5$, that the number of these atoms left will be zero. The half-life of U-238 is roughly equal to Earth's lifetime, $t/t_{1/2} = 1.0$, so the number of atoms left is

$$N = N_0 2^{-t/t_{1/2}} = \left(10^6\right) 2^{-1.0} = 5 \times 10^5$$

Lastly, the number of half-lives for K-40 is $t/t_{1/2} = 3.6$, so

$$N = N_0 2^{-t/t_{1/2}} = \left(10^6\right) 2^{-3.6} = 8 \times 10^4$$

Thus, U-238 and K-40 are suitable for dating Earth's oldest rocks.

ASSESS   Uranium is often used for radiometric dating of rocks. A geologist can determine how many U-238 atoms a rock sample originally had by estimating the excess number of lead atoms (Pb-206) in it: $N_{U\text{-}238,0} = N_{U\text{-}238} + N_{Pb\text{-}206}$. This is because Pb-206 is what becomes of a U-238 atom when it decays, as shown in Figure 38.7.

43.   **INTERPRET**   You need to determine when a house with excess radon will be safe to reenter.

DEVELOP   The half-life of Rn-222 is 3.82 days, from Table 38.1. Since the activity is proportional to the number of atoms, you can use Equation 38.3b, $N = N_0 2^{-t/t_{1/2}}$, to see how long until the activity drops from 23 pCi/L to 4 pCi/L.

EVALUATE   Rearranging the equation for the number of atoms gives

$$t = t_{1/2} \frac{\ln\left(N/N_0\right)}{\ln\left(1/2\right)} = \frac{(3.82 \text{ d})}{\ln\left(1/2\right)} \ln\left(\frac{4 \text{ pCi/L}}{23 \text{ pCi/L}}\right) = 9.6 \text{ d}$$

ASSESS   This would be the maximum time to wait before returning to the house. The time can be less if windows are opened and the air is allowed to circulate.

45.   **INTERPRET**   This problem concerns the decay of $^{232}_{90}\text{Th}$ that results in a series of short-lived nuclei. We are to find the third daughter nuclei in the decay series of thorium-232, given the decay mechanism of the second daughter nuclei, and make a chart showing the first three steps in its decay.

DEVELOP   In a-decay, the numbers of neutrons and protons both decrease by two, whereas in β^--decay, the number of neutrons decreases by one and the number of protons increases by one (see Equations 38.4 and 38.5a). Thus, after one α- and two β^--decays, the number of neutrons is decremented by four and the number of protons is unchanged. Given that $A = 232$ and $Z = 90$ for thorium-232, the third daughter nuclei must have $A = 232 - 4 = 228$ and $Z = 90$.

EVALUATE   The third daughter nucleus is therefore $^{228}_{90}\text{Th}$, which is another thorium isotope.

(b) The chart is shown in the figure below. The half-lives are given in the figure.

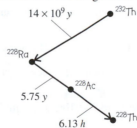

ASSESS   The reactions shown in the figure are

$$^{232}_{90}\text{Th} \rightarrow {}^{228}_{88}\text{Ra} + \alpha$$
$$^{228}_{88}\text{Ra} \rightarrow {}^{228}_{89}\text{Ac} + \beta^-$$
$$^{228}_{89}\text{Ac} \rightarrow {}^{228}_{90}\text{Th} + \beta^-$$

47.   **INTERPRET**   This is a problem about carbon-14 dating, which we are to use to deduce the age of a bone.

DEVELOP   Using Equation 38.3b, $N = N_0 2^{-t/t_{1/2}}$, the age t of the bone is given by

$$t = t_{1/2} \frac{\ln\left(N_0/N\right)}{\ln 2}$$

where N is the number of nuclei remaining when the bone is unearthed by the archeologists. From Table 38.1, the half-life of ^{14}C is $t_{1/2} = 5730$ y.

EVALUATE   From the above equation, we determine the age of the bone to be

$$t = t_{1/2}\frac{\ln(N_0/N)}{\ln 2} = (5730\text{ y})\frac{\ln(1/0.34)}{\ln 2} = 8.9\times10^3\text{ y}$$

ASSESS   As a quick check, we note that one half-life reduces the number to 50% of the original, two half-lives to 25%, three to 12.5%, and so on. In our ^{14}C dating case, we have $t = (8920\text{ y})/(5730\text{ y}) = 1.56t_{1/2}$. Therefore, we expect the fraction remaining to be between 25% and 50%. This is consistent with the 34% given in the problem statement.

49. **INTERPRET**   This problem is about the decay of ^{131}I in milk. We are given the initial activity per liter of milk, and we want to know how long it takes for the activity to decay to the level provided by the given safety guideline.

    DEVELOP   Using Equation 38.3a, $N = N_0 2^{-t/t_{1/2}}$ the waiting time t required can be written as

    $$t = t_{1/2}\frac{\ln(N_0/N)}{\ln 2}$$

    where the ratio N_0/N is the ratio of the initial activity to the final activity. The half-life of ^{131}I is $t_{1/2} = 8.04$ d.

    EVALUATE   For Poland, we find

    $$t = t_{1/2}\frac{\ln\text{ (reported level/safety guideline)}}{\ln 2} = (8.04\text{ d})\frac{\ln(2000/1000)}{\ln 2} = 8.04\text{ d}$$

    Similar calculations yield 16.2 d for Austria, and 10.0 d for Germany.

    ASSESS   As a quick check, we note that one half-life reduces the number to 50% of the original, two half-lives to 25%, three to 12.5%, and so on. For Poland, the waiting time is exactly one half-life since the initial reported level is twice the safety guideline. For Austria, $N/N_0 \approx 1/4 = 0.25$, so about two half-lives (16 days) of waiting time are required. Similar reasoning can be applied to the Germany case.

51. **INTERPRET**   This problem involves using the decay rate of ^{40}K to deduce the age of a rock.

    DEVELOP   Using Equation 38.3b, $N = N_0 2^{-t/t_{1/2}}$, the age t of the rock is given by

    $$t = t_{1/2}\frac{\ln(N_0/N)}{\ln 2}$$

    where N is the number of ^{40}K nuclei remaining and N_0 is the original number of ^{40}K nuclei. The half-life of ^{40}K is $t_{1/2} = 1.2\times10^9$ y. If 82% of the original ^{40}K decayed, then 18% remains in a rock of age t, so $N_0/N = 1/0.18$.

    EVALUATE   From the above equation and the given half-life, we get

    $$t = t_{1/2}\frac{\ln(N_0/N)}{\ln 2} = (1.2\times10^9\text{ y})\frac{\ln(1.0/0.18)}{\ln 2} = 3.0\times10^9\text{ y}$$

    Note: A type of lunar highlands rock rich in potassium (K), rare earth elements (REE), and phosphorus (P), is called KREEP norite.

    ASSESS   The age of the rock is on the same order as the age of the Earth, which is about 4.5 billion years old.

53. **INTERPRET**   The problem asks what fraction of a cancer-fighting agent will remain radioactive one year after the intervention.

    DEVELOP   The number of Ir-182 will drop relative to the initial amount according to $N/N_0 = 2^{-t/t_{1/2}}$, where the half-life is 74.2 days.

    EVALUATE   The percentage of radioactive iridium left after 365 days is

    $$\frac{N}{N_0} = 2^{-t/t_{1/2}} = 2^{-365/74.2} = 3.31\%$$

    ASSESS   Although the seeds remain radioactive for a considerable time, the gamma rays produced only travel a few millimeters through the body, so ideally only the tumor site is irradiated.

55.   **INTERPRET**   You're asked what is the long-term radioactivity of plutonium in terms of nuclear waste storage.

DEVELOP   The amount of plutonium will decrease by $2^{-t/t_{1/2}}$, where the half-life is 24,110 y.

EVALUATE   The fraction of Pu-239 after 1 million years is

$$\frac{N}{N_0} = 2^{-t/t_{1/2}} = 2^{-24,110/1e6} = 3 \times 10^{-13}$$

ASSESS   This is almost nothing, as we'd expect after 41 half-lives of plutonium. However, the plutonium itself is only part of the problem, since it's decay products will be radioactive as well. Notably, Pu-239 decays to U-235, which has a half-life of 704 million years.

57.   **INTERPRET**   For a 3.2-GW reactor, we are to determine how much ^{235}U would be needed to fuel the reactor for one year.

DEVELOP   The energy content of ^{235}U is (see Appendix C) 2.3×10^7 kWh/kg. Thus, we can calculate the amount needed to power a 3.2-GW reactor for one year.

EVALUATE   Letting the units guide us, we find that the mass m_U of ^{235}U needed to power the reactor for one year is

$$m_U = \frac{3.2 \text{ GW} \cdot \text{y}}{2.3 \times 10^7 \text{ kW} \cdot \text{h/kg}} \left(\frac{24 \text{ h}}{1 \text{ d}} \right) \left(\frac{365 \text{ d}}{1 \text{ y}} \right) = 1.3 \times 10^3 \text{ kg}$$

ASSESS   How much natural uranium do we need to obtain this amount of ^{235}U? Because natural uranium contains about 0.7% 235U, we would need $(1.3 \times 10^3 \text{ kg})/(0.007) \sim 2 \times 10^5$ kg of natural uranium. If we estimate the ratio of the densities of natural uranium (i.e., ^{238}U) and water by the ratio of their atomic masses, we can estimate that the volume of ^{238}U to be

$$V_U = \frac{m_U}{\rho_U} \sim \frac{m_U}{\rho_w (238/18)} = \frac{2 \times 10^5 \text{ kg}}{(1 \text{ kg})/(10^3 \text{ cm}^3)(238/18)} \left(\frac{1 \text{ m}}{10^2 \text{ cm}} \right)^3 = 15 \text{ m}^3$$

This would fit nicely into an average-sized truck, in agreement with Example 38.5.

59.   **INTERPRET**   This problem concerns a nonrelativistic collision between a neutron and a stationary deuteron, so conservation of momentum and energy are involved. We are to find the fraction of the kinetic energy that is transferred from the neutron to the deuteron.

DEVELOP   The particles involved in the fission reactions discussed all have nonrelativistic energies, so Equations 9.15a and 9.15b, for a head-on elastic collision between a neutron (mass m_n) and a deuteron initially at rest ($v_{d,i} = 0$) gives

$$v_{d,f} = \frac{2m_n v_{n,i}}{m_n + m_d}$$

The ratio of the initial to final kinetic energy is

$$\frac{K_{d,f}}{K_{n,i}} = \frac{m_d v_{d,f}^2}{m_n v_{n,i}^2}$$

Using the expression above for $v_{d,f}$ and the fact that $m_d = 2m_n$ allows us to calculate this ratio.

EVALUATE   The fraction of the neutron's initial kinetic energy transferred to the deuteron is therefore

$$\frac{K_{d,f}}{K_{n,i}} = \frac{m_d v_{d,f}^2}{m_n v_{n,i}^2} = \frac{4m_n m_d}{(m_n + m_d)^2} = \frac{8}{9} = 89\%$$

ASSESS   The remainder of the kinetic energy is retained by the neutron.

61. **INTERPRET**   We are to calculate the amount of ^{235}U consumed in one year in a reactor that produces 1.5 GW of power.

    DEVELOP   This problem is similar to Problem 38.57, except that we are asked to find the amount of ^{235}U required for the reactor instead of the amount of 238U. From Appendix C, we find that the energy content of pure ^{235}U is 2.3×10^7 kW·h/kg. The energy required for the reactor for a full year's operation is 1.5 GW·y, so we can calculate the mass of ^{235}U required.

    EVALUATE   The amount of ^{235}U required is

    $$m_{^{235}U} = \frac{1.5\,\text{GW}\cdot\text{y}}{2.3\times10^7\,\text{kW}\cdot\text{h/kg}}\left(\frac{24\,\text{h}}{1\,\text{d}}\right)\left(\frac{365\,\text{d}}{1\,\text{y}}\right) = 5.7\times10^2\,\text{kg}$$

    ASSESS   As expected, this is much less than the amount of natural uranium required (cf. Problem 38.57).

63. **INTERPRET**   This problem involves fission reactor dynamics. We are to find the time it takes for a 100-fold increase in power for a reactor that goes prompt critical.

    DEVELOP   This problem is similar to Example 38.6. Similar reasoning leads in this case to the equation $k^n = 10^2$, or $n \log k = 2$, where n is the number of generations. The time t required for this increase in power is the number of generations n multiplied by the generation time τ.

    $$t = n\tau = \frac{2\tau}{\log k}$$

    EVALUATE   Inserting $\tau = 100$ μs and $k = 1.001$ gives

    $$t = \frac{2(100\,\mu s)}{\log(1.001)} = 0.461\,\text{s}$$

    ASSESS   This is a very fast increase in power, which gives an indication of the care that must be taken to prevent the reaction from going prompt critical.

65. **INTERPRET**   This problem involves calculating the rate at which the Sun consumes protons given that 4 protons must combine to give 27 MeV. We are also to find how long the Sun's current proton-consuming phase will last.

    DEVELOP   The number of protons consumed per second is the power output divided by the energy release E per proton, where $E = (27\,\text{MeV})/4$ because 4 protons are required for each reaction that produces 27 MeV. The amount of hydrogen originally contained by the Sun is $(0.71)(2 \times 10^{30}\,\text{kg})/(1.67 \times 10^{-27}\,\text{kg/proton}) = 8.5 \times 10^{55}$ protons, so we can find how long it takes to consume 10% of this by dividing by the consumption rate.

    EVALUATE   (a) The proton consumption rate dN_p/dt is about

    $$\frac{dN_p}{dt} = \frac{4\times10^{26}\,\text{W}}{27\,\text{MeV/proton}}\left(\frac{1\,\text{MeV}}{1.602\times10^{-13}\,\text{J}}\right) = 3.7\times10^{38}\,\text{s}^{-1}$$

    (b) The consumption of 8.5×10^{55} protons at the rate found in part **(a)** would take about

    $$\frac{8.5\times10^{55}\,\text{protons}}{3.7\times10^{38}\,\text{protons/s}} = 2.3\times10^{17}\,\text{s} = 7.3\times10^9\,\text{y}$$

    ASSESS   The present age of the Sun is about 4.5 billion years.

67. **INTERPRET**   To continue the argument of the previous problem, you want to estimate how long nuclear fusion could supply the human population with energy it demands.

    DEVELOP   Since you have already calculated the energy density of seawater in the previous problem $(\rho_{H2O} = 4.4\times10^{10}\,\text{J/gal})$, you need now only estimate the number of gallons in the world's oceans. Assuming that 70% of the Earth's area $(A = 4\pi R_E^2)$ is covered with water to an average depth of 3 km, the volume of water is
    $$V_{H2O} = 0.7\cdot4\pi R_E^2 d = 0.7\cdot4\pi(6.37\times10^6\,\text{m})^2(3\,\text{km}) \approx 10^{18}\,\text{m}^3 \approx 3\times10^{20}\,\text{gal}$$

    EVALUATE   Using the planet's water supply, the total energy available from fusion is

$$E_{tot} = \rho_{H2O} V_{H2O} \simeq \left(4.4 \times 10^{10} \text{ J/gal}\right)\left(3 \times 10^{20} \text{ gal}\right) \simeq 1.3 \times 10^{31} \text{ J}$$

Assuming humans continue to consume energy at the current rate of 15 TW, fusion energy could last for

$$t = \frac{E_{tot}}{P} \simeq \frac{1.3 \times 10^{31} \text{ J}}{15 \times 10^{12} \text{ W}} \approx 8 \times 10^{17} \text{ s}$$

This is about 26 billion years, or roughly 20 billion years longer than the Sun will shine.

ASSESS This is a very rough calculation, but even so it should make it clear that fusion energy is effectively limitless once the technological hurdles can be overcome.

69. **INTERPRET** We want to identify the product of a nuclear fusion reaction.

DEVELOP Bismuth has atomic number $Z = 83$; chromium has atomic number $Z = 24$. Since the nuclear reaction does not produce any beta particles, we can assume that the number of protons and the number of neutrons are both conserved separately.

EVALUATE To conserve the proton number, the final product must have an atomic number equal to the sum of reactants' atomic numbers:

$$Z = 83 + 24 = 107$$

Looking in the periodic table, this corresponds to the element bohrium (Bh). To find its mass number, we need to account for the extra neutron produced in the reaction:

$$^{209}_{83}\text{Bi} + ^{54}_{24}\text{Cr} \rightarrow ^{262}_{107}\text{B} + ^{1}_{0}n$$

So the heavy nucleus is borhium-262 $\left(^{262}_{107}\text{Bh}\right)$.

ASSESS This isotope of borhium has a half-life of only about a tenth of a second. There are no stable forms of this element, so it is not found naturally in the environment. It was first synthesized in a laboratory in the 1980s.

71. **INTERPRET** This problem involves finding the products of a nuclear decay reaction. In addition, given the decay constant, we are to find the time required for the daughter nuclei to outnumber the parents by 2 to 1.

DEVELOP In a β^--decay, the atomic number of the parent nucleus increases by one while the mass number stays the same (i.e., $Z \rightarrow Z + 1$ and $A \rightarrow A$). Given the Ni has $Z = 28$ and $A = 65$, we can identify the daughter nucleus. The number of parent nuclei at time t (in a sample that was pure at $t = 0$) is $N = N_0 e^{-\lambda t}$ (see Equation 28.3a) which, when we solve for t, gives

$$t = \frac{\ln\left(N_0/N\right)}{\lambda}$$

Since a single daughter nucleus is produced each time a parent nuclei is consumed in the reaction, the number of daughter nuclei is $N_d = N_0 - N$, which allows us to find the time t when $N_d = 2N$ (recall that N is the number of parent nuclei).

EVALUATE (a) The daughter nuclei is $^{65}_{29}\text{Cu}$.

(b) When $N_d = 2N$, N0/N = 3, so the time required is

$$t = \frac{\ln 3}{\lambda} = \frac{\ln 3}{0.275 \text{ h}^{-1}} = 3.99 \text{ h}$$

ASSESS The reaction may be written as

$$^{65}_{28}\text{Ni} \rightarrow ^{65}_{29}\text{Cu} + \beta^-$$

73. **INTERPRET** This problem is about the energy released in D-T fusion. We also want to compare the fusion energy with the energy produced by a coal-burning power plant.

DEVELOP Half of the number of deuterons in one fuel pellet is

$$N = \left(\frac{1}{2}\right)\frac{2.5 \text{ mg}}{(2 \text{ u})\left(1.66 \times 10^{-27} \text{ kg/u}\right)} = 3.77 \times 10^{20}$$

since there is 2.5 mg of deuterium in a pellet and the mass of a deuteron is approximately 2 u. As given in Equation 38.10a, each D-T reaction releases $E_1 = 17.6$ MeV of energy.

EVALUATE   (a) The total amount of energy released in the D-T fusion reaction is

$$E = NE_1 = \left(3.77 \times 10^{20}\right)\left(17.6 \text{ MeV/fusion}\right)\left(1.602 \times 10^{-13} \text{ J/MeV}\right) = 1.1 \text{ GJ}$$

(b) The thermal power output, 3.000 GW, is equal to the energy release per pellet times the rate that pellets are consumed, hence this rate dN_{pel}/dt of pellet consumption is

$$\frac{dN_{pel}}{dt} = \frac{3.000 \text{ GW}}{1.06 \text{ GJ}} = 2.8 \text{ s}^{-1}$$

(c) 5.0-mg pellets, consumed at this rate for a year, have a total mass of

$$\left(5.0 \text{ mg}\right)\left(2.83 \text{ s}^{-1}\right)\left(3.156 \times 10^7 \text{ s/y}\right)\left(1 \text{ y}\right) = 450 \text{ kg}$$

A comparable coal-burning power plant uses more than 7 million times this mass for its fuel.

ASSESS   The amount of energy released by D-T fusion is enormous compared to the energy content of coal fuel. Controlled fusion devices could play a key role in future energy production.

75.   **INTERPRET**   We are to find the value of Z within the liquid-drop model that gives the minimum nuclear mass as a function of A. This value of Z will be the most stable of the nuclei for that atomic mass.

DEVELOP   The formula given us for the liquid-drop model is

$$M(A,Z) = c_1 A - c_2 Z + \left(c_2 A^{-1} + c_3 A^{-1/3}\right) Z^2$$

We shall take the partial derivative $\partial M/\partial Z$ and set it equal to zero to find Z_{min}.

EVALUATE   Taking the derivative gives

$$\frac{\partial M}{\partial Z} = 0 = -c_2 + 2\left(\frac{c_2}{A} + \frac{c_3}{A^{1/3}}\right) Z$$

$$Z_{min} = \frac{c_2}{2\left(c_2/A + c_3/A^{1/3}\right)} = \frac{A}{2\left[1 + \left(\frac{c_3}{c_2}\right) A^{2/3}\right]}$$

ASSESS   To check that this is a minimum, we take the second partial derivative, evaluate it at Z_{min}, and verify that the result is positive:

$$\frac{\partial^2 M}{\partial Z^2} = 2\left(\frac{c_2}{A} + \frac{c_3}{A^{1/3}}\right)$$

which is positive if $c_2/A > -c_3/A^{1/3}$.

77.   **INTERPRET**   When one radionuclide decays into another radioactive nucleus, the total activity of the sample becomes somewhat complicated. Here we will calculate the total activity for a two-step decay reaction, starting with the differential equations that define activity.

DEVELOP   The activity of a sample is $R(t) = \lambda N(t)$ so if we find $N(t)$ for each radionuclide, we can find the activity

$$R(t) = \lambda_A N_A(t) + \lambda_B N_B(t)$$

The equation for the parent nuclide is found in the usual way—it does not depend on the daughter at all. Thus, it is given by Equation 38.3a:

$$dN_A/dt = -\lambda_A N_A$$
$$N_A(t) = N_0 e^{-\lambda_A t}$$

The differential equation for the daughter nuclide has the usual decay term, but it also has a growth term since each decay of a parent nucleus creates a daughter nucleus. Thus,

$$dN_B/dt = -\lambda_B N_B + \left(-dN_A/dt\right)$$

where we *subtract* the rate of change in the number of parent nuclides because a decrease in the parent (i.e., dN_A/dt < 0) count is an increase in the daughter count. We will solve this second differential equation for $N_B(t)$ and then find the activity $R(t)$.

EVALUATE   The rate of change in the daughter nuclide is

$$\frac{dN_B}{dt} = -\lambda_B N_B + \left(-\frac{dN_A}{dt}\right) = -\lambda_B N_B + \lambda_A N_A = -\lambda_B N_B + \lambda_A N_0 e^{-\lambda_A t}$$

Try a solution of the form $N_B(t) = A\left(e^{-\lambda_A t} - e^{-\lambda_B t}\right)$. Inserting this into the differential equation for $N_B(t)$ to find the coeficient A gives

$$A\left(-\lambda_A e^{-\lambda_A t} + \lambda_B e^{-\lambda_B t}\right) = -A\lambda_B\left(e^{-\lambda_A t} - e^{-\lambda_B t}\right) + \lambda_A N_0 e^{-\lambda_A t}$$

$$-A\lambda_A e^{-\lambda_A t} = -A\lambda_B e^{-\lambda_B t} + \lambda_A N_0 e^{-\lambda_A t}$$

$$A(\lambda_B - \lambda_A) = \lambda_A N_0$$

$$A = \frac{\lambda_A N_0}{\lambda_B - \lambda_A}$$

so

$$N_B(t) = \frac{\lambda_A N_0}{\lambda_B - \lambda_A}\left(e^{-\lambda_A t} - e^{-\lambda_B t}\right)$$

Combining this with $N_B(t)$ to get the total activity $R(t)$ gives

$$R(t) = N_0 e^{-\lambda_A t} + \frac{\lambda_A N_0}{\lambda_B - \lambda_A}\left(e^{-\lambda_A t} - e^{-\lambda_B t}\right)$$

$$= \frac{N_0}{\lambda_B - \lambda_A}\left(\lambda_B e^{-\lambda_A t} - \lambda_A e^{-\lambda_B t}\right)$$

ASSESS   Interestingly, the activity can actually *increase* with time! This is a serious problem right now with some of the radioactive holding tanks left over from the production of bomb-making material during WWII—their activity is increasing, and is expected to continue increasing for several more decades.

79.   **INTERPRET**   You're curious whether enough of a short-lived radioisotope is being made for a medical scan planned for one of your family members.

DEVELOP   The initial mass of Tc-99* will be reduced to $M = M_0 2^{-t/t_{1/2}}$ (a variation on Equation 38.3b) during the 90 minutes that it takes to transport it to the nuclear medicine department. Although you are given the half-life of the parent isotope, molybdenum-99, the only relevant half-life is that of the excited state of technetium-99, $t_{1/2} = 6.01$ h.

EVALUATE   Writing 90 min as 1.5 h, the amount of the initial 12 mg of Tc-99* that remains is

$$M = M_0 2^{-t/t_{1/2}} = (12\text{ mg})2^{-1.5/6.01} = 10.1\text{ mg}$$

This is (barely) above the required 10 mg, so yes, the initial amount will be enough.

ASSESS   It might seem like the hospital staff is not leaving themselves much room for delay, but they presumably do not want to over-produce the amount of Tc-99*, since that would increase costs, as well as increase the risks from handling this radioactive material.

81.   **INTERPRET**   We will explore the circumstances surrounding a natural nuclear fission reaction that occurred 2 billion years ago.

DEVELOP   We want to know what the ratio between U-235 and U-238 was two billion years ago in natural uranium. In the previous problem, we showed that $N_0/N = 7.2$ for U-235. For U-238 with $t_{1/2} = 4.46\times10^9$ y,

$$\frac{N_0}{N} = 2^{t/t_{1/2}} = 2^{2/4.46} = 1.4$$

EVALUATE   Two billion years ago, a typical sample of uranium would have as its ratio of U-235 to U-238:

$$\frac{N_{235,0}}{N_{238,0}} = \frac{N_{235}}{N_{238}}\left(\frac{N_{235,0}/N_{235}}{N_{238,0}/N_{238}}\right) = (0.7\%)\left(\frac{7.2}{1.4}\right) = 3.7\% \approx 4\%$$

The answer is (b).

ASSESS   Since U-235 decays 6 times faster than U-238, there had to be relatively more of U-235 in the past.

83.   **INTERPRET**   We will explore the circumstances surrounding a natural nuclear fission reaction that occurred 2 billion years ago.

DEVELOP   There are many radioactive products from nuclear fission, and so we might expect small amounts of the three given isotopes were created at Oklo.

EVALUATE   The half-lives of Sr-90, C-14 and Pu-239 are, respectively, $29\,\text{y}$, $5730\,\text{y}$ and $24,000\,\text{y}$. Over the 2 billion years since the fission reactions ended, there have been over 60 million lifetimes of strontium-90, 350,000 lifetimes of carbon-14, and over 80,000 lifetimes of plutonium-239. Clearly, there will no longer be any trace of these elements.

The answer is (d).

ASSESS   Scientists have found measurable amounts of other fission products at Oklo, such as particular isotopes of neodymium and ruthenium. These products happen to be stable, so there's no risk of them decaying away like the ones in this problem.

FROM QUARKS TO THE COSMOS

39

EXERCISES

Section 39.1 Particles and Forces

19. **INTERPRET** This problem involves finding the lifetime of a virtual photon by applying the uncertainty principle. If the photon were to live longer than the time allowed by the uncertainty principle, conservation of energy would be violated.

     DEVELOP In order to test the conservation of energy in a process involving a single virtual photon, a measurement of energy with uncertainty less than the photon's energy $\Delta E < hc/c$ must be performed in a time interval Δt less than the virtual photon's lifetime (i.e., $\Delta t < \tau$). Thus, $\Delta E \Delta t < hc\tau/\lambda$. But Heisenberg's uncertainty principle limits the product of these uncertainties to $\Delta E \, \Delta t \leq \hbar$, so $hc\tau/\lambda > \hbar$. Knowing the wavelength allows us to find the upper bound on the lifetime of the virtual photon.

     EVALUATE With $\lambda = 633$ nm, we get

$$\tau > \frac{\lambda}{2\pi c} = \frac{633 \times 10^{-9} \text{ m}}{2\pi \left(3.00 \times 10^8 \text{ m/s}\right)} = 3.36 \times 10^{-16} \text{ s}$$

     ASSESS If the lifetime of a virtual photon of wavelength 633 nm were less than 0.336 fs, no measurement showing a violation of conservation of energy would be possible.

Section 39.2 Particles and More Particles

21. **INTERPRET** We are to write the reaction for the decay of a positive pion to a muon and a neutrino.

     DEVELOP The decay process must conserve both charge and lepton number L (see Table 39.1). The positive pion has charge $+e$ and lepton number 0. The positive muon has charge $+e$ and lepton number -1, because it is an antiparticle. To conserve change and lepton number, the neutrino must therefore have charge zero and lepton number 1. The muon neutrino satisfies these conservation requirements.

     EVALUATE The decay of the positive pion is

$$\pi^+ \rightarrow \mu^+ + \nu_\mu$$

     ASSESS Note that the lepton numbers of antiparticles are the opposite of those of their corresponding particles.

23. **INTERPRET** We are asked to verify the conservation laws for the decay of η^0 into a positive, a negative, and a neutral pion.

     DEVELOP The decay process $\eta \rightarrow \pi^+ + \pi^- + \pi^0$ must conserve charge, strangeness, and baryon number (see Table 39.1).

     EVALUATE The decay conserves charge ($0 = 1 - 1 + 0$), and all the particles have zero baryon number and strangeness, so these quantities are conserved as well.

     ASSESS The problem illustrates how conservation laws restrict the possible decay modes of a particle.

25. **INTERPRET** In this problem we are asked to apply conservation laws to decide if the give interaction is possible.

     DEVELOP The relevant properties to be considered here are charge, baryon number, and spin (see Table 39.1).

     EVALUATE The reaction $p + p \rightarrow p + \pi^+$ violates the conservation of baryon number $(0 + 0 \neq 0 + 1)$, and also angular momentum, since the proton's spin is $\frac{1}{2}$ and the spin of the pion is 0. Therefore, the process is not allowed in the standard model with electro-weak unification.

     ASSESS The problem illustrates how conservation laws restrict the possible outcomes of particle interactions.

Section 39.3 Quarks and the Standard Model

27.   **INTERPRET**   We want to know the quark composition of a baryon with strangeness −3.

   DEVELOP   A baryon is a particle that consists of three quarks. A strange quark has strangeness s = −1 and charge −e/3 (see Table 39.2).

   EVALUATE   The baryon that is composed of three strange quarks with strangeness s = −3 and charge −e is the $\Omega^- = sss$.

   ASSESS   Gell-Mann's prediction of the existence of Ω^- was confirmed experimentally and he received the Nobel Prize in 1969 for his work on sub-atomic particles.

Section 39.4 Unification

29.   **INTERPRET**   The Kamiokande experiment uses 50,000 tons of water and is designed to detect rare nuclear reactions such as neutrino interactions and hypothetical proton decays. We want to estimate the volume of the water.

   DEVELOP   The mass of 50,000 tons of water is (see Appendix C for the conversion factors)

$$M = (5 \times 10^4 \text{ tons})(2000 \text{ lb/ton})(0.454 \text{ kg/lb}) = 4.54 \times 10^7 \text{ kg}$$

   The volume is $V = M/\rho$ where ρ is the density of water.

   EVALUATE   At the ordinary density of $\rho = 1.00 \times 10^3$ kg/m^3, this amount of water occupies a volume of

$$V = M/\rho = \frac{4.54 \times 10^7 \text{ kg}}{1.00 \times 10^3 \text{ kg/m}^3} = 4.54 \times 10^4 \text{ m}^3 = 1.20 \times 10^7 \text{ gal}$$

   ASSESS   This is the volume of a cube of side length

$$L = \left(4.54 \times 10^4\right)^{1/3} \text{ m} = 35.7 \text{ m} = 117 \text{ ft.}$$

31.   **INTERPRET**   In this problem we want to estimate the temperature in a gas of particles such that the thermal energy is about 10^{15} GeV; the energy of grand unification.

   DEVELOP   The temperature corresponding to the energy where the strong and electro-weak forces unify is given by $U_{GUT} = 10^{15}$ GeV $= 10^{24}$ eV $= k_B T$, where $k_B = 8.617 \times 10^{-5}$ eV·K^{-1} is Boltzmann's constant.

   EVALUATE   The temperature is

$$T_{GUT} = \frac{U_{GUT}}{k_B} = \frac{10^{24} \text{ eV}}{8.167 \times 10^{-5} \text{ eV·K}^{-1}} \approx 10^{28} \text{ K}$$

   ASSESS   This is an extremely high temperature! According to the inflationary Big Bang Theory, at about 10^{-35} s after the Big Bang, the grand-unified interaction breaks up into strong and electro-weak interactions.

Section 39.5 The Evolving Universe

33.   **INTERPRET**   We use Hubble's law to calculate the distance to a galaxy that is receding at a given speed.

   DEVELOP   Inverting Hubble's law (Equation 39.1) gives the distance in terms of the speed: $d = v/H_0$, where the Hubble constant is $H_0 = 22.7$ km/s·Mly.

   EVALUATE   For a recession speed of $v = 2 \times 10^4$ km/s, the distance in light years is:

$$d = \frac{v}{H_0} = \frac{2 \times 10^4 \text{ km/s}}{22.7 \text{ km/s·Mly}} = 900 \text{ Mly}$$

   ASSESS   The light that we are receiving from this galaxy left it 900 million years ago. When astronomers study galaxies at such distances, they are really studying what the universe was like millions of years ago.

35.   **INTERPRET**   We are asked to find the age of the universe based on a given Hubble constant.

   DEVELOP   As discussed in Example 39.2, the age of the universe is given by $t = 1/H_0$, where H_0 is the Hubble constant.

   EVALUATE   With $H_0 = 25$ km/s·Mly, the age of the universe would be

$$t = \frac{1}{H_0} = \frac{1 \text{ Mly}}{25 \text{ km/s}} \left[\frac{\left(3 \times 10^5 \text{ km/s}\right) \cdot \left(1 \text{ y}\right)}{1 \text{ ly}} \right] = 12 \text{ Gy}$$

ASSESS   Note that we used the definition of a light year as the distance that light travels in one year.

PROBLEMS

37.   **INTERPRET**   We are asked to apply conservation laws to decide whether or not the given interactions are possible.

     DEVELOP   The relevant properties to be considered here are charge, baryon number, and spin (see Table 39.1).

     EVALUATE   **(a)** The decay, $\Lambda^0 \rightarrow \pi^+ + \pi^-$ conserves charge $[0 = 1 + (-1)]$, but violates conservation of baryon number $(1 \neq 0 + 0)$ and angular momentum because the spin of Λ^0 is $\frac{1}{2}$ and the spin of each pion is 0. Thus, this reaction is not allowed.

     (b) The decay $K^0 \rightarrow \pi^+ + \pi^-$ conserves angular momentum (at zero), charge (at zero), and lepton number (at zero), but does not conserve strangeness, which is 1 for the kaon and zero for each pion. However, strangeness is not conserved by the weak interaction, so this decay is possible.

     ASSESS   The decay of part (b) is an observed weak interaction. This problem illustrates how conservation laws restrict the possible outcomes of particle interactions.

39.   **INTERPRET**   This problem is about the conservation laws in the hypothetical proton decay suggested by the grand unification theory. We are to find if baryon number and charge are conserved in this reaction.

     DEVELOP   The relevant properties we wish to consider in the hypothetical decay $p \rightarrow \pi^0 + e^+$ are charge and baryon number (given in Table 39.1).

     EVALUATE   **(a)** The hypothetical decay $p \rightarrow \pi^0 + e^+$ does not conserve baryon number (which is 1 for the proton and 0 for mesons and leptons), nor does it conserve lepton number.

     (b) The hypothetical decay does conserve charge $(1 = 0 + 1)$.

     ASSESS   Baryon number is conserved in the Standard Model but not in the Grand Unification Theory.

41.   **INTERPRET**   We are to determine the composition of the J/ψ particle, which includes charm quarks but has charm $c = 0$.

     DEVELOP   The J/ψ particle is a meson. A meson is a particle that consists of one quark and one antiquark, so its baryon number is zero. A charmed quark (c) has charm c = +1, whereas an anticharmed quark $(\overline{c})$ has $c = -1$.

     EVALUATE   The J/ψ particle must have quark content $c\overline{c}$ in order to have zero net charm.

     ASSESS   Charmed is one of the six flavors of quarks (up, down, strange, charmed, bottom, and top).

43.   **INTERPRET**   This problem explores some properties of the Tevatron accelerator at Fermilab.

     DEVELOP   The conversion from eV to J can be done by noting that one TeV is equal to 1012 eV and $1 \text{ eV} = 1.602 \times 10^{-19}$ J (see Appendix C). To answer part (b), note that the energy of a particle of mass m falling a distance y in Earth's gravitational field is $U = mg\Delta y$ (see Equation 7.3).

     EVALUATE   **(a)** $1 \text{ TeV} = \left(10^{12} \text{ eV}\right)\left(1.602 \times 10^{-19} \text{ J/eV}\right) \approx 0.16 \text{ μJ}$.

     (b) Setting $mg\Delta y = 1$ TeV with $m = 1$ g, the height from which the mass must drop is

$$\Delta y = \frac{U}{mg} = \frac{0.16 \text{ μJ}}{\left(1.0 \times 10^{-3} \text{ kg}\right)\left(9.8 \text{ m/s}^2\right)} = 16 \text{ μm}$$

     ASSESS   This is a very small height, but remember that 1 g of protons contains about 6×10^{23} particles, whereas each proton in the Tevatron beam has this energy.

45.   **INTERPRET**   We want to estimate the time it takes for a 7-TeV proton to travel a circular path with a circumference of 27 km.

     DEVELOP   With a kinetic energy $K = 7$ TeV, the speed of a proton $(m_p = 938 \text{ MeV}/c^2)$ is very close to the speed of light (see preceding problem). Thus, the time in the lab frame that it takes to complete one circuit is given by $t = d/c$, where $d = 27$ km.

EVALUATE   Using $c = 3.0 \times 10^5$ km/s, the time required to travel 27 km is

$$t = \frac{d}{c} = \frac{27 \text{ km}}{3.0 \times 10^5 \text{ km}} = 90 \text{ } \mu s$$

ASSESS   Light travels about 3×10^5 km in one second. Since the speed of the proton is so close to the speed of light, the time it takes to travel 27 km is very short. In the proton's frame, the time t' it takes to travel the distance is

$$t' = \frac{t}{\gamma} = \frac{90 \text{ } \mu s}{7.46 \times 10^3} = 12 \text{ ns}$$

(see Equation 33.3).

47.   **INTERPRET**   We are to estimate the diameter of the Sun if it had the same density as the critical density of the universe. We will assume that the mass of the Sun is the same, and use the value for ρ_c obtained in the Problem 39.46.

DEVELOP   The mass of the Sun is M = 1.99×1030 kg (see Appendix E) and the critical density (from Problem 39.44) is

$$\rho_c = 1.03 \times 10^{-26} \text{ kg/m}^3$$

The volume of a sphere is $V = \frac{4}{3}\pi r^3$, and $\rho = m/V$. We shall solve for r.

EVALUATE

$$\rho_S = \frac{M_S}{\frac{4}{3}\pi r^3} = \rho_c$$

$$r = \sqrt[3]{\frac{3M_S}{4\pi\rho_c}} = \sqrt[3]{\frac{3(1.99 \times 10^{30} \text{ kg})}{4\pi(1.03 \times 10^{-26} \text{ kg/m}^3)}} = 3.59 \times 10^{18} \text{ m} = 379 \text{ ly}$$

ASSESS   The universe is really not a very dense place at all. It's only dense in small, very widely separated regions such as planets, solar systems, galaxies, and so on.

49.   **INTERPRET**   In this problem we want to find the size and the ground-state energy of a muonic atom, which consists of a proton at the nucleus and a muon in the place of the electron.

DEVELOP   We can use the results for the Bohr atom, with $m_\mu = 207m_e$ replacing m_e (see Equations 34.12a and 34.13). Thus, the radius of the muonic atom is

$$a_\mu = \frac{\hbar^2}{m_\mu ke^2} = \frac{\hbar^2}{207m_e ke^2} = \frac{1}{207}a_0$$

and its ground-state energy is

$$E_n^{(\mu)} = -\frac{ke^2}{2a_\mu}\left(\frac{1}{n^2}\right) = -\frac{ke^2}{2(a_0/207)}\left(\frac{1}{n^2}\right) = 207E_n$$

where a_0 and E_n are the Bohr radius and ground-state energy of the hydrogen atom, respectively.

EVALUATE   (a) For the ground state ($n = 1$),

$$r_1 = \frac{a_0}{207} = \frac{0.0529 \text{ nm}}{207} = 256 \text{ fm}$$

(b) The ground-state energy is $E_1^{(\mu)} = 207E_1 = 207 \times (-13.6 \text{ eV}) = -2.81 \text{ keV}$.

ASSESS   The size of the muonic atom is reduced by a factor of 207 with respect to the hydrogen atom, and its ground-state energy is increased (i.e., more negative) by the same factor of 207.

51.   **INTERPRET**   The problem involves finding a galaxy's recession speed from the wavelength shift in one of its emission lines.

DEVELOP   We have a moving source (the galaxy) that is receding from us the observer, so we use Equation 14.14b for the Doppler shift:

$$\lambda' = \lambda\left(1 + \frac{u}{v}\right)$$

where u in this case is the recession speed of the galaxy, and $v = c$ is the speed of the electromagnetic waves that we are receiving. We are told that the hydrogen-β spectral line $(\lambda = 486.1\ \text{nm})$ is shifted to $\lambda' = 495.4\ \text{nm}$.

EVALUATE   (a) Solving for the recession speed, we find

$$u = c\left(\frac{\lambda'}{\lambda} - 1\right) = (3 \times 10^5\ \text{km/s})\left(\frac{495.4\ \text{nm}}{486.1\ \text{nm}} - 1\right) = 5.740 \times 10^3\ \text{km/s}$$

(b) Plugging this speed into the Hubble law, we find

$$d = \frac{u}{H_0} = \frac{5.740 \times 10^3\ \text{km/s}}{22.7\ \text{km/s} \cdot \text{Mly}} = 253\ \text{Mly}$$

ASSESS   The speed in part (a) is less than 2% of the speed of light, so we are justified in using the nonrelativistic Doppler formula.

53.   **INTERPRET**   This problem explores using the width of the measured energy distribution of a particle to estimate its lifetime using the energy-time uncertainty principle.

     DEVELOP   In the energy-time uncertainty relation (Equation 34.16), $\Delta t = \tau$ can be taken to be the lifetime of the particle (i.e., the time available for the measurement) and $\Delta E = \Gamma$ to be its width (i.e., the spread in measured rest energies). Thus, $\tau = \hbar/\Gamma$ (this is, in fact, the definition of the width).

     EVALUATE   For the particle Z^0, a width $\Gamma = 2.5\ \text{GeV}$ implies a lifetime

$$\tau = \frac{\hbar}{\Gamma} = \frac{6.582 \times 10^{-16}\ \text{eV} \cdot \text{s}}{2.5\ \text{GeV}} = 2.6 \times 10^{-25}\ \text{s}$$

     ASSESS   These extremely short-lived particles are called "resonances."

55.   **INTERPRET**   We are to compare the age of the universe obtained from the two different values of the Hubble constant, using the procedure given in Example 39.2.

     DEVELOP   In Example 39.2, we see that the age of the universe is $t = H_0^{-1}$. The two values of the Hubble constant given are $H_0' = 17\ \text{km} \cdot \text{s}^{-1} \cdot \text{Mly}^{-1}$ and $H_0 = 22.7\ \text{km} \cdot \text{s}^{-1} \cdot \text{Mly}^{-1}$.

     EVALUATE   We use $t = H_0^{-1}$. For $H_0 = 22.7\ \text{km} \cdot \text{s}^{-1} \cdot \text{Mly}^{-1}$, t = 13.2 Gy (as shown in the example), and for $H_0' = 17\ \text{km} \cdot \text{s}^{-1} \cdot \text{Mly}^{-1}$, $t' = 18$ Gy.

     ASSESS   The larger our estimate of the Hubble constant, the lower the age of the universe.

57.   **INTERPRET**   We are to find the value of the Hubble constant for a 60-Gy-old universe, assuming the expansion of the universe is constant.

     DEVELOP   We shall use $t = H_0^{-1}$ for the age of the universe but solve for H_0 using $t = 60$ Gy as the age of the universe instead of the accepted value of $t = 13.2$ Gy (see Problem 39.55).

     EVALUATE   The new value of the Hubble constant would be $H_0 = t^{-1} = 5.0\ \text{km} \cdot \text{s}^{-1} \cdot \text{Mly}^{-1}$.

     ASSESS   The Hubble constant is not really a constant—it just seems like it is because it changes rather slowly at the current age of the universe.

59.   **INTERPRET**   We consider an earlier-tested cancer treatment using pions to kill cancer cells.

     DEVELOP   Recall that the nucleus is positively charged, so there should be a Coulomb attraction between it and negatively-charged particles.

     EVALUATE   The only negatively charged particle in the list is the negative pion.
The answer is (c).

     ASSESS   There's a second advantage to using negative pions over neutral pions (π^0). Since π^- mesons are charged, they can be deflected with magnetic fields. Thus, a beam of negative pions can be focused with magnets onto a tumor site.

61.   **INTERPRET** We consider an earlier-tested cancer treatment using pions to kill cancer cells.

DEVELOP The negative pion has charge $q = -e$ and mass $m = 139.6 \text{ MeV}/c^2$. We're asked what quark combination could give this.

EVALUATE From Table 39.2, a particle with quarks uud would have a charge of $q = +\frac{2}{3}e + \frac{2}{3}e - \frac{1}{3}e = +e$. This therefore can't be the negative pion (in fact, this is the quark combination for the proton, see Figure 39.4). We could have also ruled this choice out because pions, like all mesons, are by definition made up of 2 quarks, not 3. Moving on, a particle with quarks $\overline{d}u$ would have a charge of $q = -\frac{1}{3}e - \frac{2}{3}e = -e$, where we have used the fact that an antiquark of a particular flavor will have the opposite charge of the quark with the same flavor. This could be the negative pion. The next choice ud has a charge of $q = +\frac{2}{3}e - \frac{1}{3}e = +\frac{1}{3}e$. This is not possible, since no free particles are known to have fractional charges. The last possibility is $c\overline{c}$ with a charge of $q = +\frac{2}{3}e - \frac{2}{3}e = 0$. This doesn't match the negative pion, so by elimination π^- is $\overline{d}u$.

The answer is (b).

ASSESS Notice that this result is similar to the quark content of the π^+ meson: $u\overline{d}$, shown in Figure 39.5. We didn't consider the masses, but it turns out that most particles have more mass than just the sum of their quark masses. From Table 39.2, we might assume a particle made up of d quark and an anti-u quark would have a mass of around $9 \text{ MeV}/c^2$, but the negative pion actually is 15 times more massive than that. The extra mass comes from the internal energy between the strongly-interacting quarks.